The Evolution of
NORTH AMERICAN INDIANS

A 31-volume series of outstanding dissertations

Edited by

DAVID HURST THOMAS
American Museum of Natural History

A GARLAND SERIES

THE EVOLUTION OF NORTH AMERICAN INDIANS

Cultural Diversity Among Aboriginal Cultures of Coastal Virginia and North Carolina
Lewis R. Binford

Prehistory of the Central Maine Coast
Bruce J. Bourque

Aboriginal Cultural Adaptations in the Midwestern Prairies
James A. Brown

PostColumbian Culture History in the Northern Columbia Plateau A. D. 1500–1900
Sarah K. Campbell

The Effect of European Contact and Trade on the Settlement Pattern of Indians in Coastal New York, 1524–1665
Lynn Ceci

Late Prehistoric and Early Historic Chiefdoms in the Southeastern United States
Chester B. DePratter

Prehistoric Exchange and Sociopolitical Development in the Plateau Southwest
Amy A. Douglass

Playing with Tradition among the Mescalero Apaches
Claire R. Farrer

An Ecological Analysis Involving the Population of San Juan Pueblo, New Mexico
Richard Irving Ford

Cherokee Folk Zoology: The Animal World of a Native American People
Arlene Fradkin

Preserving Diversity: Ethnoarchaeological Perspectives on Culture Change in the Western Canadian Subarctic
Robert R. Janes

The Evolution of Chumash Society: A Comparative Study of Artifacts Used in Social System Maintenance in the Santa Barbara Channel Region Before A. D. 1804
Chester D. King

Pre-Seminole Florida: Spanish Soldiers, Frars, and Indian Missions
Robert A. Matter

The Regional Organization of the Hohokam in the American Southwest: A Stylistic Analysis of Red-on-Buff Pottery
Jill E. Neitzel

Spanish-Indian Relations in Florida: A Study of Two Visitas, 1657–1678
Fred Lamar Pearson, Jr.

Seasonal Agricultural Circulation and Residential Mobility: A Prehistoric Example from the Pajarito Plateau, New Mexico
Robert Washington Preucel, Jr.

Emergence of Native American Nationalism in the Columbia Plateau
Jeffrey C. Reichwein

The Ceramic Sequence in the Chaco Canyon, New Mexico, and Its Relation to the Cultures of the San Juan Basin
Frank H. H. Roberts

Prehistoric Dimensions of Status: Gender and Age in Eastern North America
Nan A. Rothschild

Aging and Osteoporosis in Native Americans from Pecos Pueblo, New Mexico: Behavioral and Biomechanical Effects
Christopher Ruff

The Acoma Culture Province: An Archaeological Concept
Reynold J. Ruppé

Long Term Adaptations among Arctic Hunter-Gatherers: A Case Study from Southern Baffin Island
George Sabo, III

Interdependence in the Prehistoric Southwest: An Ecological Analysis of Plains-Pueblo Interaction
Katherine A. Spielmann

Clown in Native North America
Julian Haynes Steward

The Mikasuki Seminole Medical Beliefs and Practices
William C. Sturtevant

Rock Art of the Southern Black Hills: A Contextual Approach
Linea Sundstrom

Hunting by Prehistoric Horticulturalists in the American Southwest
Chrisatine R. Szuter

In the Maelstrom of Change: The Indian Trade and Cultural Process in the Middle Connecticut River Valley, 1635–1665
Peter A. Thomas

Prehistoric Tewa Economy: Modeling Subsistence Production on the Pajarito Plateau
Nicholas Trierweiler

The Dentition of Arctic Peoples
Christy G. Turner, II

Prehistoric Warfare on the Great Plains: Skeletal Analysis of the Crow Creek Massacre Victims
P. Willey

Aging and Osteoporosis in Native Americans from Pecos Pueblo, New Mexico

Behavioral and Biomechanical Effects

CHRISTOPHER RUFF

Garland Publishing, Inc. New York & London 1991

Library of Congress Cataloging-in-Publication Data

Ruff, Christopher.
Aging and osteoporosis in native Americans from Pecos Pueblo, New Mexico : behavioral and biomechanical effects / Christopher Ruff.
p. cm. — (The Evolution of North American Indians)
Revision of the author's thesis (Ph. D.)—University of Pennsylvania, 1981
Includes bibliographical references.
ISBN 0-8240-2515-6 (alk. paper)
1. Pueblo Indians—Diseases. 2. Pueblo Indians—Aged. 3. Pueblo Indians—Anthropometry. 4. Osteoporosis—Epidemiology. 5. Femur—Measurement. 6. Tibia—Measurement. 7. Human mechanics. 8. Pecos National Monument (N.M.)—Antiquities. 9. New Mexico—Antiquities. I. Title. II. Series.
E99.P9R85 1991
616.7'16—dc20 91-2397

Printed on acid-free, 250-year-life paper.
Manufactured in the United States of America.

Design by J. Threlkeld

Dedicated to my wife, Frances

PREFACE

The following manuscript was submitted in 1981 as a Ph.D. thesis in the Department of Anthropology, University of Pennsylvania. Most of the work was carried out in 1979 and 1980 in the Peabody Museum, Harvard University, and the Biomechanics Laboratory, Beth Israel Hospital, Boston. Because this was just prior to the dawn of the word-processing age, all of the manuscript was typewritten. Thus, it has been left largely unedited, with the exception of a few sections deleted to reduce its total length by approximately 20 percent. These include in particular the original analyses of polar second moments of area, 2 and 3-way analyses of variance, an appendix listing all data broken down into six age-group divisions, the index, and three short sections dealing with technical issues or other information not necessary for understanding the rest of the study. The complete thesis with these sections intact is available through University Microfilms (Ann Arbor, Michigan). I attempted to eliminate most references to these passages in what remains, although a few were left in in the chapter summaries and Discussion section. Again because this was not done on a word-processor, there are a few instances where editing has resulted in some blank spaces or lines within the text. Finally, because like most theses this was originally produced in sections for independent review by different faculty advisors, there are numerous instances throughout the text, not really necessary now, where parenthetical reference is made to previous or future sections of the thesis. I hope that the reader will bear with these minor obstacles in reading the manuscript.

Osteoporosis, or bone loss leading to an increase in fracture incidence in old age, is a serious medical and societal problem in much of the world today. The annual direct and indirect medical costs of treating age-related fractures in the United States alone have been estimated to be in excess of 7 billion dollars, and are expected to rise in the years ahead as the population ages (Kelsey 1987). While much has been learned in recent years regarding the etiology of osteoporosis (Riggs and Melton 1986, Parfitt 1987, Raisz 1988, Stini 1990), the specific risk factors involved and the most effective treatments for this condition are still matters of controversy. Part of the problem is that certain key aspects of the skeletal aging process are still little understood, in particular structural changes in the skeleton beyond simple alterations of bone mass or volume that may be critical in determining fracture risk (Ruff and Hayes 1988, Beck et al. 1990).

The primary objective of the present study was to determine what structural changes in the femur and tibia occurred with aging in a well-defined archaeological population sample from Pecos Pueblo, New Mexico. The structural properties examined included geometric properties and bone mineral content of 11 cross sections in the shafts and near the hip (femoral neck), as well as several angles and measures of curvature of the whole bones and bone core densities of a subset of individuals. These properties were chosen because they have direct *biomechanical* significance, that is, they directly reflect the strength of bones and their ability to withstand applied forces without fracturing. Technological advances newly available at the time of the study, particularly a semi-automated technique for calculating cross-sectional geometric properties, allowed the analysis of a much larger sample than had ever before been possible. A total of 119 individuals and over 1300 bone cross sections were included in the study. Because the size of the sample, this was the first true demographic analysis of these biomechanical properties that had been carried out, and still today represents the largest single population sample to have been so analyzed.

The most important finding of the age analysis was that despite cortical thinning and some loss of bone density with age, geometric remodeling of bone (subperiosteal expansion) maintained bone strength, in both sexes. This contrasted with a much smaller study of some modern Euroamericans (Martin and Atkinson 1977), in which only males showed such geometric "compensation" with aging. The difference between the studies was provisionally attributed to higher levels of physical activity among Pecos Pueblo females, which could have stimulated more compensatory bone remodeling throughout life. It was suggested that a larger study of a modern population sample was needed to further test this possibility.

In addition to the aging analysis, comparisons between the sexes and between left and right sides were also carried out. The most interesting finding here was that males and females at Pecos appeared to have engaged in activities that placed systematically different mechanical stresses on the lower limb. The sex differences in cross-sectional geometry of the femur and tibia were consistent with more running and long distance travel among males, possibly combined with sexual dimorphism in pelvic shape related to childbirth requirements in females. These apparent differences in behavior and basic body morphology between the sexes also had subtle but important effects on patterns of skeletal remodeling with age. Some evidence of differential use of the right and left lower limbs, partly sex-specific, was also found. Again, the need for more comparative data from other population samples with known differences in behavior was stressed.

The results of this study have appeared in a series of papers—on the re-aging of the Pecos sample and its implications for demographic reconstructions (Ruff 1981), the bone cross-sectional geometric findings (Ruff and Hayes 1982, 1983a, 1983b), and the bone mineral content and density findings (Ruff and Hayes 1984a, 1984b). In the last two papers it was re-emphasized that simple measures of bone mass are not adequate to fully document or understand the skeletal aging process.

In the years following the completion of the thesis, a study of a large (n=100) sample of femora and tibiae collected from U.S. White autopsy material was carried out to further test one of the major hypotheses suggested by the original Pecos study—that females from modern industrial societies may suffer large increases in fracture rate in old age partly because they do not undergo optimal geometric remodeling of bone with aging. In fact, the hypothesis was strongly supported by the results of this study (Ruff and Hayes 1988). The difference in bone geometric remodeling between Pecos and U.S. White females matches differences in osteoporotic fracture incidence observed among living populations, where women in developing countries show much lower incidences (e.g., Barss 1985). Dietary or other factors do not appear to explain the discrepancy, while increased physical activity in non-mechanized societies very well might (*ibid.*).

The general sex difference in lower limb bone structure observed in the Pecos sample was further explored in both the new U.S. White sample as well as a number of other samples, archaeological and modern, collected by myself and others (Ruff 1987a). It was found that there is a strong correlation between the degree of sexual dimorphism in certain structural properties and subsistence strategy, with hunter-gatherers showing a high degree of dimorphism, agriculturalists a moderate degree, and industrial societies virtually none. The particular properties exhibiting this variable dimorphism, centered near the knee, are precisely those that would be expected, from a biomechanical standpoint, to reflect different levels of long-distance mobility. The ethnohistoric literature clearly documents such differences in mobility associated with subsistence strategy due to changes in the sexual division of labor (*ibid.*). The link between morphology and behavior was greatly strengthened by the breadth of the comparison (geographic and temporal) and the consistency of the trends. Even Upper Paleolithic and Neanderthal samples, despite some overall differences from recent samples, showed the same sexual dimorphism in bone shape as modern hunter-gatherers, suggesting a similar sexual division of labor. In contrast, a sex difference in bone structure near the hip, probably reflecting pelvic sexual dimorphism, is relatively constant between different groups, probably reflecting the relatively constant differential demands of childbirth on females, regardless of subsistence strategy.

This study is perhaps the clearest example to date of a fairly subtle but widespread behavioral change being detected through structural analysis of long bone shafts. Although the comparison was carried out mainly using cross-sectional geometric analysis, it was emphasized that even simpler traditional measurements, such as bone external breadths, can provide useful information, as long as they are interpreted within a biomechanical framework. As an example, differences reported by various authors in the efficacy of sexing techniques for archaeological material based on long bone shaft breadths were shown to be most likely due to the systematic behavioral differences outlined above. Thus, these differences must be taken into account when using or interpreting such (forensic) techniques.

As implied in this discussion, the last decade has seen a large increase in the number of different population samples studied using cross-sectional geometric analysis. I have recently reviewed these studies in detail elsewhere (Ruff in press a). They include investigations of the effects of subsistence strategy (Ruff et al. 1984, Brock and Ruff 1988, Bridges 1989a,b, Robbins et al. 1989, Ruff and Larsen 1990, Fresia et al. 1990, Larsen and Ruff 1991, Ruff in press b), growth and development (Sumner 1984, VanGerven et al. 1985), and skeletal aging (Martin et al. 1985) in archaeological samples, as well as structural analyses of earlier human paleontological samples (Kimura and Takahashi 1982, Trinkaus and Ruff 1989, Trinkaus et al. 1990) and nonhuman primates (Burr et al. 1981, 1982, 1989; Schaffler et al. 1985, Ruff 1987b; see Ruff 1989 for a review of nonhuman primate studies). As a result it is now becoming possible to carry out some fairly broad comparative analysis of recent human populations (e.g., Ruff 1987a, in press b, Larsen et al. 1991). However, there is still a need to increase the geographic range of such samples, particularly from regions other than North America and East Asia. There is also a need to extend such studies to more juvenile samples to better understand the ontogenetic development of skeletal form. I am currently analyzing a juvenile sample from Pecos (30 individuals between 6 and 19 years). Initial results indicate, among other things, that the sex difference in lower limb bone structure related to sexual division of labor appears at least as early as mid-late adolescence. I am also currently carrying out a study of upper limb bone (humerus, radius, ulna) cross-sectional geometric properties in matched samples of the same juveniles and 60 adults from Pecos. Results will be used to help interpret patterns of limb use and locomotion in earlier hominids.

Skeletal structural studies including cross-sectional geometric analysis have been greatly aided in recent years by the increasing availability of noninvasive imaging techniques, in particular computed tomography, or CT (Ruff and Leo 1985). These techniques have not only permitted the study of internal bone structure without damaging specimens, but have also eliminated much of the laborious manual labor in a study of this kind, e.g., as reflected in the description of methodology in the present work. New structural information can also be extracted by modifying existing technology. For example, photon absorptiometry, originally designed to measure bone mineral content, as reported in the thesis, has recently been extended to determine structural properties of the proximal femur important in determining hip fracture risk (Beck et al. 1990). Work is continuing to further automate such analyses to make them even more widely available to both researchers and clinicians.

Finally, the value of *combined* analyses of cross-sectional geometry of long bone shafts with other structural features such as articular geometry (Ruff 1988), patterns of arthritis (Larsen et al. in press), and bone microstructure (Burr et al. 1990) has recently been demonstrated. Bone adaptations to mechanical and other factors can take place on many different structural levels, and thus biobehavorial reconstructions of past populations from their skeletal remains should take advantage of every preserved structural feature. It can be anticipated that many future studies will employ such a multiple technique approach, particularly as technical advances increase the feasibility of these studies.

In closing, I would like to thank Garland Publishing for including my work in this exemplary series on the American Indian. An appreciation of historical context is necessary to better understand and develop treatments for medical and other problems in modern societies. The Pecos sample, being late prehistoric and protohistoric, but also preindustrial, may provide a good baseline with which to compare modern populations in this regard. The apparent value of increased physical activity in preventing osteoporosis, and the unanticipated problems brought about by increasing mechanization and sedentism in industrialized societies, are lessons that have been learned in part through such studies.

Christopher Ruff

REFERENCES

Barss, P.
1984 Fractured Hips in Rural Melanesians: A Nonepidemic. *Tropical and Geographic Medicine* 37: 156–159.

Beck, T.J., C.B. Ruff, K.E. Warden, W.W. Scott, and G.U. Rao
1990 Predicting Femoral Neck Strength from Bone Mineral Data: A Structural Approach. *Investigative Radiology* 25:6–18.

Bridges, P.S.
1989a Changes in activities with a Shift to Agriculture in the Southeastern United States. *Current Anthropology* 30:385–394.

1989b Bone Cortical Area in the Evaluation of Nutrition and Activity Levels. *American Journal of Human Biology* 1:785–792.

Brock, S.L. and C.B. Ruff
1988 Diachronic Patterns of Change in Structural Properties of the Femur in the Prehistoric American Southwest. *American Journal of Physical Anthropology* 75:113–127.

Burr, D.P., G. Piotrowski, R.B. Martin, and P.N. Cook
1982 Femoral Mechanics in the Lesser Bushbaby (Galago senegalensis): Structural Adaptations to Leaping in Primates. *Anatomical Record* 202:419–429.

Burr, D.B., G. Piotrowski, and G.J. Miller
1981 Structural strength of the Macaque Femur. *American Journal of Physical Anthropology* 54:305–319.

Burr, D.B., C.B. Ruff, and C. Johnson
1989 Structural Adaptations of the Femur and Humerus to Arboreal and Terrestrial Environments in Three Species of Macaque. *American Journal of Physical Anthropology* 79:357–368.

Burr, D.B., C.B. Ruff, and D.D. Thompson
1990 Patterns of Skeletal Histologic Change through Time: Comparisons of an Archaic Native American Population with Modern Populations. *Anatomical Record* 226:307–313.

Fresia, A., C.B. Ruff, and C.S. Larsen
1990 Temporal Decline in Bilateral Asymmetry of the Upper Limb on the Georgia Coast. *Anthropological Papers of the American Museum of Natural History* 68:121–132.

Kelsey, J.L.
1987 Epidemiological Research. Paper presented at the Scientific Workshop on Research Directions in Osteoporosis, National Institutes of Health, Bethesda, Md., February 9–11.

Kimura, T. and H. Takahashi
1982 Mechanical Properties of Cross Section of Lower Limb Long Bones in Jomon Man. *Journal of the Anthropoligical Society of Nippon* 90:105–117.

Larsen, C.S., C.B. Ruff, and M.C. Griffin
Implications of Changing Biomechanical and Nutritional Environments for Activity and Lifeway in the Eastern Spanish Borderlands. In *Disease and Demographic Collapse in the Spanish Borderlands*, B.J. Baker and L.L. Kealhofer, eds. (in press).

Larsen, C.S., C.B. Ruff, and R.L. Kelly
1991 Skeletal Structural Adaptations in Prehistoric Western Great Basin Hunter-Gatherers. Presented at the annual meeting of the American Association of Physical Anthropologists, Milwaukee.

Martin, R.B.and P.J. Atkinson
1977 Age and Sex-Related Changes in the Structure and Strength of the Human Femoral Shaft. *Journal of Biomechanical Engineering* 10:223–231.

Martin, R.B., D.B. Burr, and M.B. Schaffler
1985 Effects of Age and Sex on the Amount and Distribution of Mineral in Eskimo Tibiae. *American Journal of Physical Anthropology* 67:371–380.

Parfitt, A.M.
1987 Bone Remodeling and Bone Loss: Understanding the Pathophysiology of Osteoporosis. *Clinical Obstetrics and Gynecology* 30:789–811.

Raisz, L.G.
1988 Local and Systematic Factors in the Pathogenesis of Osteoporosis. *New England Journal of Medicine* 318:818–827.

Riggs, B.L. and L.J. Melton, III
1986 Involutional Osteoporosis. *New England Journal of Medicine* 314:1676–1686.

Robbins, D.R., K.R. Rosenberg, and C.B. Ruff
1989 Activity Patterns in Late Middle Woodland, Delaware. *American Journal of Physical Anthropology*78:290–291.

Ruff, C.B.
1981 A Reassessment of Demographic Estimates for Pecos Pueblo. *American Journal of Physical Anthropology* 54:147–151.

1987a Sexual Dimorphism in Human Lower Limb Bone Structure: Relationship to Subsistence Strategy and Sexual Division of Labor. *Journal of Human Evolution* 16:391–416.

1987b Structural Allometry of the Femur and Tibia in Hominoidea and Macaca. *Folia Primatologica* 48:9–49.

1988 Hindlimb Articular Surface Allometry in Hominoidea and Macaca, with Comparisons to Diaphyseal Scaling. *Journal of Human Evolution* 17:687–714.

1989 New Approaches to Structural Evolution of Limb Bones in Primates. *Folia Primatologica* 53:142–159.

Biomechanical Analyses of Archaeological Human Material. In *The Skeletal Biology of Past Peoples.* S.R. Saunders and A. Katzenburg, eds. New York: Alan R. Liss (in press).

Biomechanical Analyses of Northern and Southern Plains Femora: Behavioral Implications. In *Skeletal Biology in the Great Plains: A Multidisciplinary View.* D.W. Owsley and R.L. Jantz, eds Washington, D.C.: Smithsonian Institution Press (in press).

Ruff, C.B. and W.C. Hayes
1982 Subperiosteal Expansion and Cortical Remodeling of the Human Femur and Tibia with Aging. *Science* 217:945–948.

1983a Cross-sectional Geometry of Pecos Pueblo Femora and Tibiae—A Biomechanical Investigation. I. Method and General Patterns of Variation. *American Journal of Physical Anthropology* 60:359–381.

1983b Cross-sectional Geometry of Pecos Pueblo Femora and Tibiae—A Biomechanical Examination. II. Sex, Age, and Side Differences. *American Journal of Physical Anthropology* 60:383–400.

1984a Bone Mineral Content in the Lower Limb: Relationship to Cross-sectional Geometry. *Journal of Bone and Joint Surgery* 66A:1024–1031.

1984b Age Changes in Geometry and Mineral Content of the Lower Limb Bones. *Annals of Biomedical Engineering* 12:573–584.

1988 Sex Differences in Age-Related Remodeling of the Femur and Tibia. *Journal of Orthopedic Research* 6:886–896.

Ruff, C.B. and C.S. Larsen
1990 Postcranial Biomechanical Adaptations to Subsistence Changes on the Georgia Coast. *Anthropological Papers of the American Museum of Natural History* 68:94–120.

Ruff, C.B., C.S. Larsen, and W.C. Hayes
1984 Structural Changes in the Femur with the Transition to Agriculture on the Georgia Coast. *American Journal of Physical Anthropology* 64:125–136.

Ruff, C.B. and F.P. Leo
1986 Use of Computed Tomography in Skeletal Structural Research. *Yearbook of Physical Anthropology* 29:181–196.

Schaffler, M.B., D.B. Burr, W.L. Jungers, and C.B. Ruff
1985 Structural and Mechanical Indicators of Limb Specialization in Primates. *Folia Primatologica* 45:61–75.

Stini, W.A.
1990 "Osteoporosis": Etiologies, Prevention, and Treatment. *Yearbook of Physical Anthropology* 33:151–194.

Sumner, D.R.
1984 Size, Shape, and Bone Mineral Content of the Human Femur in Growth and Aging. Unpublished doctoral thesis from the University of Arizona.

Trinkaus, E., S.E. Churchill, and C.B. Ruff
1990 Neanderthal Post-traumatic Humeral Assymetry and the Interpretation of Fossil Diaphyseal Morphology. Paper presented at the annual meeting of the American Association of Physical Anthropologists, Miami, Fla.

Trinkaus, E. and C.B. Ruff
1989 Diaphyseal Cross-sectional Morphology and Biomechanics of the Fond-de-Foret 1 Femur and the Spy 2 Femur and Tibia. *Bulletin Société Royale Belgique Anthropologie et Préhistoire* 100:33–42.

VanGerven, D.P., J.R. Hummert, and D.B. Burr
1985 Cortical Bone Maintenance and Geometry of the Tibia in Prehistoric Children from Nubia's Batn el Hajar. *American Journal of Physical Anthropology* 66:275–280.

TABLE OF CONTENTS

TABLES

ILLUSTRATIONS

Figure		Page
15b	Age differences in female femoral cross section centroid positions.	187
16a	Tibial cortical areas: male/female.	198
16b	Femoral cortical areas: male/female.	200
17a	Change with age in bone areas of tibial 65% cross section: males.	208
17b	Change with age in bone areas of tibial 65% cross section: females.	209
18a	Change with age in bone areas of femoral 50% cross section: males.	210
18b	Change with age in bone areas of femoral 50% cross section: females.	211
19a	Change with age in %CA of tibial 65% cross section: males and females.	219
19b	Change with age in %CA of femoral 50% cross section: males and females.	220
20a	Tibial cross section outlines in one individual, showing angle of orientation of major axis, phi.	234
20b	Femoral cross section outlines in one individual, showing angle of orientation of major axis, phi.	235
21	Change with age in femoral angles: females.	252
22a	Tibial I_{max} and I_{min}: male/female.	264
22b	Femoral I_{max} and I_{min} : male/female.	265
23a	Sex differences in tibial I_{max}/I_{min} ratio.	267
23b	Sex differences in femoral I_{max}/I_{min} ratio.	270
24	Change with age in I_{max} of femoral 50% cross section.	278
25a	Change with age in tibial I_{max}: 40+/20-39 year ratio.	281
25b	Change with age in femoral I_{max}: 40+/20-39 year ratio.	283

Figure		Page

ACKNOWLEDGMENTS

I would first like to thank Dr. Francis Johnston, my principal advisor for this project, for his support, general guidance, critical comments, and encouragement throughout the course of this dissertation. The concept of continuing growth beyond the end of the adolescent "growth" period, and the willingness to use somewhat non-traditional approaches to explore traditional physical anthropological issues, I owe to him as important stimuli in the formulation of my thinking throughout this project and my graduate career. I would also like to thank the Department of Anthropology, University of Pennsylvania, for a grant to help defray the costs of the project, particularly the initial setting up and implementation of the programs used in the digitizing and computer analysis of cross sections.

To Dr. Wilson Hayes, Director of the Biomechanics Laboratory, Beth Israel Hospital, I owe a debt of gratitude extending over several phases of the study: bringing me through basic engineering concepts, constantly helping in defining the implementing the project, and most importantly, making available all of the equipment and computer software used in analyzing cross section geometric properties. It is no exaggeration to say that without his assistance, this project could never have been completed as planned. I would also like to thank Mark Nagurka for making available program SLICE and for helpful suggestions in setting up the program, and others in the biomechanics lab, particularly David Black, who helped in putting up and debugging the system, and answering questions when they arose.

I deeply appreciate the generosity and foresight of Dr. Erik Trinkaus for making available the Pecos collection for this kind of study, and for several valuable discussions during its course. To him and to others on the Harvard Peabody Museum staff, I extend thanks for free use of storage and work space during the project, use of the circular saw and other equipment, and generally being made to feel at home.

I would also like to th ank Dr. William Laughlin and Dr. Al Harper for use of the ^{125}I Bone Mineral Analyzer in the Department of Biobehavioral Sciences, University of Connecticut, and Dr. David Thompson for generously allowing me to use his preliminary data on directly measured bone core densities.

Finally, I thank my wife, Fran, for typing much of the manuscript, and for all the sacrifices she made during the gathering and writing of this dissertation.

Chapter 1

Introduction

The study of human morphological variability has been a central concern of biological anthropologists since the founding of the discipline two hundred years ago (Blumenbach, 1776). Initially mainly descriptive in character, more recent studies have attempted to explain human structural differences in terms of functional adaptations (Washburn, 1951). A relatively new approach to the study of structure-function relationships has been through the use of biomechanical theory. Biomechanics, the study of the application of engineering or mechanical principles in biological contexts (Frankel and Burstein, 1970), is particularly attractive for these kinds of analyses because of its ability to give quantitative expression to complex functional relationships. Biomechanical theory has been found to be applicable to many issues of traditional anthropological concern, for example, structural changes in early hominids (Lovejoy et al., 1973; McHenry, 1975), nonhuman primate locomotion (Wells and Wood, 1975; Vilensky, 1979), and comparative primate anatomy (Jungers and Minns, 1979). Together with the anthropologist's interest in and knowledge of intra- and interpopulational variability, biomechanics provides a potentially powerful new tool for the interpretation of human morphological variation and adaptation.

To date, however, few studies have attempted to apply this approach on a large-scale population basis. Investigations of human biological form using biomechanical priniciples have tended to be primarily methodological in intent (Piotrowski and Wilcox, 1971; Rybicki et al., 1972; Piziali et al., 1976), limited to very small sample sizes (Kimura, 1974; Minns et al., 1975, 1977; Lovejoy et al., 1976; Miller and Purkey, 1980) and/or concentrated on one particular structural feature such as the femoral neck (Backman, 1957). A major obstacle to conducting detailed structural analyses of large samples has been the difficulty in determining relevant biomechanical parameters without resorting to time-consuming manual measurement techniques (e.g., Lovejoy et al., 1976).

One purpose of the present study is to demonstrate a rapid automated method for analyzing complex biological structures (Nagurka and Hayes, 1980). Inputting data through an electronic digitizer linked directly with computer programs of analysis eliminates many of the time constraints and sources of measurement error inherent in manual inputting of data. Using this system, in the present study it has been possible to analyze a very large number of bone cross sections--1320--within a relatively short length of time (one year). The data output by this system included all geometric parameters critical in evaluating bone strength and rigidity, including cross-sectional areas, principle second moments of area and their orientations (see chapter 4). Thus, both precise measurements of important structural characteristics, combined with a large enough sample size to clearly show central tendencies in these characteristics, were available in this study.

A second aim of the present study is to show how biomechanical analysis can be used to investigate sources of within-population structural variation. In the few previous studies of structural variation where moderate sample sizes (20-40 individuals) have been available (Amtmann, 1971; Martin and Atkinson, 1977), specimens were obtained from autopsy or dissection room samples. Such collections are not random population samples, and allow little control over many environmental and genetic factors which may influence structural variation in unpredictable ways. In the present study, 120 individuals from a single archaeologically-defined population sample (Pecos Pueblo, New Mexico) were used. To the extent to which Pecos Pueblo represents a relatively homogeneous closed population unit (see chapter 3), the effects of factors such as sex and age on intrapopulational structural variation may be defined more clearly without the additional statistical "noise" inherent in heterogeneous, mixed population samples. In addition, the size of the study sample is large enough to permit subdivision along several axes, allowing comparisons between various effects while maintaining adequate subsample sizes within each treatment cell.

In the present study, the most important effect considered as a source of intrapopulational structural variation is that of age. Bone loss with age and increase in fracture incidence among the elderly constitute a major medical problem of concern to both biological and social scientists. As described in chapter 2, previous attempts to predict fractures or to evaluate bone strength on the basis of sample measures of total bone volume or bone mass have been largely unsuccessful. This may be due in part because of lack of understanding of some of the more subtle aspects of bone structural change with age,in particular changes in bone geometry. Geometrical changes which redistribute bone and alter stress distributions could significantly affect the functioning of a bone and its success or failure (fracture) under physiological loading (Klenerman et al., 1957). Although measures of total bone mass, volume, and density were obtained for all specimens included in this study, the novel aspects of the study involve the description and analysis of variations in bone shape (as measured by cross-sectional geometry). These data should aid in the formulation of more sophisticated and accurate models of skeletal aging. The practical benefit of such information in various clinical applications are discussed further in chapter 2.

The final purpose of this study is to provide comparative data for a particular human population--the inhabitants of Pecos Pueblo--which may be used in future studies involving other population samples. By extending this type of analysis to interpopulational comparisons, effects on structural variation of differences in factors such as activity levels, nutrition, culturally defined sex or age roles, and genetic relationships could be evaluated with more precision. The long-term goal of these kinds of biomechanical studies should be to provide a sounder basis for evaluating observed morphological differences within and between human populations, especially as these regard functional adaptations to various stresses. The approach is not limited to present or recent populations, but may be extended into the past to provide an evolutionary perspective to the study of human structural variation (Lovejoy and Trinkaus, 1981).

The two bones described in the present study--the femur and tibia--were chosen for analysis for several reasons. First, both show age-related increases in fracture incidence (Buhr and Cooke, 1959; Alffram, 1964; Nilsson and Westlin, 1977). Both also present problems in fracture treatment and fixation, both in older and younger adult age groups (Per Linton, 1949; Rockwood and Green, 1975; Watson-Jones, 1976; Minns et al., 1977; Carter and Vasu, 1980). Both the tibia and femur have been previously implicated as undergoing age-related remodeling in adults (Smith and Walker, 1964; VanGerven et al., 1969; Garn, 1970; Carlson et al., 1976; Ruff, 1977; Mensforth and Lovejoy, 1978; Ericksen, 1979). Physiological loading conditions for both bones have been described (e.g., Paul, 1966, 1967, 1971; Morrison, 1968, 1969, 1970). The femur and tibia are among the best preserved material in archaeological collections, assuring adequate sampling from the Pecos Collection. In addition, being large bones (i.e., calcium reservoirs), changes in bone volume and mineral mass in the femur and tibia may better reflect changes in the total skeleton than smaller bones such as the second metacarpal (Garn, 1970). This is also a reason why two bones from each individual were used--this should represent a better sampling of the skeleton as a whole than one bone alone. A final reason for analyzing two different bones is that the femur and tibia are subjected to somewhat different patterns of loading under normal physiological conditions (Morrison, 1970; Paul, 1971; Carter, 1978). Thus, a biomechanical study of the femur and tibia should illustrate different bone-specific functional adaptations to the aging process, while also indicating more global, whole-skeleton changes.

Previous approaches to the problem of skeletal aging are outlined in chapter 2, while chapter 3 describes the characteristics of the Pecos sample and the techniques used to sex and age individuals included in the study. The methodology employed to derive biomechanical parameters of the bone cross sections is described in chapter 4, and results of the analysis are presented in chapter 5. Chapter 6 presents a discussion and summary of the results of the study.

Chapter 2

Skeletal Aging

Osteoporosis

Loss of bone--bone mass, volume, and density--with aging of the human skeleton is a phenomenon which has been recognized for many years by clinicians and human biologists (Ingalls, 1931). Albright and coworkers (Albright et al., 1941; Albright and Reifenstein, 1948) originally defined "osteoporosis" as "a quantitative reduction in bone mass without a qualitative change in its composition". Since loss of bone with aging normally does not involve a change in chemical composition or degree of mineralization of organic matrix (Melick and Miller, 1966; Mueller et al., 1966; Dequeker, 1975), osteoporosis, whether defined as "post-menopausal", "senile", or simply "involutional" (Doyle, 1972) is an adequate term to describe the changes which occur with aging of the human skeleton. Although many pathological conditions, such as Paget's disease, hyperparathyroidism, and osteomalacia can cause loss of bone (Jowsey, 1977), age-related osteoporosis is probably best viewed not as a disease, but as a normal part of the aging process (Newton-John and Morgan, 1970; Dequeker, 1975). Bone loss with aging probably occurs as a function both of decreased intestinal absorption of calcium and reduction in estrogenic hormone levels, which have been shown to exert a protective effect against the bone resorbing action of parathyroid hormone (Dequeker, 1975). As a result, bone resorption rates exceed bone formation rates in old age, upsetting the normal resorption-formation balance (Harris and

Heany, 1969) and leading to overall loss of bone tissue.

The decline in skeletal mass and volume with aging has been demonstrated using a variety of techniques, both *in vivo* and *in vitro*. Trotter and coworkers (Trotter et al., 1959, 1960) investigated changes with age in the density (measured as bone weight over whole bone volume) of vertebrae, ribs and long bones obtained from an anatomical collection, and found an average decline of about 20% from the fourth through the ninth decade. They concluded that there was no significant difference in the rate of bone loss in different areas of the skeleton or in different sex or race groupings (although American Blacks had significantly denser bone than Whites over all age groups). Arnold et al. (1960, 1966) using autopsy material, found marked differences in the rate of femoral cortical thinning in males and females, although both sexes exhibited comparable declines with age in vertebral and rib density (ash wt./cc). Many other authors have documented sex differences in timing and rate of bone loss with aging, with females beginning somewhat earlier and proceeding at a faster rate than males (e.g., Garn, 1970).

A simple radiographic index for the detection of osteoporotic changes (sum of two cortical breadths/total subperiosteal breadth) applicable *in vivo* was introduced by Barnett and Nordin in 1960. Nordin and coworders (1966) later went on to show a clear relationship between age of menopause and onset of bone loss in women, a finding substantiated by most other studies. They also suggested a difference in timing and rate of bone loss in different areas of the skeleton, with vertebral bone loss beginning earlier than bone loss in the second metacarpal, followed by the femoral diaphysis (see below). Radiographic cortical thickness measurements and derived indices have continued to be very popular as methods for measuring bone loss with age, despite inherent problems in their interpretation (Dequeker, 1972; Morgan, 1973). Studies of age changes in the metacarpals (Garn et al., 1967a; Garn, 1970; Exton-Smith et al., 1969; Adams et al., 1970; Horsman and Simpson, 1975), radius (Meema and Meema, 1969), humerus (Bloom and Laws, 1970), tibia (Garn, 1970), and femur (Smith and Walker, 1964) have been carried out in this way. Radiographically measured bone density using an aluminum or hydroxyapatite step wedge has been used extensively by Meema and coworkers (Meema and Meema, 1965) and Doyle (1972). Meema (Meema and Meema, 1969) concluded that radiographic bone density is a more sensitive technique for measuring osteoporotic changes than cortical thickness, since some loss in density and increased porosity at the endosteal surface may occur before being detected in a cortical breadth measurement.

A better technique for the *in vivo* assessment of bone mass utilizing low energy photon absorption was introduced by Cameron and Sorenson in 1963. Photon absorptiometry has been shown to be an accurate and reproducible method for determining bone mineral content (BMC), and gives results which are well correlated with *in vitro* measurements of total bone weight and bone ash weight (Cameron and Sorenson, 1963; Mazess et al., 1964; Cameron et al., 1968; Mazess, 1971). Many clinical studies using the photon absorptiometric technique followed Cameron and Sorenson's report (Johnston et al., 1968; Goldsmith et al, 1971; Mazess et al., 1972; Smith, 1972; Boyd et al.,

1974). These document an increase in BMC through the fourth decade of life, followed by a loss of BMC after about age 50, more rapid in women, changes which parallel those observed for bone volume measurements of cortical breadths. Average BMC and BMC/bone width in individuals in their eighth and ninth decades may reach values one-half or less those of younger adults (Johnston et al., 1968).

Demonstration of bone loss with aging is not limited to living or recent populations. Changes in bone volume (Dewey et al., 1969; VanGerven et al., 1969; VanGerven et al., 1969; VanGerven, 1973; Carlson et al., 1976; Ericksen, 1976) and BMC (Perzigion, 1973) have been documented in archaeological samples. Age-related trends in these past populations follow patterns similar to those of living populations.

Fracture Incidence

The clinical significance of age-related bone loss lies in its effects on bone strength and susceptibility to fracturing. As Klenerman et al. (1967) have said, "Viewed from the functional standpoint, the important mechanical properties of a bone are its strength and rigidity; osteoporosis is clinically important only in so far as it interferes with these properties, especially the strength." An increase in fracture incidence with aging has been shown for many areas of the skeleton, including the vertebrae, proximal femur, distal radius, proximal humerus, pelvis, and tibial plateaus (Stewart, 1955; Buhr and Cooke, 1959; Alffram and Bauer, 1962; Alffram, 1964; Chalmers and Ho, 1970; Nilsson and Westlin, 1977). Fractures among the elderly are a major medical problem, with age-specific rates for some fractures such as those of the proximal femur rising 500 times or more over younger adult levels (Chalmers and Ho, 1970; Lewinnek et al., 1980). Many of these fractures also pose difficulties in treatment and fixation, often causing serious debilitation (Rockwood and Green, 1975; Watson-Jones, 1976).

While various factors, including impaired muscular coordination, reduced sense of balance, and arthritis have been suggested as contributory to the observed rise in fracture incidence among the elderly (Buhr and Cooke, 1959), by far the most investigated factor has been loss of bone, or age-related osteoporosis. Highly significant correlations between bone tensile and compressive strength and bone ash content (Currey, 1969), density (Carter and Hayes, 1976), and photon absorptiometric BMC (Foltz et al., 1976) have been shown. Somewhat lower but still significant correlations ($r = .5$ to $.7$) between simple bone volume measures such as breadths or areas extrapolated from breadths and bending strength of whole bones have also been demonstrated (Klenerman et al., 1967; Foltz et al., 1976). Based on these data, a direct cause-and-effect relationship between loss of bone, "bone fragility", and the rise in fracture incidence with aging has often been implied or assumed (e.g., Saville, 1972).

However, despite these empirical *in vitro* findings and continued intensive research in the area, to date no study has succeeded in distinguishing clearly between fracture (or fracture-prone) patients and control groups on the basis of simple measures of bone mass or volume. The various techniques attempted for this purpose include photon absorptiometry (Smith et al., 1972; Boyd et al., 1974; Meema and Meema, 1974; Fredensborg and Nilsson, 1977, Nilsson and Westlin, 1977), radiogrammetry (Stevens et al., 1962; Meema and Meema, 1974; Fredensborg and Nilsson, 1977), subjective rating of vertebral densities (Iskrant and Smigh, 1969), and quantitative histology (Aaron et al., 1974). In the only longitudinal study published to date which attempted to link a bone parameter with fracture incidence, Adams et al. (1970) found no relationship between changes in cortical thickness of the second metacarpal over an eleven year period and fracture occurence. They concluded:

> No differences could be found between subjects who had had fractures and those who did not with respect to bone mass, bone loss, the presence of back pain, or loss of height...Furthermore bone mass, as judged by the thickness of the second metacarpal cortex, was distributed normally in both sexes at both surveys; a distinct population of individuals with less bone than others was not found. (p. 614)

Although significant mean differences in bone mass, volume, or density between fracture and non-fracture groups have been found, the differences are always relatively small when compared to the overlap between groups. Fracture patients rarely fall outside the "normal" range of variablity, usually defined as ±2 S.D. of the control mean, making diagnosis of increased fracture risk on an individual basis impossible for all but a very few cases (Doyle, 1972; Saville, 1972; Dequeker, 1975).

One reason why simple bone measures such as BMC or cortical thickness may not be effective in discriminating between high and low fracture-risk patients is that other factors besides total mass or volume may have significant effects on bone strength. Bone strength is dependent upon many factors including material properties (chemical composition, density), microstructure (e.g., orientation and number of Haversian systems), macrostructure (trabecular orientation, geometry of compact cortical bone), and total bone mass and volume. Most studies of the living have concentrated on the last of these parameters--total mass and volume--while *in vitro* studies of skeletal aging have examined material properties (Mueller et al., 1966; Dequeker and Merlevede, 1971), microstructure (Jowsey, 1960; Currey, 1970), and trabecular patterns (Hall, 1961; Arnold et al., 1966). However, very few studies have attempted to measure macrostructural changes in shape and orientation of cortical bone with aging, even though changes of this type significantly affect the strength of a bone under loading and consequently its susceptibility to fracturing (Klenerman et al., 1967).

Cortical Bone Remodeling with Aging

Noninvasive techniques.

Smith and Walker (1964) were the first to point out the potential importance of age-related remodeling or redistribution of cortical bone. Basing their study on femoral A-P radiographs of 2030 women aged 45 to 90 years, they found a significant increase in subperiosteal diameter with aging, particularly near the midshaft. Endosteal diameter increased at a faster rate than did subperiosteal breadth, leading to a reduction in cortical breadth. However, because of the overall expansion of the diaphysis, cortical area increased 11% between the youngest and oldest age groups, and calculated section moduli increased 32%. The authors posited a biomechanical explanation for these geometrical changes, based on hypothetical stresses in the femur:

> This suggests that flexural stress with bowing, which is maximal at about midshaft, activates the periosteal accretion of bone...For older femurs at midshaft, gains in section moduli mean increased resistance to flexural forces. This has more significance for subjects in whom skeletal ingredients are diminished, since the resistance of the shaft to flexure can be maintained even with less bone provided it has been remodeled into a cortex of larger diameter. (p. 157)

Following Smith and Walker's initial report, other studies documented significant gains in subperiosteal or outer bone dimensions with aging among adults in several other skeletal areas, including the ribs (Epker and Frost, 1966), second metacarpal (Garn et al., 1967b), tibia (Garn, 1970), and craniofacial region (Israel, 1973; Ruff, 1980). Thus, adaptation to bending or "flexural" stress similar to that found in the femur may not account for all of the observed increases with age in bone dimensions. Expansion of outer bone dimensions during aging may serve as partial compensation for loss of bone in many areas of the skeleton under a variety of different physiological loadings (Garn et al., 1967b.).

The interpretation of these findings has been open to some debate, however. First, some of these studies are cross-sectional in design, sampling from age groups which may have grown and matured under different environmental conditions. Thus, secular effects on size and possibly shape could produce spurious results similar to those reported by Smith and Walker and others. In fact, Trotter and coworkers (Trotter et al., 1968) working with the Terry Skeletal Collection, which spans over a hundred years in birth dates, were able to show just such a secular trend in the subperiosteal diameter of the femoral midshaft diaphysis. From the mid-1800's to the mid-1900's, femora in their sample not only became longer (part of the oftnoted trend in height during the last century) but also narrower. Thus, although there was a positive correlation between mediolateral subperiosteal breadth (measured directly) and age at death, partial correlations of breadths on age holding birth year constant were not significant in three out of the four race/sex groupings tested.

Since Smith and Walker's study sample was drawn from another population, and several other studies, longitudianl in design (e.g., data quoted in Garn, 1970; Israel, 1973) have also shown increases in outer bone dimensions with aging, the general conclusions drawn from these studies are not necessarily invalidated by Trotter et al.'s reappraisal. However, Trotter's study does demonstrate the caution which must be exercised in using cross-sectional data to support hypotheses with regard to age-related changes. In addition to secular trends, selective survival must also be considered when interpreting cross-sectional "growth" data (see chapter 3).

Another problem common to all radiographic studies involves the extrapolation from one-dimension radiographic breadths to two-dimensional areas and geometric properties such as the section modulus. Calculations of this kind assume circular or cylindrical cross sections, an assumption rarely justified by the complex cross-sectional geometry of most long bone sites. VanGerven and coworkers (VanGerven et al., 1969) found errors as high as 25% or more in femoral midshaft areas extrapolated from A-P radiographic breadths when compared to direct measurement of cut bone cross sections. Furthermore, because of complex changes in the shape of the femoral cortex with aging (see below), the degree of error changed with aging, and was different in males and females. Horsman and Leach (1974) studied variations in cross-sectional shape of the radius and ulna and found that radiographic breadths varied in the accuracy of predicted two-dimensional areas depending upon location of the cross section. The present author (Ruff, n.d.) found errors as high as 40% in calculating cross-sectional areas from radiographic breadths at several sites on the tibia, humerus, femur, and radius. Combining radiographic breadths from two planes of measurement helped to reduce error somewhat, although differences between radiographic and direct measurement values still averaged 15%. Errors in extrapolating second moments of area from A-P and M-L radiographic breadths averaged 20-25% (the greater errors here were expected, since areal moments are expressed as fourth powers of linear dimensions and are thus more sensitive to inaccuracies in breadth measurements).

A further problem in using radiographic measurements to infer morphological changes is the presence of radiographic "shadowing" in many areas, due to the overlap of bony structures when viewed in the A-P or M-L plane. For example, much of the lateral tibial cortex is obscured by the normal curvature of the anterior tibial crest, limiting cortical and subperiosteal breadth measurements to a relatively small area of the distal tibial diaphysis (Ruff, 1977). Thus, many skeletal areas of interest cannot be investigated using this technique, and it must be assumed that the measurements which can be made are representative of the whole bone, or in some cases, the total skeleton.

Since there is evidence that site-specific bone remodeling occurs with aging, i.e., bone loss and remodeling occur at different rates at different levels of the diaphysis of long bones (e.g., Carlson et al., 1976), this assumption is probably not justified.

Thus, great care must be exercised in extrapolating from radiographic data to two- or three-dimensional models of skeletal aging. At best, radiographic data may be used as general indicators of age-related size and shape variation in long bone diaphyses. More precise descriptions of bone morphology are necessary if biomechanical interpretations of observed age trends are to be made.

The only noninvasive technique currently capable of reconstructing two-dimensional cross sections of body segments in sufficient detail to allow biomechanical analysis is computerized tomographic scanning (Ledley et al., 1974). However, because of the expense involved in carrying out analyses and the limited availability of CT scanners, biomechanical studies of the skeletal system using this technique have been limited primarily to methodological demonstrations (Lovejoy and Burstein, 1977; Jungers and MInns, 1979). Consequently, the most detailed studies to date of skeletal aging have been accomplished through direct measurement of cut bone cross sections, obtained either from autopsy material or archaeological collections.

Direct measurement techniques.

Investigations using the direct measurement approach to determine bone cross-sectional properties have been of three general types. First, several researchers have conducted detailed biomechanical analyses of diaphyseal cross sections (all of the tibia) in small samples (n = 4-10) with no attempt to distinguish age or sex differences (Kimura, 1974; Minns et al., 1975, 1977; Lovejoy et al., 1976; Miller and Purkey, 1980). With the exception of Lovejoy et al.'s study, these investigations were designed not primarily to measure interindividual variation, but rather to provide normative data against which hypotheses regarding function under physiological loading (Kimura, 1974), stress distributions, and proper fixation of fractures (Minns et al., 1977) could be tested. Lovejoy and coworkers' study was significant in that it attempted to quantify in biomechanical terms the functional significance of "platycnemic" (mediolaterally flattened) and "eurycnemic" (rounder) tibial diaphyses. They concluded that platycnemic tibiae are better adapted to resist bending moments developed during "active locomotion", while eurycnemic tibiae are more adapted to "generalized" loadings including torsion as well as bending. This conclusion has evolutionary significance in that earlier Mesolithic and Paleolithic populations are characterized by more platycnemic tibiae than later populations, suggesting a change in activity patterns through time (Lovejoy, 1970).

In the second type of direct measurement study, larger samples have been used to characterize age and sex differences in cortical bone thickness or cross-sectional area, but without investigating more complex geometrical properties such as moments of inertia (Dewey et al., 1969; VanGerven et al., 1969; VanGerven and Armelagos, 1970; VanGerven, 1973; Carlson et al., 1976; Mensforth and Lovejoy, 1978). All of these studies were of archaeological samples, and with the exception of VanGerven's (1970) report which included the tibia and humerus, all were carried out on the femur, usually limited to the femoral midshaft. Most of the results simply corroborate radiographic studies of the living, showing bone loss with aging through cortical thinning (measured at several sites around the cross section perimeter), more rapid among women. Subperiosteal expansion with age has also been documented (VanGerven et al., 1969; Carlson et al., 1976). In addition, more complex shape changes in the femoral mid-diaphysis with aging have been suggested by VanGerven et al. (1969) and Mensforth and Lovejoy (1978). Both studies suggest greater endosteal resorption and subperiosteal apposition of cortical bone in anterior and posterior quadrants, leading via "cortical drift" to a more mediolaterally flattened shaft with aging among males (e.g., VanGerven et al., 1969: 36, figure 17). Carlson et al. (1976) found differential bone loss at five cross section locations along the length of the femoral diaphysis, with a greater loss of cortical bone area in the more distal sites, and different patterns of cortical thinning with aging in males and females. In addition, they reported a possible association between degree of cortical thinning and subperiosteal expansion at different sites, with sites exhibiting the greatest degrees of cortical thinning also showing the greatest increase in subperiosteal diameters.

Thus, redistribution of compact cortical bone in long bones with aging, both within transverse cross sections and along the length of the diaphysis, has been suggested by several of these authors. However, except in terms of simple area or breadth measurements, these complex remodeling or "shape" changes have not been quantified.

The last type of study utilizing the direct measurement technique has included both large enough sample sizes and detailed enough measurements to be able to quantify some remodeling changes with aging. These include the studies of Amtmann (1968, 1971) and Martin and Atkinson (1977; see also Atkinson and Weatherall, 1969), all of autopsy material and all of the femur. Amtmann investigated sex and age differences in radiographic density and breaking strength (of dry bone samples) in each of four quadrants of five diaphyseal cross sections. Several geometric properties, including second moments of area were also claimed to have been calculated (Amtmann, 1971: 18), however, with the exception of cross-sectional area, the sex and age distributions of these parameters are not reported in the results of the study. Perhaps the most relevant reported finding for the purposes of the present study was the observation that locations of maximum density shifted somewhat along and around the diaphysis with aging in females, becoming more distally and dorsally oriented (*ibid.*: 31, figure 12). Again, as in the archaeological studies quoted above, a shape change or redistribution of bone volume with aging is suggested, although not specifically quantified: "The enlargement of the medullary canal with increasing age is possibly reached at the cost of the tissue parts lying in the sagittal plane (a significance for this cannot be proved) whereby the femoral crest is apparently not spared" (*ibid.*: 53).

Martin and Atkinson (1977) have produced the most detailed and complete study of geometrical changes with aging in cortical bone published to date. Several important section properties, including area, second moments of area about A-P and M-L axes, and principal moments and their orientations with respect to reference axes, were computed for four diaphyseal cross sections. In addition, bone densities (measured directly) in four quadrants of each cross section were available from an earlier study (Atkinson and Weatherall, 1967). The major finding of these studies was that cortical bone density of the femoral shaft (and derived ultimate bending strength) declined on average in both sexes, but that only males compensated geometrically for this decline through an increase in area moments. Both cortical area and second moments of area about A-P and M-L axes decreased with age along with density in females. Thus, the calculated mean bending moment for failure remained approximately constant in males, while declining significantly in females. The results suggest that geometrical changes in males may be "biomechanically induced, in keeping with Wolff's Law", but that "female femora are apparently denied this advantage". The authors note that while no explanation currently exists for this apparent sex difference, the results may "possibly explain the special problems that women have with osteoporosis" (Martin and Atkinson, 1977: 229).

Although providing important data suggesting additional avenues of research, there are several problems with both Amtmann's and Martin and Atkinson's studies. Samples, although apparently adequate in size to give some indication of interindividual variation (n = 24 and 37, respectively), are not equally distributed throughout the adult age range, and in particular lack younger adults. Amtmann's study includes only individuals 56 years of age and older, a range which completely excludes premenopausal women and younger men. Thus, his findings are based on just the tail end of the adult aging process, after significant age-related changes have already begun. Martin and Atkinson's sample composition is better,including 18 individuals younger than 50 years. However, if immature individuals (under 20 years) are excluded and the sample is broken down by sex, only five adult men and seven women under the age of 50 are included. Given the large range of normal variablility <u>within</u> sex and age groupings in both material properties and cross-sectional geometry (e.g., see their figure 8, p. 228), these numbers are not sufficient for distinguishing clearly between sex and age categories, and small sample size may account for some of the more unusual findings of this study (for example, no significant increase in subperiosteal breadth with aging in either males or females).

The importance of sampling the entire adult age range in studies of skeletal aging becomes apparent when one considers that the bone remaining in older adulthood is dependent not only on the rate of bone loss with age, but also on the amount of bone present to begin with in younger age groups. Doyle (1972) pointed this out in his critique of Albright et al.'s original (1941) paper: "In his thinking about osteoporosis, Albright left one factor completely out of account--the mass of bone present before the 'disease process' began". As Garn (1970) has shown, cortical thickness of the second metacarpal continues to increase through the third and fourth decades of life, a finding he interpreted as preparatory for later bone losses. Increases in bone mass and remodeling of bone <u>before</u> the age of 50 may very well have important effects on subsequent loss of bone and susceptibility to fracturing, and must be considered in any complete model of skeletal aging.

Another problem with the studies of Amtmann and Martin and Atkinson, mentioned earlier in chapter 1, is one common to all studies utilizing autopsy material: these collections do not in any sense represent a random sample of a single population. Individuals of various ethnic and national origins, living under a wide assortment of different environmental conditions are very likely included in such samples. The lack of control over these variables, together with the small sample sizes, may explain why clearer sex and age differences in many of the geometrical properties examined in these two studies were not found.

A final problem with these two studies is that for the most part they attempt no functional interpretation of their results. Although the principle embodied in "Wolff's Law"--that bone will be remodeled in order to reduce physiological stress--is acknowledged by both authors, neither applies it in any strict sense to his data. Observed geometrical and material property changes with age should be viewed in terms of the differences in stress patterns and fracture susceptibility in different age groups that such changes bring about. Knowledge of the *in vivo* physiological forces which act upon the skeletal system throughout life is still far from complete, however, several studies (e.g., Minns et al., 1977) have demonstrated the possibility of constructing at least approximate models of stress distributions in long bones under simulated loading conditions. Such interpretations are necessary if the true functional significance of observed age-related changes is to be understood.

Physical Activity and Bone Remodeling

One factor frequently suggested as important in determining rates and patterning of bone remodeling throughout life is activity level. It has been shown that complete inactivity due to immobilization leads to significant bone loss and changes in bone geometry, for example, angle of the femoral neck in growing children (Amtmann, 1971: 8). Conversely, experimental studies with animals have demonstrated gain of bone mass and thickness under increased physiological stress (ibid.: 9; Woo et al., 1980). It is more difficult to demonstrate significant changes in bone mass or volume under less extreme conditions; however, several lines of evidence indicate that variation in physical acitivity levels may play an important role in determining rates of bone gain or loss.

Two studies (Nilsson and Westlin, 1971; Dalen and Olsson, 1974) have used photon absorptiometry to show higher bone mineral content at several skeletal sites in amateur atheletes compared to nonathletic control groups. Dalen and Olssson, comparing male cross country runners and controls, all between the ages of 45 and 60, found differences in BMC in the wrist, humeral head, femoral shaft, and calcaneus of 13% or more. Other sites, including the femoral neck and third lumber vertebra, exhibited less significant differences, although all differences between the groups were positive (runners higher). Nilsson and Westlin measured BMC of the distal femur in four groups ranging from "top rank" amateur athletes to nonexercising controls, and demonstrated significant activity-related differences. Interestingly, the greatest (absolute and percentage) differences occurred between "nonathlete" controls who engaged in some form of physical fitness training,and controls who participated in little, if any training. The authors note that "This finding indicates that even moderate physical activity may increase the bone mass" (Nilsson and Westlin, 1971: 182). Another study relevant in this context is that of Doyle et al. (1970), who found a significant correlation between post mortem ash weight of the third lumber vertebral body and weight of the psoas muscle, a relationship which remained even after weight, height, and age were partitioned out through partial correlation.

Because of the possibility of pre-selection of athletes, however (i.e., larger individuals with denser bones may be predisposed to enter athletics), it is difficult to prove an activity-bone mass relationship using this type of study design. Another kind of study which controls for this factor is the comparison of right and left sides, or dominant versus nondominant limbs, _within_ individuals. Bilateral asymmetry in length and weight of human upper limb bones is well documented (e.g., Ingalls, 1931; Latimer and Lowrance, 1965) and may also exist in lower limb bones (_ibid_.). Several studies have shown very marked bilateral asymmetry in dominant/nondominant limb bone prameters such as BMC or cortical thickness in athletes (King et al., 1969; Nilsson and Westlin, 1971; Watson, 1973; Jones et al., 1977). Although some evidence contradictory to purely environmental (exercise-related) explanation for bilateral asymmetry has been found (Schultz, 1926; Garn et al., 1976; Plato et al., 1980), on the whole it appears that degree of bilateral asymmetry is related to activity levels and consequent differential stress in dominant and nondominant limbs.

This finding takes on added significance in view of the reported difference in rate of occurence of femoral neck fractures on right and left sides (Chalmers and Ho, 1970). A previous study by the present author (Ruff and Jones, 1981) showed bone to be lost with age at different rates in right and left tibiae and humeri, with more bone being lost from the side with more bone in younger adulthood (left tibiae and right humeri). Side-related differences in bone loss and bone remodeling will be explored further in chapters 5 and 6.

The best way to demonstrate an activity-bone gain relationship, and the most difficult to carry out in practice, is through longitudinal studies of changes in bone mass in exercise and non-exercise control groups. Two studies have reported such findings. Smith and coworkers (Smith and Babcock, 1973; Smith and Reddon, 1975) examined bone mineral content in the mid-radius of 39 subjects in a nursing home (29 in the later report) aged 55 to 94 years (69 to 95 years in the later report). In the initial report, changes in BMC over an eight month period were compared in three groups: a group of patients undergoing physical therapy treatment, a "physical activity" group (participating in an exercise prpgram three times a week), and a non-exercise control group. The control group showed no change in BMC, the "physical activity" group increased 2.6%, and the physical therapy group increased 7.8%. In the second report, subjects had been followed at three month intervals for a period of 30 months. Participants in the exercise program exhibited an average gain in BMC of 3.0%, while non-exercise controls <u>lost</u> an average of 2.25%. The authors conclude that "the physical activity program appeared to slow bone mineral loss in the nursing home population" (ibid.: 84).

Dalen and Olsson (1974) in another part of the study quoted above, could demonstrate no significant change in BMC in a group of men put on a daily exercise routine when compared to a nonexercise control group. However, this study lasted only three months, a time period in all probability too short for significant changes in bone mineral content to be detected. The age range of the subjects included (25 to 52) years) is also different from that used by Smith and co-workers, and may have affected the results of the study. As the authors note, "It would thus seem impossible to obtain a rapid increase in the amount of bone mineral in the skeleton of healthy normal male subjects by means of physical training. No conclusions can be drawn from these results regarding the effect of physical training under pathological conditions" (ibid.: 174).

Thus, very short-term therapy regimens have little if any effect on bone gain, at least in younger males, while longer-term exercise programs may significantly increase bone mass in older adults. Unfortunately, given the nature of currently available *in vivo* bone measurement techniques (other than CT scanning, which exposes patients to relatively high radiation doses), it has not been possible to measure longitudinal changes in skeletal parameters other than total bone mass in studies of this kind. Given indirect evidence from radiographic studies (e.g., Jones et al., 1977), it may be expected that significant remodeling of bone takes place concurrently with overall changes in bone mass under exercise programs.

Of course, the most important single criterion for assessing the efficacy of osteoporosis treatments is the effect such treatments have on the incidence of fractures themselves. While no longitudinal studies relating the incidence of fractures to exercise have as yet been reported, some indirect evidence suggests that physical activity levels throughout life may be important in determining later susceptibility to fracturing. Chalmers and Ho (1970) compared the incidence of proximal femoral fractures in five population samples from Great Britain, Sweden, Hong Kong, Africa (Bantu) and Singapore. While the Western European groups, particularly females, showed dramatic increases in fracture rates after the sixth decade, Bantus increased very little and Singapore females only moderately. In addition, the European groups were the only samples in which females significantly outnumbered males (corrected for age-specific population numbers) in fracture incidence. The relative importance of genetic factors ("race") in determining this distribution is unknown, and must be considered a possible influence (e.g., see Trotter et al., 1960). However, as the authors note, "With regard to environmental influences, it is hard to reconcile the two most widely accepted causal factors--the menopause and malnutrition--with the observed racial variations", since there is no evidence for earlier menopause in any of the samples tested, and dietary calcium and other nutrients would be expected to be more plentiful in the European groups. On the other hand, physical activity levels show an almost perfect inverse relationship with fracture incidence, with less affluent populations engaged in more manual labor--work shared by the women (Bantu, Chinese)--showing the lowest incidence.

"It is therefore reasonable to postulate that the skeletal atrophy of age may be modified by physical stress to which the individual has been exposed over prolonged periods. Senile osteoporosis may therefore be to some degree an undesirable consequence of the easier work load which the more advanced societies appear to regard as a major objective and desirable attainment" (ibid.: 673).

This conclusion may find support in other evidence from dietary supplementation programs. Garn (1970) could demonstrate no relationship between dietary calcium, vitamin D, or protein intake in several population samples, including some from developing countries (Central America), and rate of bone loss in the second metacarpal. In general, the success of various dietary supplementation programs designed to inhibit or reverse bone loss with aging (e.g., Jowsey et al., 1972) has been very limited. In addition, the wisdom of long-term calcium supplementation or estrogen replacement in treating osteoporotic bone loss has been questioned since the effects of such treatments on other organs, e.g., the blood vessels, are not fully understood (for a discussion of these points, see Dequeker, 1975: 109-110).

A critical distinction to bear in mind when discussing bone loss with aging (or under any circumstances) is that between overall loss of bone from the total skeleton and localized bone loss and remodeling. Harris and Heaney (1969) have made the important observation that whereas the overall magnitude of bone loss from the human skeleton may be determined to a large extent by hormonal and possibly other internal influences, the location of such bone loss may be mediated by biomechanical factors. In other words, some reduction in bone mass may indeed be an almost inevitable consequence of the aging process, but the actual pattern of bone loss and remodeling may be variable and dependent upon the localized physiological forces acting upon the bone. Other authors (Frost, 1964; Villaneuva et al., 1966) have also stressed the importance of biomechanical factors in determining the patterning of bone loss and remodeling. Under this hypothesis, differences in fracture incidence between populations such as those reported by Chalmers and Ho could be explained on the basis of localized remodeling and strengthening due to characteristically higher stress levels during life in those skeletal areas (it would be interesting to see if other fractures--e.g., those of the vertebrae or wrist--were also lower in these populations). This might also explain why bone mass in some areas of the skeleton seems more affected by certain kinds of exercise than in other areas (Watson, 1973; Dalen and Olsson, 1974), and why some authors have found different timing and rates of osteoporotic bone loss in different skeletal areas (Nordin et al., 1966).

Because stresses under physiological loading are distributed throughout entire bones (and the body as a whole), the effects of skeletal remodeling in one area will influence the stresses felt in another area. The sum total of the interactions between generalized skeletal changes in bone mass and localized remodeling in different skeletal regions must be appreciated if we are to predict the likelihood of fracture under loading. Until we better understand these complex interactions, it will be difficult to evaluate the true effectiveness of any osteoporosis preventive therapy. Other benefits of a study of bone remodeling with aging include more precise interpretation of already existing data on skeletal aging (e.g., subperiosteal expansion), and possibly better treatment of fractures and design of prosthetic devices. The implications of the various factors discussed above in terms of both general models of skeletal remodeling and skeletal aging, and specific clinical issues, will be discussed in chapter 5 within the context of the results of the present study.

Summary

1. Decline in bone mass, volume, and density of the human skeleton after the age of 50 is well-documented in both living and past populations.
2. Fracture incidence, particularly of the vertebrae, proximal femur and humerus, and distal radius, also increases after about age 50. While a simple relationship between bone loss, "bone fragility", and fracturing has often been assumed, no study to date has demonstrated a clear difference between fracture and non-fracture groups in terms of bone mass or volume measurements generally employed--i.e., simple total mass or volume determinations, with no considerations of bone shape.
3. Changes in bone shape, or remodeling with aging have been documented in several studies. These include subperiosteal expansion, "cortical drift" in long bone cross-sectional geometry, and differential bone loss at different long bone diaphyseal sites. Outside of one or two studies, however, except for simple breadth or

area measurements these shape changes have not been biomechanically quantified.

4. Changes in bone mass and remodeling throughout life are probably important in preparing for later bone loss in older adulthood. A critical factor in determining rate of bone loss or gain in all age groups may be activity level.
5. Generalized changes in total skeletal mass must be distinguished from localized remodeling due to particular biomechanical stresses. Only by understanding the relationships between overall loss of bone and local bone remodeling will it be possible to predict fractures and evaluate fracture prevention therapies.

Chapter 3

Pecos Pueblo Sample

The Pecos Pueblo Site

Pecos Pueblo, located in north-central New Mexico some 15 miles southeast of Sante Fe and one mile from the Pecos River (Kidder, 1958), was first excavated on a scientific basis by Kidder beginning in 1915 (Kidder and Kidder, 1917). Over the next 15 years a wealth of material, including literally millions of pottery sherds, the remains of at least six distinct large architectural structures, and over 1900 human burials were recovered (Kidder, 1958). The site of Pecos Pueblo is of signal importance in the history of Southwestern archaeology, since it was here that pottery sequences were described which have served as one of the bases of Southwestern chronology (Kidder and Kidder, 1917; Kidder, 1931; Kidder and Shepard, 1936). Eight pottery types from Pecos were originally described, from earliest to latest: Black-on-white, Glaze I, Glaze II, Glaze III, Glaze IV, Glaze V, Glaze VI, and Modern (Kidder and Kidder, 1917). The particular value of this chronology in the present context is that most burials can be associated, directly or indirectly, with a pottery type (Hooton, 1930: 16, Table II-1). This allows analysis of secular trends in the Pecos population (*ibid.*), as well as some control over sample selection in the present study with respect to temporal periods (see below).

Pecos Pueblo was occupied until well into historic times, being first discovered by Coronado's party in 1540 and not abandoned until 1838 (Kidder, 1924: 4-15). The dwelling was originally thought to have been founded at about 800 A.D. (Hooton, 1930: 332). As dendrochronological dating for the American Southwest became available, however, this date was continually revised upward (i.e., to more recent): an addendum to Hooton's 1930 monograph (p. 342) based on new dendrochronological dates, gives 1100 A.D. as the earliest possible date for the founding of Pecos; this was later revised to 1200 A.D. (Kidder, 1931), 1250 A.D. (Kidder, 1932), and finally, 1300 A.D. (Kidder and Shepard, 1936). Thus, the total period of occupation of the site lasted only some 500 years, about half of which was within the historic period.

The Pecos site, on the top of a flat, rocky ridge or "mesilla" may have been chosen for defensive purposes due to the arrival of nomadic eastern tribes (e.g., Apache) in the area during the fourteenth and fifeenth centruies (Kidder, 1958: 43, 308-309). The first structure built was a "straggling one-story community" occupied until the beginning of the Glaze III period, about the middle of the fifteenth century (ibid.: 107-108). At this time, a larger structure, the Quadrangle, was constructed—a four-storied building some 250 by 550 feet, in the center of which was a spacious plaza, the whole surrounded by a low defense wall (ibid.: 63). It was this pueblo which was first sighted and described by the Spanish in 1540. Despite sporadic contact throughout the next 60 years, it was not until 1598 that the real Spanish colonization period began, with the setting up of an administrative government and Christian Mission at Pecos. This arrangement lasted until 1680, when the Pecos inhabitants revolted, killing 400 Spaniards and beginning a long series of conflicts which lasted until the abandonment of the pueblo. The next

150 years document a period of gradual decline in the size of Pecos Pueblo, hastened by subsequent Spanish retaliation, wars with neighboring tribes, constant harassment and attack by Comanches, and a smallpox epidemic (Kidder, 1924: 14-15).

The population of Pecos at its peak (Glazes II through V) has been estimated at anywhere from 1000 to 4000 inhabitants (Hooton, 1930: 331-341; Howells, 1960: 167-168). Only approximately 12 to 15 per cent of the site was ever excavated (Kidder, 1932: 4). Burials were discovered in all parts of the site--under room floors, under the plaza, and in refuse piles around and within the pueblo:

> That interments were made almost everywhere is clearly shown by the dots on Figure 20. The great east midden was crowded from bottom to top with the dead (Kidder, 1924: Pl. 11). On the <u>mesilla</u>, graves were dug where-ever there was sufficient earth; most, except the earliest, in accumulations of refuse. There, as was noted in discussing conditions on the west side of the Quadrangle, there was a strong tendency to bury just outside the outer walls of groups of inhabited living rooms...This often enabled us to determine the phase of occupancy of such groups. (Kidder, 1958: 280).

Many burials were not preserved by the excavators, those in very poor condition being reburied after noting age, sex, and any features of special interest (Hooten, 1930: 16). Somewhat over 900 individuals are now housed at the Harvard Peabody Museum; these served as the source from which the present study sample was derived.

The origin and historical affinities of the Pecos Pueblo inhabitants to other Southwestern groups are unknown, although Kidder believed that the town was most likely settled by sedentary agricultural "immigrants from the west" (Kidder, 1958: 3). Conditions for the cultivation of maize, beans, squash, and sunflower seeds (Kidder, 1932:

303) were good, with sufficient rainfall for some dry farming in addition to irrigated crops grown along either the Pecos River or in a smaller arroyo with a year-round spring running next to the pueblo (ibid.: 5-6; Kidder, 1924: 4; 1958: 131). Irrigation streams are described by Spanish explorers, who also noted large stores of grain in each house (quoted in Kidder, 1924: 12). Kidder felt, barring large-scale warfare and social disruption (characteristic of Pecos Pueblo during its final stages of occupation), that "there is no reason to believe...they were ever faced by a serious shortage of their staple food" (Kidder, 1932: 6). Judging from the numerous animal bones found in the rubbish heaps of the site (ibid.), hunting of wild animals, which would have been available in the nearby mountains and plains to the east, also constituted an important element in the diet (see also the discussion of pathologies below).

Social class distinctions among the Pecos Pueblo inhabitants appear to have been almost unknown, judging from the lack of difference in grave goods recovered from the site: "Nothing more clearly illustrates the classlessness of the Pecos that the uniform simplicity of their graves. Not one of the nearly two thousand we opened was outstanding in construction or...in wealth of offerings" (Kidder, 1958: 289). This type of egalitarian social structure appears to have been typical of most if not all Southwestern societies of this time period (Willey, 1966: 186). No significant in-migration of other populations not native to the area appears to have occurred prior to the Spanish colonization period, although there is some evidence of consolidation of other small local communities into the pueblo during its occupation

(Kidder, 1958: 59).

Hooton's Study of the Pecos Skeletal Material

One advantage to using the Pecos skeletal sample in the present study is that it has already been the subject of an earlier very extensive investigation. Hooton's 1930 monograph, "The Indians of Pecos Pueblo: A Study of Their Skeletal Remains", remains a classic in the field of human osteology. No aspect of skeletal analysis as it was then practised was overlooked by Hooton: demography (including age and sex distribution and estimates of population size), cranial and postcranial osteometrics, morphological "types" and their historical affinities, and occurence of pathologies and traumas were all explored with great thoroughness. While certain of the methodologies employed and conclusions reached are today outmoded or subject to revision (such as the typological analyses), in breadth of treatment and care in execution the monograph still serves as an exemplary model for studies of this kind.

While close to 2000 human burials were discovered during the course of the Pecos excavations, only a relatively small subset of these was actually used in most of Hooton's analyses. Since his study was begun in 1925, none of the 300+ burials discovered after the 1924 field season were included (Kidder, 1958: 283). Also, beginning in 1920, as mentioned above much of the skeletal material in particularly poor condition was reburied after noting sex and age. Of the 973 individuals sent to the Harvard Peabody Museum, 532 were not included in osteometric or typological analyses due to either immaturity or the fragmentary nature of the material. Almost half of those remaining were still missing significant portions of the skeleton or were of "doubtful" sex or archaeological provenience, leaving only 250 in "good condition" for the study (*ibid*.: 18).

For the purposes of the present study, the most important aspects of Hooton's investigation were his analysis of secular trends in physical characteristics and his description of pathologies. I discuss secular changes at Pecos here, while pathologies are treated in a later section of this chapter.

Because of the small number of usable individuals from some of the archaeological strata represented at Pecos, Hooton grouped skeletons into four pottery-defined periods for the analysis of secular trends at the site: Black-on-white and Glaze I, Glazes II and III, Graze IV, and Glazes V and VI (Modern period skeletal remains were not analyzed). After an extended osteometric analysis of the cranial and postcranial remains of the 250 individuals in the study sample, Hooton could find little evidence of any significant change through time. For example, with respect to cranial measurements, he wrote:

> An inspection of these tables shows that the deviations of the whole series are less than would be expected if the Glaze groups were replaced by random samples. This is true both in measurements and in indices and for every subgroup, except the earliest (Black-on-white and Glaze I). In the latter subgroup the deviations are large enough to be of possible significance both in measurements and in indices, although the deviations even in this group are not sufficiently marked so that it could be said that the crania of this group are certainly differentiated from those of the entire series. (Hooton, 1930: 74)

Postcranial measurements also showed little variation from period to period, for example, in stature (reconstructed from tibial and femoral length using Pearson's formula): "None of the Glaze subgroups can be said to differ significantly in stature from the mean of the [male]series...The females also present an exceptional constancy in stature" (ibid.: 178). Hooton did find a slight tendency for cranial characteristics (both metric data and morphological "types") to become more "homogeneous" through time, although postcranial measurements became somewhat less so (ibid.: 181). To explain this apparent discrepancy he hypothesized that "craniometric features are more largely controlled by genetic factors than other skeletal characters", and thus reflect more accurately the "continuous process of amalgamation through inbreeding" occurring during the Pecos Pueblo occupation (ibid.: 181, 345). This explanation appears somewhat unlikely, however, given the facts that a) the pueblo was occupied only a few hundred years, too short a time period for any significant inbreeding to have occurred, b) cranial features are not necessarily under more genetic control than postcranial features, and c) if anything, the Pecos Pueblo population might be expected to have become more physically heterogeneous through time, since there is some evidence that smaller local communities were absorbed by the pueblo in later periods (see above).

The temporal trend towards more "heterogeneous" postcranial measurements was explained by Hooton to be the result of greater "fluctuations in the prosperity and health of the community" due to increased environmental stress, e.g., food shortages and enemy attacks, in the later occupation periods (ibid.: 181-182). He also discerned a "slight but perceptible decrease in the gross dimensions during these[last] periods" (*ibid*.), also presumably because of increased environment stress. In fact, inspection of Hooton's own summary tables of data (*ibid*.: 132-184) fails to support either of these contentions. In none of the postcranial dimensions analyzed--lengths and breadths of long bones, shape indices, stature estimates--is there any consistent trend through successive chronological periods in either mean dimensions or coefficients of variation (my calculations). Apparently what Hooton was referring to by more or less "homogeneous" or "heterogeneous" was the number of measurements falling inside or outside twice the "probable error" (analogous to a standard error) range around the grand mean over all period groupings (*ibid*.: 70 ff.), and not to relative dispersion within groups. However, the pattern of such "excesses" (deviations from the overall mean) in each period group is inconsistent, some dimensions deviating in one direction, others in another, and none in a linear fashion through time (e.g., summary Table V-75, pp. 182-183). More sophisticated univariate and multivariate statistical techniques might still identify significant temporal trends in physical characteristics at Pecos; however, on the basis of the evidence presented, and especially given the small sample sizes involved (for some dimensions less than 20 individuals per sex/period grouping), what relatively minor differences between groups there are would seem to be best explained on the basis of sampling error.

Sample Selection Criteria

However, in order to avoid any of the possible effects of the extreme social and economic disruption characteristic of Pecos Pueblo during its final years, and also the possibility of Spanish interbreeding, only individuals recovered from strata through Glaze V were used in the present study, eliminating any skeletons from the Glaze VI or Modern periods. While precise dates for the beginning and ending of most of the pottery-defined periods at Pecos are not given in Kidder's publications, it is evident from the appearance of European trade goods at the site that Spanish contact first occurred during the Glaze V period (Hooton, 1930: 332-333). This is supported by Kidder in his last communication regarding Pecos: "These [years of occupation of one of the pueblo structures] span a period beginning in Late Glaze IV and extending into Glaze V...say from about 1500 to 1600" (Kidder, 1958: 121). Glaze V may have lasted from about the early sixteenth to the middle seventeenth centuries (Hooton, <u>loc. cit</u>.; Kidder, 1932: 4). Therefore, eliminating skeletons from all periods past Glaze V restricts the present study sample to precontact and early Spanish contact periods. While some of the individuals from Glaze V burials must have lived during the Spanish Colonial era, relatively few from this period were <u>born</u> during it, and any event, the most serious effects on the physical health of the Pecos inhabitants occurred after the early colonization period, in the late 1600's and 1700's (see above). This restriction also limits the time span of the study sample to no more than 350 years, or about 12-15 generations.

Agreement was reached with Dr. Erik Trinkaus, Curator of Physical Anthropology at the Harvard Peabody Museum, to section a sample of long bones from the Pecos Collection. Only individuals with both right and left intact femora and tibiae were used, in each case taking the <u>less</u>

well-preserved side for sectioning. This procedure leaves one bone of each type for each individual intact for future scientific study.

After surveying the entire collection, it was decided to include tibiae and femora from either right or left sides (not restricting the study sample to a single side) for the following reasons: a) this significantly increased the available sample sizes, particularly in the rarer older adult age groups, b) bilateral asymmetry is generally fairly small in the lower limbs, c) if significant bilateral asymmetry in lower limb bone structure does occur, it should be investigated. It has been traditional in physical anthropology to use only one side in morphological analyses (usually the left among American anthropologists), since this eliminates an extra source of variability (side differences) and makes interpopulational comparisons easier. However, side differences are potentially informative in biomechanical studies, since bilateral asymmetry in bone structure may reflect differential use of right and left sides, different mechanisms of bone loss on the two sides, and different side-related susceptibility to fracturing (see chapter 2). Also, in functional terms it may not be justifiable to generalize from one side to the other if the two sides perform somewhat different functions in certain contexts. In any case, since the focus of the present study is on structural variation and its functional interpretation, it seemed reasonable to include bilateral asymmetry as another source of variability worthy of study. True within-individual bilateral comparisons cannot be made using the present experimental design, since in each case one side must be left intact. However, because of the large number of individuals involved (see below), sampling bias in choosing right or left sides (e.g., from larger or smaller individuals) should be largely canceled out, allowing reasonable comparisons between overall right and left side means, even though not derived from the same individuals.[1]

For these reasons, some care was taken in sampling right and left sides to ensure proper representation in the final sample composition. An equal number of individuals represented by right or left sides was selected in each of the 12 sex/age groupings analyzed (see below). This distribution allows comparisons across side (with equivalent sex/age composition within each side), and between sex, age, or sex/age groups (holding number of right and left sides constant within each group). In addition, within individuals, the tibia and femur were always taken from the same side (i.e., each individual is represented by either two right bones or two left bones). This reflects the fact that the two lower limb bones (femur and tibia) form a functional complex in weight-bearing and locomotion; analyzed together they provide data on structural remodeling of the right and left lower limbs as a whole.

The other restriction on sample selection was that each individual included have enough intact skeletal areas for definite assignment of sex and age. The minimum requirement for sex assignment was the presence of at least one fairly intact innominate, with a measurable sciatic notch angle and enough of the pubis and ischium preserved to derive the ischium-pubis index (for details see below). In almost all

---

[1] See also discussion in chapter 5 of side differences in cross-sectional areas.

cases substantial additional portions of the skeleton, e.g., the skull, were also available to aid in making sex determinations. The minimum requirements for making an age determination were the presence of at least one intact pubic symphyseal face, and the molar regions of the upper and lower dentitions for use in the Miles method (see below). Again, as with the sex assessments, usually many other features of the skeleton were available for making additional age estimates. The only exception to this last requirement were very young adults in which skeletal areas with late epiphyseal union, particularly the medial clavicle, were preserved, allowing an independent estimate of age to about age 26. Also, one individual without a "readable" pubic symphysis but obviously very old from all other skeletal indicators--dentition, cranial sutures, vertebral osteophytosis--was included in the sample. It might be remarked that probably the most limiting requirement in deriving an adequate sample size was the presence of an intact pubic symphysis. Many otherwise almost perfect skeletons had to be disgarded because of worn or missing symphyseal faces, this being a very fragile area of the skeleton particularly vulnerable to rapid disintegration under archaeological conditions. Still, even with this massive elimination of potential subjects, it was felt very important to maintain as strict controls as possible over age determinations and ensure consistency in assignment of age to all individuals in the sample, since so many of the future analyses of the study depend upon accurate age estimates. Only adults 20 years and over were included in the study sample. More details concerning the sexing and aging techniques employed are given below.

Every effort was made _not_ to exclude any "unusual" or apparently "abnormal" individuals from the sample, except in the cases of pathologies resulting in obvious bone distortion. In particular, arthritis, even if quite severe with marked effects on articular surfaces, was not considered sufficient cause for excluding an individual. At least moderate arthritic involvement of many limb bones is common among older adults from Pecos (see below), and eliminating arthritic individuals might have severely biased the sample in these age groups. Also, since arthritis has been mentioned as a possible contributory factor in the etiology of age-related fracturing (Buhr and Cooke, 1959), it is important that these individuals be included.

Several individuals were eliminated from the study sample because of gross pathological changes in skeletal structure. These included several cases of apparent periostitis affecting the tibial or femoral diaphyses, and a possible case of tuberculosis involving the vertebral column and proximal femur. One individual was originally placed within the study sample but was later removed after sectioning and analysis of geometric properties (chapter 4). It was originally thought that the changes in bone structure in the tibia and femur of this individual were simply manifestations of extreme arthritis, however, as explained more fully in the section on pathologies below, the degree and patterning of skeletal distortion was such that it could not be considered to fall within the "normal" range of variation, and so was excluded.

With the criteria described above in mind--presence of intact femora and tibiae of both sides, appropriate sex and age indicators, adult--the entire Pecos Skeletal Collection housed in the Harvard

Peabody Museum ($n > 900$) was surveyed. During the survey, preliminary estimates of age and sex were made for all individuals meeting the criteria. Two hundred and nine individuals were identified as possible subjects for study. Of these, 12 had to be eliminated because they had been recovered from Glaze VI or Modern periods at Pecos (see above),[2] leaving just under 200 individuals from which to select the study sample. An additional sample of about 30 juveniles, aged 6 to 19 years, was also identifed at this time for later use in "calibrating" the Miles method of aging (see below).

It was apparent after reviewing the initial sample that only 40 to 50 "old" adults (over 40 years) were available for study. Desiring to obtain as even an adult age distribution as possible, and given the limits of accuracy of skeletal aging techniques (see below) a "target" final sample of 120 individuals,[3] 10 males and 10 females from each of the following six age groups--20-24, 25-29, 30-34, 35-39, 40-49, and 50+ years--was decided upon. In addition, as discussed above, equal numbers of right and left lower limb bones (each individual represented by a femur and tibia of the same side) in each sex/age

---

[2]Period designations for all individuals could not be obtained from Hooton's monograph or the Peabody Museum grave cards. However, since only 4% of all stratigraphically placed burials were recovered from post-Glaze V periods (Hooton, 1930: 338, Fig. XI-3), the probability of inadvertantly including a Glaze VI or Modern individual in the study sample is quite small.

[3]As discussed above, one individual was later ommited from the study leaving a final study sample of 119 individuals.

group were needed. Fortunately, individuals in the initial survey sample were found to fulfill all of these requirements. The final study sample is large enough to permit sex and age comparisons of various kinds while still maintaining adequate sample sizes within each subgrouping. As discussed in chapter 2, the "normal" range of variation in long bone cross-sectional geometry even within similar sex and age groupings is large; because of this, a sample size of at least 10 individuals in each sex/age treatment cell was considered necessary if clear sex and age differences were to be delineated.

Sexing Techniques

The primary skeletal area used in sexing the Pecos study sample, as in all skeletal samples where adequate preservation allows its use, was the pelvis. Several morphological features of the pelvis were examined; the most important of these were the sciatic notch angle and the ischium-pubis index (Washburn, 1948; Krogman, 1962). As previously mentioned, all specimens included in the study had enough of the relevant areas of an innominate preserved to derive at least these two indices. As a test of the reliability of association of these two indicators, a bivariate plot of sciatic notch angle and ischium-pubis index, similar to that illustrated in Krogman (op. cit.: 141, fig. 40), was constructed using the study sample data. As predicted, the points formed two fairly distinct clusters (males and females), separated by a diagonal. There was some difference in the average values obtained for the ischium-pubis index with those of the published graph (obtained from a study by Hanna and Washburn, 1953, of Eskimo pelves), mine being somewhat higher--this was almost certainly due to differences in measuring technique (locating the exact location of the junction of the ischium and pubis in the acetabulum is difficult). What is more important, however, is consistency in locating landmarks and taking measurements, so that while average values may vary from observer to observer, within-observer sex discrimination should be constant. Using just these two indicators, I felt confident in sexing the great majority of the study sample, although other measurements and observations were also made whenever possible.

Other pelvic sex indicators found to be useful were the subpubic angle (especially when both innominates had been preserved), shape of the subpubic cavity (Bass, 1971: 157-159), breadth of the medial aspect of the ischiopubic ramus (*ibid*.), and size of the acetabulum relative to the innominate as a whole. I specifically avoided using such features as overall size of the pelvis or robusticity of muscle markings, since these may be related to physical activity levels and could have biased the study sample (e.g., only males are "robust"). On the basis of the pelvis alone, Krogman (1962) estimated that 95% of all skeletons could be sexed correctly.

The skull was used a secondary sex indicator, always examined when adequately preserved, but never outweighing or "overruling" the pelvis as to sex determination. While a number of features were examined, e.g., relative size of mastoids and nuchal crests, roundedness of orbit margins, slope of forehead, and development of chin, the only feature which seemed to be really well associated with the pelvic determinations was development of the supraorbital tori (Krogman, 1962: 115). In a few individuals in the study sample this feature helped decide sex when pelvic indicators were equivocal. Size or robusticity of long bones or other areas of the skeleton were not used in making sex determinations, for the reasons given above for the pelvis (possible sampling bias). No individual in the initial survey sample of about 200 had to be eliminated from consideration because of inability to determine sex.

Hooton claimed to have had considerable trouble in sexing the Pecos skeletal material:

> The Pecos skeletal material offered more difficulty in the determination of sex than any other series I have studied. Of course a large number of the skeletons are clearly defined as to sex, whether male or female, but the doubtful specimens are extraordinarily puzzling and quite numerous. (Hooton, 1930: 26)

(In spite of these problems, it should be noted that "only about 1.5 per cent" of the sample was rejected by Hooton because of inability to determine sex--*ibid.*: 18.)

It is interesting to examine why Hooton had such problems in sexing the Pecos skeletal remains, since, with all due modesty, I found really very few individuals of equivocal sex (this view is shared by Dr. Trinkaus). First, Hooton used some unreliable visual characteristics to guide him in his sex assessments. For example, the presence of a preauricular sulcus on the ilium was given some weight in sex assessment; however, this trait is not a foolproof indicator of sex, being more frequent in, but by no means limited to females (Krogman, 1962: 129; and my own personal observations). Second, Hooton relied *only* on visual nonmetric characteristics in sexing, scoring even potentially metric traits like the subpubic angle as discreet subjective categories rather than measuring them (Hooton, 1930: 29, Table II-9). The emphasis since Hooton's time has been on developing more objective, quantifiable techniques of sex determination (e.g., Krogman, op. cit.: chapter 5), the ischium-pubic index being a case in point; these newer metric techniques provide a finer discrimination between male and female categories, especially when several are used in conjunction with purely visual nonmetric traits.

An additional factor leading to problems in the determination of sex of some individuals may have been that Hooton was unprepared for the degree of female robusticity in the Pecos sample, especially as compared to modern samples. For example, at one point he complains "We do not...expect males to have broad, low, feminine-looking pelves. Yet not infrequently in the Pecos series skeletons with exaggeratedly masculine crania and long bones have features which one expects to see only in the female pelvis" (Hooton, 1930: 27). It must be remembered that this was one of the first large-scale systematic studies of its kind, and that little comparative archaeological material was as yet available (see also the discussion of pathologies below). Finally, as a general rule, Hooton may have placed too much emphasis on skeletal features other than the pelvis, particularly the skull, in making sex determinations. This emphasis probably in part reflects the then current great interest in cranial features in general, e.g., in developing racial typologies. Cranial characters are not, however, as accurate as pelvic morphology in assessing sex (Krogman, 1962: chapter 5); this is especially true in the Pecos sample, where almost universal cranial deformation due to cradle boarding makes some cranial sex indicators more difficult to apply (my personal experience).

In order to test the correspondence of my sex estimates with previous assessments of the Pecos sample, after completing the sexing and aging of the study sample I compared the individual assignments with those listed in the Peabody Museum grave cards for Pecos. Of the 120 individuals in the study sample, previous sex assignments could be found for only 101. The results of this comparison are shown in Table 1. It should be noted that this comparison is based on Todd's sex estimates for Pecos, not Hooton's. Todd visited the Peabody Museum in 1927 and independently resexed and aged 587 individuals in the collection (Hooton, 1930: 18). Since Todd's age estimates for Pecos published in Hooton's monograph (ibid.: 19, Fig. II-1) have been used extensively by other authors for demographic comparisons, I used them in my comparisons of present and previous age estimates. To be consistent, I also used Todd's individual sex assignments for comparison with my data. Using Hooton's individual sex assessments yield almost precisely the same results, which is not surprising since overall Hooton and Todd's sex assignments differed by less than one half of one percent (ibid.: 30, Table II-11).

Table 1 shows that sex assignments using present and previous estimates are significantly different ($\chi^2 = 3.92$, $p < .05$). Using the previous estimates, a 61/40 male/female ratio for the 101 individuals compared is obtained; using my estimates, this same sample yields a 51/50 male/female ratio. Nearly one-quarter of the individuals sexed by me as female were originally sexed as male. This difference is probably the result of both lack of

Table 1

Comparison of Previous and Present Sex Assignments for the Pecos Study Sample

Present Sex Assignment	n	Previous Sex Assignment[1] Male	Female
Male	51	49	2
Female	50	12	38
Totals	101	61	40

[1] Hooton, 1930: Todd's estimates (see text).

experience with robust female skeletons at the time of the earlier study (see quote above), as well as the noted tendency among past investigators to assign male sex to individuals showing moderate or intermediate expression of sex-related traits (Weiss, 1972). Hooton himself noted the possibility that some females in the sample were being sexed as males (Hooton, 1930: 31). The 60/40 male/female adult ratio obtained by Hooton and Todd for the Pecos sample as a whole (*ibid*.: 16, 30) is among the most unbalanced of any large skeletal sample ever reported (Weiss, 1972: 241, Table 1). Since other more recent analyses of archaeological material (Ubelaker, 1974; Lovejoy et al., 1977; Owsley and Bass, 1979) have all obtained close to a 50/50 adult sex ratio, it seems reasonable that the new sex estimates for the study sample, with more skeletons sexed as female, are more accurate than the original sex assessments. This also suggests that the original published figures for the Pecos sample should be revised to reflect more equal sex proportions.

Aging Techniques

General issues.

As stated above, each individual in the study sample was placed into one of six age groups: 20-24, 25-29, 30-34, 35-39, 40-49, and 50+ years. Becuase the age estimates for the sample form such a critical part of the study, a few preliminary comments regarding the applicability and accuracy of the aging techniques employed are in order.

First it should be noted that all ages derived from skeletal remains are only approximations of true chronological age. Skeletal age changes indicate biological age only, and not (directly) chronological age at death. Also, the aging techniques used here were derived principally from modern U.S. White and Black samples, thus their applicability to a Native American sample living several hundred years ago requires some extrapolation. It must be assumed that the same approximate time schedule of biological (skeletal) events during life holds true for all recent human populations if skeletal aging methods are to be considered universally applicable. This may be a fairly safe assumption to make however, at least within certain limits, since similar skeletal changes in childhood and adolescence have been documented in many different population samples (Krogman, 1962: chapter 2), and at least one adult aging technique--use of the pubic symphysis--has been shown to be applicable not only to American samples (Todd, 1920, 1921; McKern and Stewart, 1957), but to Hungarians (Acsadi and Nemeskeri, 1970), and Japanese (Hanihara, 1952). Also, one of the aging methods used in the present study--the Miles method of functional dental wear--is derived on a population-specific basis, using dental wear in juveniles of the same sample for "calibration" of the aging standards. Still, it must be expected that individuals will vary somewhat from the "ideal" or "normal" pattern of biological aging, and that some overlap between age groups will occur, particularly among older adults, where idiosyncratic age effects have had more time to accumulate. This is another reason why adequate numbers of individuals in each sex/age group are essential: if enough subjects are included, even if total age ranges overlap between groups, the mean age differences between groups should be about correct. In other words, variation in individual patterns of growth and development may occur, but if the sample is large enough, these variations will tend to cancel each other out, possibly somewhat blurring, but not distorting true age trends.

It should also be noted that even if absolute chronological age at death derived from skeletal indicators varies somewhat between populations, relative age, i.e., the sequence of indicators, almost surely stays the same. Thus, since the primary aim of the present study is to examine the aging process, that is, the changes which occur from age group to age group, slight deviations in exact chronological ages derived from skeletal aging techniques may not be too significant here.

A second point to be made concerns the importance of using multiple, rather than single age estimators. Many investigators in the field of forensic medicine have stressed this point (Todd, 1923; Brooks, 1955; Krogman, 1962; Acsadi and Nemeskeri, 1970). Acsadi and Nemeskeri found frequent misagings of 10 or more years using single age estimators, while a multiple aging technique using four different indicators yielded an average margin of error of only ±2.5 years (Acsadi and Nemeskeri, 1970: 122-135). It is particularly important to include aging techniques based on different, relatively independent areas of the skeleton, e.g., the pubic symphysis and the teeth. Also, experience with age changes in different skeletal features in the sample as a whole is important, since this helps in identifying individual anomalies not fitting into the total aging pattern. As the above authors have noted:

> The final age diagnosis should be established in a most circumspect manner, considering all available factors. When dealing with biological phenomena, an over-simplified, mechanistic employment of the methods must be avoided by all means. (*ibid*.: 135)

Certain skeletal aging techniques were specifically not used in the present study. These included changes in the cancellous bone of the proximal humerus and femur (Acsadi and Nemeskeri, 1970), and arthritic changes in the long bones. Use of these methods could have produced systematic biases and involved circular reasoning in a study of cortical bone remodeling with aging. While all skeletal age indicators are obviously functionally linked to some extent with age changes in the skeleton as a whole, the features used in the present study for aging were chosen to be as independent as possible from the object of study--structural changes in the lower limb bones.

Younger adults.

Adults 20 years and over were distinguished from juveniles on the basis of long bone epiphyseal union, union of the iliac crests and third molar eruption and wear. Pubic symphyseal age changes and the Miles method (see below) were also used for this purpose. The most important of these age indicators was epiphyseal union of the long bones, since fully grown (in length) tibiae and femora were required for the study. Epiphyseal union of all centers of the femur and tibia is complete by about 18-19 years of age (Krogman, 1962: chapter 2; McKern and Stewart, 1957). The last long bone epiphysis to unite is the proximal humerus, or humeral head: this occurs at about 20 years of age, and is therefore a useful skeletal feature in distinguishing between "under 20" and "20 years and over" age categories. Complete union of the iliac crests occurs at about 20-21 years (ibid.), thus, several of the very youngest individuals included in the sample were just completing union of this element. Third molar eruption and slight polishing were also taken as a sign of having reached at least 20 years of age, although lack of M3 eruption did not necessarily lead to elimination, since third molar eruption times can be quite irregular (Legoux, 1966) and M3 agenesis (or noneruption) was occasionally observed in the Pecos sample.

The aging of younger adults, 20 to 40 years of age, was accomplished using the pubic symphysis and Miles methods of dental aging, with epiphyseal union of the medial clavicle also helpful in the lower part of this range. The pubic symphysis aging technique was originally described by Todd (1920, 1921), although McKern and Stewart's (1957) and Gilbert and McKern's (1973) modified standards, for males and females, respectively, were used here. The method as revised depends upon the assignment of scores from 1 to 5 to each of three symphyseal components (dorsal plateau, ventral rampart, and symphyseal rim), adding the three scores to form a total score, and comparing this against the standard tables to arrive at an age (or age range). As various authors have noted (e.g., Stewart, 1957), because of the pitting and scarring of the symphyseal faces which frequently occurs, presumably as a result of childbearing, females are more difficult to age using this technique (this increased difficulty is reflected in the larger standard deviations for ages using the Gilbert and McKern female standards). In particular, distinguishing between the build-up and breakdown stages of the ventral rampart and symphyseal rim can present a problem (Gilbert and McKern, 1973 ; Suchey, 1979) (this holds true to a lesser degree for males as well). It is here that the value of a multiple aging approach is most apparent. Fucntional dental wear usually distinguishes easily between relatively older and younger adults: this frequently provides the clue necessary for interpreting an intermediate expression of a pubic symphyseal component as either "developing" or undergoing breakdown and disintegration.

The Miles method of functional dental wear (Miles, 1963) is based upon the extrapolation of tooth wear in juveniles of well-defined age to dental attrition in adults. Assuming average eruption times for the first, second, and third molars of 6, 12, and 18 years of age, respectively, one would expect about 6 years of "functional dental wear" on the M1's at the time of M2 eruption, and about 6 and 12 years of wear on the M2's and M1's, respectively at the time of M3 eruption. (Miles felt, based on his own and others' observations, that the second and third molars wear at a slightly slower rate than the M1's, in an approximate M1: M2: M3 ration of 6: 6.5: 7 years; therefore the wear standards are adjusted to reflect this fact.) By progressive extension of this principle, e.g., a 32 year-old will have M3 wear equivalent to M1 wear in an 18-year-old, ages of adults in the third, fourth, and later decades of life may theoretically be determined, although accuracy of the technique progressively declines as one moves further from the baseline wear standards established by the juveniles. Scores for each of the three molars are averaged for a combined "functional age"; other teeth may also be used, although the molars show the most consistent pattern of wear with age and are easiest to score (I used the other teeth, e.g., the premolars, as a check for the overall pattern of wear, but did not develop specific standards for them.) As stated above, a baseline sample of 30 juveniles was obtained from the Pecos Collection for "calibration" of the functional wear standards. The practical upper age limit for application of this technique in the Pecos sample was about 40 years; beyond this most individuals had lost many teeth and showed such irregular wear on those remaining that accurate functional ages were impossible to determine. In general, pubic symphysis ages and functional dental wear ages were well associated, although several cases of apparent unusually heavy or light dental wear were encountered.

As mentioned earlier, the late uniting of some skeletal epiphyses allows an additional age estimate in the early and mid-20's. The medial clavicular epiphysis begins to fuse at about 21-23 years, and completes fusion by 25-28 years (Krogman, 1962: chapter 2; McKern and Stewart, 1957). Thus a partial fusion of this epiphysis placed an individual within the 20-24 year age group, while completed fusion indicated an age of at least 25 years. Other late-uniting epiphyses completing fusion during the early 20's include the ramal epiphysis of the pelvis (about 21 years) and the heads of the ribs (20-22 years).

Using these three types of age indicators—pubic symphysis, functional dental wear, and epiphyseal union—I felt very confident in assigning individuals to the first two five-year age groups, and somewhat less confident of the 35 and 40 year "cut-off" points for the second two five-year groupings. I would estimate that age determinations in the 20's are accurate to within about 1 to 2 years, with this range of accuracy about doubled in the 30's. (Krogman (1962: 111) ventured an accuracy of about ±2 years in the third and fourth decades of life using essentially the same techniques.) Thus, the overlap between age groups increases with each subsequent age category. However, as discussed above, while the total true age ranges of adjacent age groupings hould be expected to overlap somewhat, given adequate sample sizes the mean age at death for each grouping should be approximately correct.

Older adults.

Beyond age 40 age indicators in the skeleton become less precise. Recent pubic symphyseal age standards are available for females through the sixth decade (Gilbert and McKern,1973), and Todd's original standards (1920, 1921) range through 50 years for both sexes. However, later stages in pubic symphyseal change are often difficult to distinguish from one another, due to the somewhat erratic nature of the break-down process in older adults. Todd himself noted of his standards that they were "a much more reliable age indicator from 20 years to 40 than after the latter age" (Todd, 1920: 313). Standard deviations for Gilbert and McKern's scores for older females average more than 8 years (Gilbert and McKern, 1973: 34 Table 2). Dental wear, as noted above, is also erratic and difficult to interpret in older adults, particularly in samples such as Pecos, where most teeth are lost or completely worn down by the fifth decade. Thus, additional age indicators are needed to assess age beyoond 40 years. The techniques available for this age range are not as accurate as those for younger adults, but it was felt they were adequate for placing individuals into one of two age categories: under 50 years, and 50 years and over.

Aging of older adults in the Pecos study sample was accomplished through use of endocranial suture closure, the terminal phases of pubic symphyseal age changes, and development of vertebral osteophytosis. Cranial suture closure, originally one of the most relied upon age indicators in the human skeketon (Todd and Lyon, 1924), has since fallen into some disrepute because of the wide range of variability in its age-related progress (Singer, 1953; Brooks, 1955; McKern and Stewart, 1957). Because of frequent ectocranial "lapsed union", endocranial suture closure is a more reliable age indicator (Todd and Lyon, 1924), however, even here estimates should probably only be made to the nearest decade (Krogman, 1962: 89). Acsadi and Nemeskeri (1970) recently studied endocranial suture closure in a sample of 285 skeletons of known age. Their chart (*ibid.*: 119, Table 29), along with Todd and Lyon's original observations, was used to establish the following guidelines in the present study, based on the average degree of endocranial suture closure in the three cranial vault sutures (frontal, sagittal, and occipital): a) less than 1/2 closed--no more than 49 years, most probably under 40 years; b) 1/2 to 3/4 closed--30-49 years, most probably 40-49 years; c) more than 3/4 closed--at least 40 years, most probably 50+ years. No attempt was made to score sutures other than those of the vault, and no subdivision of suture areas was made. It should also be noted that **cranial** suture closure was not referred to in making age estimates under 40 years; this is due to the very erratic and unreliable nature of the early stages of suture closure (e.g., ibid.), and also because better aging techniques were available for younger adults.

Vertebral osteophytosis (Stewart, 1958) was used more as supportive evidence for age assignments than as a specific age indicator. The guidelines here were: a) slight lipping--under 40 years; b) moderate lipping, usually localized mainly in the lumbar region--40-49 years; c) moderate to heavy lipping in most areas of the vertebral column--50+ years. As has been the experience of most other investigators working with archaeological material, I found the highest incidence of osteophytosis in the lumbar region, in contrast to modern populations, where it is more commonly found in the cervical vertebrae (Steinbock, 1976: 302). As noted earlier, arthritic changes in the long bones were specifically avoided in making age determinations, since it was felt that changes of this kind might be directly related to cortical bone remodeling with age, and would thus involve circular reasoning in the present study. For the same reason,other observations, such as lightness and texture of bones, which have been used as age indicators (e.g., Hooton, 1930: 18), were not used in the present study.

Comparison with previous estimates.

As with the new sex estimates, age assignments for the Pecos study sample were later compared with the original assignments. Todd's individual assignments, obtained from the Peabody Museum grave cards, were used in lieu of Hooton's, becuase it is Todd's ages, presented in Hooton's monograph as Figure II-1 (p. 19), which were later used by Goldstein (1953) and others for comparative purposes. The results of the present comparison are shown in Table 2 and Figure 1 (as in sex assign-

Table 2

Comparison of Previous and Present Age Estimates for the Pecos Study Sample (Sexes Combined)

Present Age Assignment		Previous Age Assignment[1]					
Age Group (yrs)	n	Age Group (yrs)					
		20-24	25-29	30-34	35-39	40-49	50+
20-24	18	10	4	1	3	---	---
25-29	17	---	2	2	5	4	4
30-34	19	---	---	1	3	13	2
35-39	13	---	---	---	4	6	3
40-49	18	---	---	---	---	5	13
50+	16	---	---	---	---	2	14
Mean Differences: Previous-Present (yrs)[2]		0	3.3	5.0	7.3	8.7	13.7

[1] Hooton, 1930: Todd's estimates.

[2] Mean differences between previous and present age estimates for each "previous estimate" age group, calculated using midpoints of age groups. Midpoint of "present estimate" 50+ year age group assumed to be 60 years; midpoints of "previous estimate" 40-49 and 50+ year age groups calculated using Todd's 5-year age groups assignments to age 80-84 years.

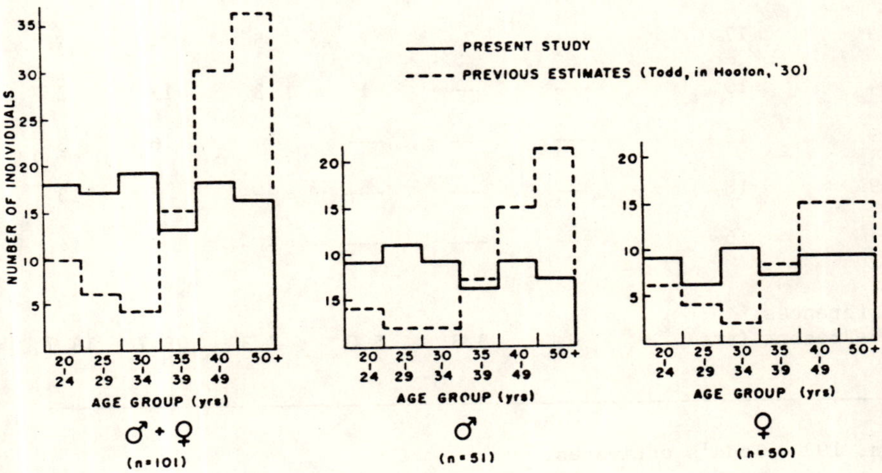

Fig. 1. Age distributions for the Pecos study sample derived from previous and present age estimates.

ments, ages for only 101 individuals were available from the previous study).

While a general correspondence between my age assignments and those of Todd is evident (Table 2), my age estimates average consistently lower over the entire adult age range. In order to quantify this difference, individuals were assigned to the midpoints of their respective age groups and the difference between average ages using present and previous assignments was calculated (bottom of Table 2). Since Todd presented his results by five-year group categories through age 80-84, to make his data comparable to mine his 40-44 and 45-49 year categories were combined, as were his categories from 50 years on (means for these larger groupings were determined on the basis of his original five-year groupings); the mean age at death for my 50+ year category was assumed to be 60 years. Average differences between previous and present age estimates increase steadily from younger to older age groups; however, relatively fewer "mismatches" occur in the first (20-24 year) and last two age groups (40+years). In other words, very young and old adults were more likely to be aged as such by both observers, while middle range adults (25-39 years) were very often aged as much older by Todd (e.g., four individuals presently aged as 25-29 were previously assigned ages of 50+ years). The comparison is broken down by sex in Figure 1, which shows the discrepancy between studies to be more marked among males than among females (sex assignments here are mine). The average difference between present and previous age estimates for all 101 individuals is 9.0 years, for females 6.2 years, for males 11.7 years.

Most of the individual discrepancies in aging are probably due to the differences in aging techniques which were applied in the two studies. Brooks (1955) has shown Todd's original pubic symphysis standards to provide too high ages on average when compared to known age. Krogman also noted "They [Todd's 1920, 1921 standards] work, but they do tend to over-age, especially in the later decades of life" (Krogman, 1962: 97). Cranial suture closure (ectocranial or endocranial not specified in Hooton's monograph) is particularly inaccurate in the younger adult age range (e.g., see Singer, 1953). Paradoxically, the more frequent occurrence of "lapsed union" in females, which leads to some underaging using this technique (Brooks, 1955), may explain why the overall previous age estimates for females were closer to my estimates (Figure 1), since this would tend to preferentially lower the age determinations of females in the sample. Another contributory factor to the 9 year average difference between the present and past age estimates is Todd's assignment of relatively precise ages to very old adults. Since cranial suture closure is essentially complete by the seventh decade (Todd, 1920, 1921; Gilbert and McKern, 1973), it is difficult to see how Todd arrived at his five-year age groupings in the seventh through ninth decades, except by purely subjective judgment. If age estimates for these very old individuals are reduced somewhat to fall only in the sixth and seventh decades, the relatively large average differences between his and my age estimates for the last two age groups (see Table 2) would be reduced by several years.

Hooton did not classify the Pecos sample into five-year age groupings as Todd did, using only three age categories: 21-35, 35-55, and

55+ years, thus, precise quantification of differences between his estimates and those of the present study is more difficult. However, Hooton and Todd were very close in their overall estimates of the age distributions for the Pecos sample (Hooton, 1930: 20-24). (It is interesting to note that Hooton, who apparently relied more on cranial suture closure than Todd, came up with a slightly higher percentage of younger adults and fewer old adults--see _ibid._: 20, Table II-3.) If my age estimates are grouped into categories similar to those Hooton used, the same pattern of previously higher ages is apparent.

Goldstein (1953) used Todd's age profile to calculate a mean age at death for the Pecos sample of 42.9 years. If the new age estimates for the study sample are applied as "correction factors" to the Pecos sample as a whole, the mean age at death for adults would have to be lowered by almost 10 years. There is another reason why this earlier figure should be revised downwards: juveniles are grossly under-represented in the age profile (15% under age 20, with only 2% of all burials 0-4 years of age--Hooton, 1930: 19, Fig. II-1). Todd himself noted that most infant burials from the site were disgarded (_ibid._: 24, 334). In a much later tabulation of demographic data for Pecos, including many burials not previously analyzed, Kidder arrived at more realistic figures for juvenile deaths (39% under age 20, 24% 0-3 years of age--Kidder, 1958: 284, Table VI). When these data are included along with the new age corrections for adults, the average age at death for the total sample falls to about 25 years. This figure accords much better with reported figures for other pre- and proto-historic archaeological samples from North America, which range from 13.2 to 22.9 years

(Johnston and Snow, 1961; Ubelaker, 1974; Lovejoy et al., 1977; Owsley and Bass, 1979).

Pathologies

All skeletons in the study sample were routinely examined for evidence of pathology. As discussed earlier, every attempt was made not to exclude possibly "anomalous" individuals from the study sample. In particular, older adults frequently manifested one or more skeletal disorders (e.g., arthritis, osteophytosis, dental abcesses): these were usually treated as simply part of the "normal" range of variation inherent in samples of this age range. However, several individuals were eliminated from the study sample because of obvious pathological distortions which could not be considered "normal". These included several cases of periostitis of the femoral or tibial diaphysis which changed the normal contour of the bone, a possible case of tuberculosis which affected the structure of the vertebral column and proximal femur (El Najjar, personal communication), and one individual with distortion of the tibial and femoral diaphyses due apparently to very severe arthritis in this limb. This last individual, a male 35-39 years of age, showed no signs of arthritis in any other limb bones or in the preserved portion of the vertebral column (lumbar region). Only the knee joint of one limb was affected, and affected in a way not seen in any of the other individuals in the Pecos sample: "thick" lipping around the distal femoral and proximal tibial contact surfaces, flattening of the femoral condyles (but not eburnation), and hollowing of the tibial plateau surfaces. As subsequent geometrical analyses demonstrated, the diaphyses of both bones were very unusual in shape, particularly the distal femur, which was extremely A-P flattened. In some geometrical properties this individual fell more than 5 standard deviations from his sex/age group mean (and 4 S.D. from the total male mean). No bowing of the shaft, lesions, healed fractures, or other evidence of pathology of traumatic injury could be detected. It might be interesting and instructive at some point in the future to do a functional analysis of this individual, trying to determine why this peculiar morphology developed (Pauwels carried out a similar functional analysis of femoral morphology in a rachitic individual, showing how the cross-sectional shape was adaptive to the distorted form of the bone--quoted in Amtmann, 1971: 10).

In my survey of the Pecos material, I was particularly interested in pathologies which might reflect dietary deficiencies, since these deficiencies could affect long bone growth and development and thus the results of the present study. Hooton conducted an extensive investigation of pathologies in the Pecos sample (1930: 118-122, 306-330). He was seemingly convinced of the overall poor health of the Pecos inhabitants, and of a general decline in health from the earliest to the latest periods of occupation:

> In general the picture of Pecos pathology from period to period is one of a struggling population apparently afflicted by the ordinary ailments of life which were neither ameliorated by a well-balanced and adequate diet nor by any effective knowledge of medicine or surgery. While the end of the pueblo's life was probably caused by epidemics and warfare, it is yet possible to trace from the skeletal remains an increased prevalence of certain diseases and a general deterioration in physique during the period of decline. (Hooton, 1930: 349)

This poor level of health Hooton apparently attributed largely to dietary factors. For example, with regard to the dentition, he concluded:

> One's impression is that the Pecos people had very poor teeth for primitive people and that the quality of the teeth seemed to deteriorate slightly but steadily throughout the archaeological periods studied.... Probably the inferior quality of these Indian's teeth is to some extent referable to the diet of maize and vegetables which must have been the rule throughout the period of occupation. Some game and fish may have been added to this diet occasionally and buffalo meat may have been an exceptional luxury. It is fairly certain from a study of the teeth that the Pecos Indians were in most cases undernourished. (*ibid*.: 119)

H.Y. Rihan (D.M.D.) concurred with this view in an appendix to Hooton's monograph dealing with the dentition, stating that the caries, loss of teeth, and pyorrhea observed in the Pecos sample were due to "the limited variety of food" (*ibid*.: 369).

Osteoporosis symmetrica, a pitting of the cranial vault bones, and cribra orbitalia, a similar condition affecting only the orbits, was said by Hooton to have been found in "a considerable number of the crania of infants and children from the Pecos collection", with traces in the orbits and porotic thickenings of the skull vault... common among the adult specimens" (*ibid*.: 318-319). Since these particular skeletal manifestations are most probably brought about in native North American populations by iron deficiency anemia, possibly compounded by protein and vitamin C deficiencies (Steinbock, 1976: 236-248), these observations, if confirmed, would also lend weight to an hypothesis of dietary insufficiency in the Pecos population.

Having presented all of the evidence to this effect, however, I would contend that Hooton's conclusions regarding the dietary standards and general health of the Pecos population are in a large part unfounded or at least questionable. First, we have Kidder's opinion, based on material remains in the refuse deposits at Pecos (and published after Hooton's monograph, when all of the site had been excavated), that malnutrition was rare and the diet reasonably varied:

> Animal and bird bones fill the rubbish heaps and show that game was of much importance. The nearby mountains are heavily forested and must at all times have provided abundance of deer, elk, and wild turkeys. Rabbits, of course, were everywhere. Antelope ranged in the more open country, the buffalo herds were only a few days foot-journey to the east. These animals supplied both meat and hides; their bones, horns and sinews were used for many purposes. (Kidder, 1932: 6)

As mentioned earlier, year-round water supplies allowed the cultivation of irrigated crops, and large stores of grain were apparently kept as a safeguard against food shortages. Thus, malnutrition and dietary deficiencies may not have been as marked as Hooton believed among the Pecos inhabitants.

Second, despite Hooton's claims that pathologies increased through time at Pecos due to environmental stress, I see very little evidence for this in his published tables. As was the case with gross body dimensions (see earlier discussion above), neither dental pathologies (Hooton, 1930: 120, Table IV-48) or postcranial pathologies (<u>ibid</u>.: 307-309, Tables X-1 through X-5) show any consistent temporal trends in incidence. Again, sample sizes here are very small with respect to the occurrence of particular skeletal disorders (e.g., absence of third molars, noted as undergoing a temporal trend at Pecos--ibid.: 118--varies in incidence within sex/Glaze period groupings from 2 to 7 individuals), and any period differences could easily be accounted for by sampling error.

As for osteoporotic pitting of the cranial vault or orbits, or thickening of the cranial vault, I observed no definite instances of this in my survey of the Pecos Collection, at least among adults (juveniles were not examined as closely). El-Najjar, who surveyed the Pecos sample during a visit to the Peabody Museum while the present study was in progress, also found no manifestations of this kind in adults, and only a few relatively minor examples among the children of the sample (El-Najjar, personal communication). Hooton himself does not provide good evidence that the pathology was "prevalent" among either juveniles or adults at Pecos: in his table of individual pathologies (pp. 320-330), he lists only 12 cases of osteoporosis symmetrica or cribra orbitalia, 8 of them juveniles, out of a sample of 503. Later he remarked "Osteoporosis symmetrica, probably a deficiency disease, occurred in a number of immature specimens, but it seems to have been much less common than among the early Mayas and Peruvians" (ibid.: 348). His one published photograph of the disease in an individual from the Pecos sample, a portion of the parietals in a 2-3 year old child (Plate X-2, p. 310), shows only a relatively minor manifestation of the condition, especially when compared to other examples illustrated in Hooton's monograph which were obtained from other sites (e.g., Plates X-9, 10, p. 316).

In a reanalysis of the Pecos dentition, Nelson (1938) presented evidence that the relatively high proportion of individuals showing dental abscesses was due primarily to mechanical stresses (attrition), rather than general malnutrition or specific dietary deficiencies, as Hooton and Rihan had believed. He also pointed out the difficulty in diagnosing periodontal disease (which Rihan had found in 72% of the individuals examined) and in distinguishing "what is pathological and what is not" (a dentist visiting the Peabody Museum during the course of the present study also has this problem). As for caries incidence, Nelson concludes:

> Despite the fact that approximately one-half the Pecos skulls are affected by caries, the actual number of teeth involved in each dentition is not great. The extent of this disease among the Indians was much less than it is among modern civilized races. (*ibid*.: 292)

In my observations on the Pecos study sample, caries were fairly common, but as Nelson observed, did not generally involve many teeth in each individual. Part of this may have been due to rapid dental attrition, which may have worn away many caries--this is supported by the fact that I found fewer instances of caries among older adults. I found definite instances of dental abscesses in 18 of the 120 individuals in the study sample, with 10 of these in the two 40+ year age groups (N=40). Alveolar resorption was commonly observed, but again, as Nelson noted, it is difficult to say whether this constitutes a true "pathology" (periodontal disease, or pyorrhea) in many cases, and in any event may be brought about by factors other than a faulty diet, e.g., attrition, lack of dental hygiene, and calculus (which was observed in some of the Pecos individuals in the study sample (Brothwell, 1972: 149)).

There are several reasons why Hooton may have exaggerated or been unduly impressed with the pathologies present at Pecos. It should be remembered that his monograph, a pioneering effort in the field, was written during the early years in the development of paleopathological analysis; thus, there were few other skeletal samples available for comparative purposes. For example, both Hooton and Todd remarked upon the relatively "low" age at death among adults at Pecos (Hooton, 1930: 18-25), yet their original age estimates, which probably should be modified as discussed above, yield an average age at death some 20 years higher than any other subsequently reported archaeological sample from North America! The number of pathologies observed in the Pecos sample, particularly in the dentitions, may have seemed excessive to someone with only modern American autopsy material for comparison. As another example, Rihan reported "marked attrition of the occlusal surfaces of the teeth" in over 97% of the Pecos sample (ibid.: 369), and Hooton graded a majority of the sample as showing "pronounced" or "very pronounced" dental wear (ibid.: 118-119). However, while dental attrition in the Pecos sample is certainly very marked when compared with that found in modern industrialized populations, it is my personal experience that it is by no means unusual when compared with other prehistoric Native American groups (for example, Central Californian populations) (Ruff and Jones, 1981).

Hooton may also have been unduly influenced by the graphic accounts of war and pestilence during the final years of Pecos' existence, perhaps neglecting the fact that fully three-quarters of the total period of occupation, and the vast majority of burials, occurred prior to the period of extreme social and economic disruption which began in the late 1600's. Kidder believed, in fact, that the general prosperity of the pueblo was not seriously disrupted until even later, about the middle of the eighteenth century (Kidder, 1932: 4).

From an overall perspective, then, considering both archaeological and osteological evidence, there appears to be no reason to believe that the population inhabiting Pecos Pueblo, prior to the last 100 or 150 years of its occupation, suffered from serious malnutrition of any kind, or an unusually high incidence of pathologies in general. My own personal impression of the Pecos skeletal sample was of a generally healthy population, with the usual age-related pathologies, such as arthritis, and a few cases of more specific disorders such as periostitis.

As mentioned earlier, osteoarthritis of some degree was very common among older adults at Pecos, in its most extreme from leading to eburnation, or a polishing and pitting of the joint surface due to complete breakdown of the articular cartilage and direct contact between adjoining bone surfaces. Four cases of eburnation were observed, three involving the elbow joint; all occurred in males aged as 50 years or over. Jurmain (1977) studied the incidence of arthritis in several skeletal samples, including Pecos, Alaskan Eskimos, and modern Black and White Americans (Terry Collection). Somewhat surprisingly, Pecos showed the lowest overall incidence of arthritis, including the two modern samples. Also, the Pecos sample showed the latest age of onset of moderate or severe arthritis (Jurmain used Hooton's and Todd's original age estimates, "spot-checked for consistency"--Jurmain, 1975: 162). He rejected mechanical factors (lower activity levels) as an explanation for these findings, preferring to attribute the lower incidence of arthritis in the Pecos sample to genetic factors "leading to a lower susceptibility to all degenerative stimuli" (Jurmain, 1977: 363). On the basis of the present study results it is difficult to evaluate this hypothesis, however, my observations indicate the presence of at least moderate and often marked arthritic changes in almost all older adults in the study sample, so it would appear that osteoarthritis was not a particularly rare occurence among the Pecos inhabitants. The original age estimates for Pecos relied upon by Jurmain may have misled him in his analysis of age trends in this sample to some extent; using the new age estimates presented here would have the effect of lowering his ages at onset somewhat, making them more comparable to those observed in his other study samples.

Partial or total vertebral collapse or "wedging" of one or more vertebrae was observed in 10 individuals in the study sample, all aged as 40 years or over, with the most extreme cases occurring in the 50+ year age group (no sex difference in incidence). These symptoms have frequently been used as clinical indications of osteoporosis in the living (see chapter 2). Partial or complete ankylosis of the sacrum and pelvis was observed in three individuals, all males, one 35-39 years, the other two 50+ years of age. One case of sacral spina bifida was noted in a 50+ year female. A humeral head fracture, partial collapse, and refusion was observed in a 50+ year male. In this same individual (originally numbered 60242, illustrated in Hooton, 1930:

307, Plate X-1,c) perforations in the pelvis are present, apparently not the result of infection, which Hooton interpreted as evidence of a healed fracture of the ilium and acetabulum (*ibid*.). It is worth noting that both these types of fractures--proximal humeral and pelvic--have also been reported as age-related fractures (chapter 2). (Hooton also noted that postcranial fractures were common in older adults--*ibid*.: 312-313.)

Several other localized pathologies were observed in the study sample. Periostitis of the distal radius and ulna were present in a 50+ year old female (also recall that several individuals were eliminated from the study sample because of periostitis of the femur or tibia). "Button" osteomas, small benign tumors found on the external surfaces of the cranial bones (Steinbock, 1976: 327-328), were found in two individuals in the study sample, females 30-34 and 40-49 years of age. As noted earlier, one individual was eliminated from consideration because of possible tuberculosis (although it is difficult to rule out other diseases, e.g., coccidioidomycosis, as causing the skeletal features observed in this individual--Steinbock, 1976: 170-188). No dental hypoplastic lines were observed in the study sample, although it is possible that a closer examination, including sectioning of teeth, might show some. Since radiographs of all tibiae and femora included in the study were taken before sectioning (see chapter 4), it will be possible to examine these at a later date for evidence of tranverse lines of growth arrest, which have been used as an index of health in prehistoric populations (Wells, 1967).

The Study Sample as a Population Sample

Before proceeding to the description of techniques used to analyze structural properties in the study sample, several assumptions underlying the use of such a sample in an investigation of this kind need to be justified. First, there is the question of how well my specific sample represents the adult segment of the population at Pecos Pueblo. Next, there is the issue of how well skeletal samples in general approximate true "population" samples, in the biological (evolutionary) sense. Finally, there are the potential problems or limitations common to all cross-sectional design studies of "growth" where longitudinal data are not available.

As noted earlier, the excavators of Pecos Pueblo felt that only about 12-15% of the site had been excavated, leaving a large proportion of the cultural (and skeletal) remains untouched. This fact in itself could constitute a problem in selecting a random sample of the Pecos population, if either nonrandom burial patterns (e.g., with respect to social class or age) or nonrandom excavation procedures, or both, were practised. Also, since much of the most poorly preserved skeletal material was disgarded and reburied after excavation, it is possible that some post-excavation selection (e.g., for the most robust individuals) took place. In addition, my sampling procedure, which required intact lower limb bones of both sides, as well as an intact pelvis, could have biased the selection of the study sample in terms of the structural properties I was investigating.

Fortunately, Kidder, being one of the preeminent archaeologists his generation, and also because of his keen interest in documenting cultural (pottery) sequences through time in this region, was particularly careful to sample all areas and levels of the Pecos site:

> Pecos is so extensive that the six field seasons spent upon it sufficed to clear only about twelve to fifteen percent of its total area. While the collections, therefore, fall far short of comprising the whole relic content of the site, care was taken to sample all major deposits, and considerable sections were dug completely...Excavations were made in all parts of the rubbish, ruined houses of each period were cleared,as well as a number of subterranean ceremonial chambers or kivas. Nearly two thousand graves were encountered. Specimens emanated from every part of the diggings...(Kidder, 1932: 4)

> Our digging was habitually carried down to bedrock or barren clay...Thus, in by far the greater part of our operations no graves were missed; none at all in such work as was accomplished on the North and West Terraces or in the midden trenches...we took special care to examine all burials below or in rooms. (Kidder, 1958: 280)

In addition to the above, it should be noted that, as mentioned earlier, no class distinctions of any kind were indicated by the burial remains of the more than 1900 graves excavated. Also, no particular association of burials by sex or age was reported for any excavated area of the site. It thus seems unlikely that any serious nonrandom sampling of the total skeketal sample present at Pecos took place, since the burials themselves seem undifferentiated and randomly distributed throughout the site, and the excavators were careful to sample all sections of the site.

As for the disgarding of skeletons recovered in poor condition, this may somewhat bias the Pecos skeletal sample in favor of more robust individuals. Hooton believed that both immature and very old individuals, as well as those "who have died of wasting and chronic diseases" probably disintegrated faster, as did females in comparison to males (Hooton, 1930: 15). It should be noted, however, that many otherwise well-preserved, sound younger adult skeletons had to be eliminated from the present study sample either because a pubic symphysis was missing or one of the femora or tibiae had been damaged through post-mortem crushing (or the least well-preserved side--the one to be removed for sectioning--was not the same for the tibia and femur, and thus could not be used). Conversely, several extremely fragile skeletons of older adults were, fortunately, possessed of all necessary skeletal criteria to be included in the study sample even though the skeleton as a whole was not in good condition. In general, the effects of differential site preservation on sample selection would seem to be less of a problem than other factors such as selective survival.

The use of skeletal samples as "population" samples has come under some criticism in recent years (e.g., Cadien et al., 1974). Because skeletal collections usually span much more than a single generation, they cannot be said to represent evolutionary units in the same sense that living contemporaneous populations do (leaving aside for the moment the problems in defining a contemporary population, or gene pool). In particular, changes through time in physical characteristics may be impossible to detect; therefore, two skeletal samples from different sites may appear very similar in both the mean and range of morphological variability present, yet represent populations undergoing completely opposite evolutionary or secular trends.

Fortunately, due to the careful excavation of Pecos, the depth of the deposits, and the pottery seriation technique which can be used to provide relative chronological positions to most of the burials, the possibility of temporal trends in physical characteristics is directly testable at this site. Excluding the Modern period (from which very few burials were recovered), the length of occupation of the site can be placed at about 400 years (1300-1700 A.D,--see above, also Hooton, 1930: 333). Using Hooton's four pottery-defined groupings and assuming an equal length of time for each grouping, this leaves about 100 years, or three to five generations on average per grouping. Thus, by comparing his four subdivisions of the Pecos skeletal sample, Hooton was actually comparing samples from relatively short, circumscribed time periods, longer than a single generation, but not spanning many generations. The fact that he was unable to show any significant differences between these groups for the vast majority of physical characteristics examined, as discussed above at length, is good evidence that no important temporal changes in morphology occurred during the occupation of Pecos Pueblo. In addition, no in-migration of foreign populations from outside the Pecos Valley is indicated by either skeletal or cultural remains at the site. In fact, Pecos Pueblo was one of the few habitations in the region to maintain its integrity beyond the fourteenth century (Kelley, 1952), and may have been the largest single settlement of the American Southwest during the fifteenth through seventeenth centuries (Kidder, 1958: 313). Thus, several of the potential pitfalls in extrapolating from a skeletal sample to a true population sample may be avoided in the Pecos sample, which apparently is derived from a stable, homogeneous group, relatively unchanged through time (prior to the late 1600's), and closed to foreign immigration.

The potential problems in using a cross-sectional sample to investigate longitudinal "growth" are somewhat more difficult to deal with. One problem--secular trends which may produce spurious "age differences" (e.g., Trotter et al., 1968--see chapter 2)--is apparently not an issue at Pecos, since no temporal changes in morphology were noted. Another factor, however--selective survival--cannot be ruled out as an influence on the results of the present study. As Hooton noted, in the great majority of cases it is not possible to determine the cause of death from skeletal remains alone (1930: 15). It is entirely plausible that certain morphological characteristics, e.g., size or robusticity, might influence the probability of survival in certain situations, e.g., enemy attacks, disease outbreaks, etc. The effects of such factors on cross-sectionally measured age differences may be small relative to those resulting from true ontogenetic age changes, however. Given the wide range of factors responsible for fatalities, and the probability that different causes of death are associated with different body builds or sizes, the correlation between age and a particular morphological type is probably low. It is encouraging in this regard that several studies of osteoporosis in archaeologic material have obtained results similar to those of longitudinal studies of the living (chapter 2). At worst, selective survival may partially obscure true longitudinal age changes, however, the general outline of these longitudinal changes should still be visible.

Summary

1. Pecos Pueblo, a large archaeological site in north-central New Mexico, has been dated to about 1300-1823 A.D. The pottery seriation technique developed by Kidder allows relative chronological placement of burials within the site. Almost 2000 burials were recovered, although only a fraction of these are in good condition and available for study.
2. A past extensive investigation of the Pecos skeletal remains (Hooton, 1930) failed to find any evidence of significant change in morphology through time at the site. There is also no evidence of any migrations of foreign populations into the area prior to the Spanish Colonial period, beginning about 1600.
3. Only individuals from pottery-defined periods before the late 1600's were used in the present study. Tibiae and femora were selected from individuals with both sides preserved, removing the least well-preserved side for sectioning. The final sample consisted of 120 adults, equally divided between males and females and six age groups, with equal numbers of right and left sides represented in each sex/age grouping (one individual was later removed due to pathology, leaving a final study sample of 119).
4. Sexing was accomplished primarily through visual observation and metric evaluation of the pelvis. Aging of younger adults (20-40 years) depended on changes in the pubic symphysis and functional dental wear (Miles method). Older adults were aged using endocranial suture closure, the terminal phases of pubic symphysis change, and degree of vertebral osteophytosis.

5. Despite Hooton's claims to the contrary, based on my own observations of the skeletal material and the archaeological and ethnohistoric record of Pecos, the diet and health standards of the Pecos population were probably fairly good. The Pecos people were subject to the usual array of pathological disorders, but do not appear to have suffered from any serious malnutrition or an unusually high incidence of disease.
6. Most of the requirements for extrapolating from a skeletal sample to a true population sample are well met by the Pecos remains. The only potential problem in using the Pecos sample to study "growth" changes--selective survival--may not be too serious, since true ontogenetic longitudinal changes are probably more important in determining age differences in morphology.

Chapter 4

Analytic Technique

Preliminary Whole Bone Measurements

Although the chief aim of the present study is to document the structural properties of a series of 11 femoral and tibial cross sections, defined below, many measurements were also taken of the "whole bones" prior to sectioning and cross-sectional analysis. These measurements were of two general types: a) measurements designed to establish a set of standardized reference axes allowing each cross section to be located in three-dimensional space after sectioning, and b) measurements of other "whole-bone" structural features of interest, for example, total bone length, femoral neck angles, and centers of articulation and muscle insertions.

Reference axes.

Standardized x, y, z reference axes have not always been clearly defined in past studies of skeletal structure and function. Because of the sensitivity of many of the structural properties examined in the present study to even very small changes in the location or positioning of the cross sections, precise and consistent definition of reference axes is vital. There are at least four considerations in establishing any set of anatomical reference axes. First, landmarks by which axes are defined should be easily recognizable, able to be precisely located, and available on every specimen. Second, these landmarks, whenever possible, should be *homologous* points, that is, points which in every specimen refer to a common structural-developmental feature (e.g., the point of origin or insertion of a muscle, articulation with another bone, or position of an anatomical feature such as a nutrient foramen). Third, reference landmarks should be as "unbiased" as possible, in other words, not based on structural features which themselves may vary with age, sex, or other factors under investigation. For example, if a longitudinal axis of the femur were defined as passing through the middle of the femoral head, and the femoral neck angle were to change with age, altering the relative position of the head, spurious "age changes" in structural properties of diaphyseal cross sections could be measured. This factor was especially important with respect to the axis along which "bone length" was measured, since in the present study cross section locations were determined as a percentage of bone length; therefore, some care was taken to obtain as "unbiased" an estimate of bone length along this axis as possible. Finally, reference axes should be defined, as far as practicable, to correspond with, or at least be referable to, those previously used by other authors. This allows not only precise comparisons between similar kinds of structural data collected by different investigators, but also the use of other anatomical and physiological data, such as position and direction of *in vivo* muscle forces, which have been reported in reference to a particular set of axes (e.g., Paul, 1967; Morrison. 1970).

In the present study, reference axes were defined insofar as possible with respect to the centers of the articular surfaces at the bone ends. This follows from the general approach to lower limb structure used in this study, in which the femur and tibia are modeled as beams, articulated at their ends to form the two major skeletal links of the lower limb. In addition to this, reference axes need to be defined in such a way as to allow x, y, z coordinates of any point on the bone surfaces to be easily determined.

Accordingly, the following general procedure was carried out for each specimen. After cleaning and making any repairs necessary, each bone was placed, dorsal side down, on a plexiglass board covered with millimeter graph paper upon which a set of perpendicular axes had been drawn (see Figure 4, p. 98). After various adjustments to bone position, described in the following paragraphs, the measuring board surface became a plane parallel to the frontal plane of the bone, with mediolateral and proximal-distal coordinates obtained directly from the graph paper on the board, and anteroposterior coordinates measured as vertical heights above this surface with the help of a small triangular wedge covered with graph paper (Figure 4). The longitudinal (proximal-distal) axis is assigned the z axis, the mediolateral the x axis, and the anteroposterior the y axis. The positive z sense is proximal, the positive y anterior, and the positive x lateral (see Figures 2 and 3) (cross sections from left limb bones were photographically reversed--made mirror images of--during analysis to maintain a constant "right hand rule" and the same positive x axis sense). Thus, the xz plane is the frontal plane, the yz the sagittal plane, and the xy the transverse plane.

The exact method of positioning the tibia with respect to reference axes is illustrated in Figure 2. The centers of the talar articulation surface and the two tibial plateau articulation surfaces were determined

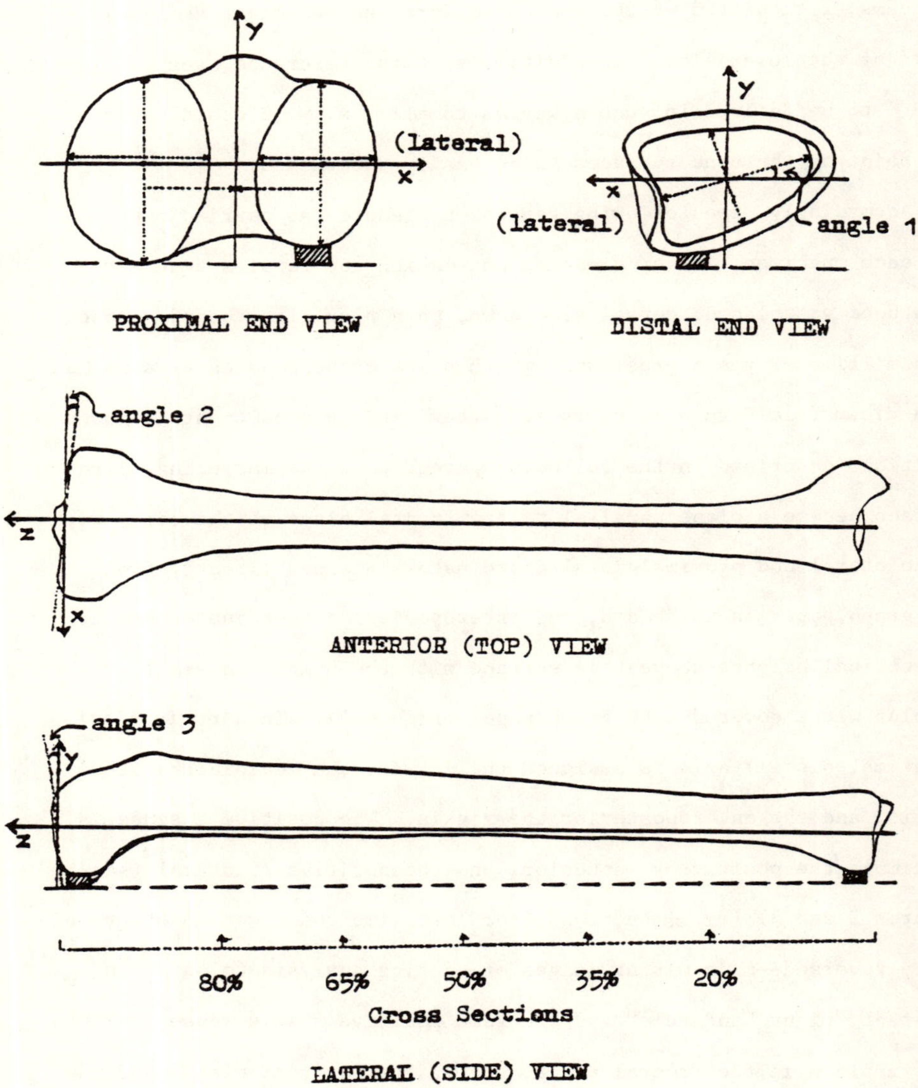

Fig. 2. Location of references axes, angles, and cross sections in the tibia (right side illustrated). See text and Table 3 for explanation of features. (Note: proximal and distal end views not drawn to same scale--x (and z) axes equidistant above measuring board surface at both ends.)

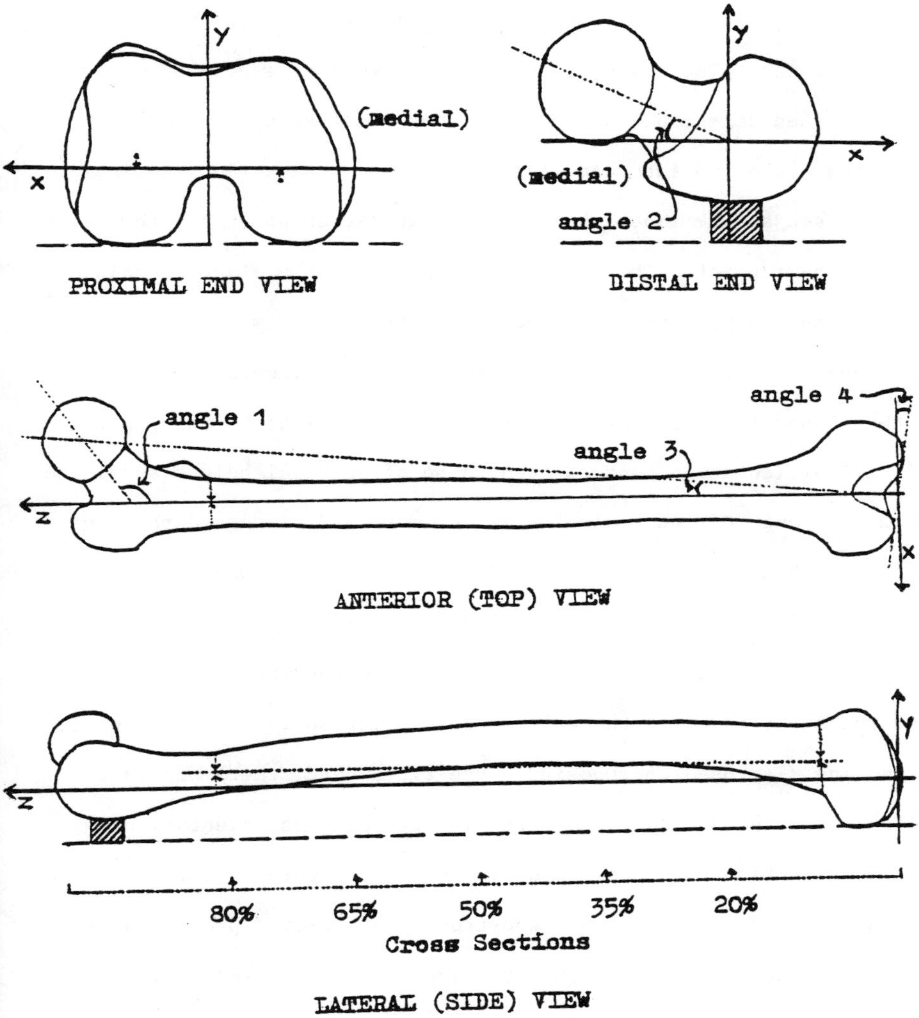

Fig. 3. Location of references axes, angles, and cross sections in the femur (right side illustrated). See text and Table 3 for explanation of features. (Note: proximal and distal end views not drawn to same scale--x (and z) axes equidistant above measuring board surfaces at both ends.)

using a pair of sliding calipers on a millimeter rule, and were marked in indelible ink. The midpoint between the two tibial plateau center points was then determined and marked (this point always fell in the notch between the two intercondylar tubercles). The z (longitudinal) axis was taken as the line intersecting this latter point and the talar articulation midpoint. The xz (frontal) plane is the plane containing the talar and two plateau center points, with the yz (sagittal) plane perpendicular to this plane and containing the longitudinal axis. To make the plane of the measuring board surface parallel to the frontal plane, all articulation centers need to be of an equal height above the surface. Because of the unequal size and more A-P elongated shape of the medial tibial plateau, this necessitated placing a small piece of modeling clay under (on the distal aspect of) the lateral plateau. Similarly, a small piece of clay was usually needed under the distal end of the tibia to bring the talar center point up to the same level. All modeling clay used to reposition the tibia (and femur) was left adhering to the bone surface, maintaining proper axes orientation, until after sectioning. Because of the concave nature of some of these surfaces, I found a simple tool, constructed of a metal probe mounted and able to slide up and down on a vertical support marked in millimeters, to be useful in determining heights of articulation centers. The probe itself was marked in millimeters and could be slid back and forth parallel to the measuring board surface--this was useful in obtaining 3-dimensional coordinates on concave surfaces (for example, in measuring bone length of the tibia, as described below).

Using the tibial plateau articulation centers to orient the mediolateral (and consequently, anteroposterior) axes of the tibia follows the procedure employed by Morrison (1970), Kimura (1974), and Minns *et al*. (1977), but differs from that of Lovejoy et al (1976) and Miller and Purkey (1980). These latter authors used the distal rather than the proximal end of the tibia to establish reference axes, fitting a cylinder into the concave talar articulation surface and using this to orient the M-L (coronal) axis. The proximal end was used in the present study for two reasons: a) Morrison's studies of physiological loadings of the tibia are the best available, and since future stress analyses of the tibia under simulated physiological loadings are planned, it seemed advisable to use his system of orientation; b) using the knee joint allows approximately the same reference points to be used in establishing femoral coordinate axes (since the tibial plateaus and femoral condyles articulate), thus, relative orientation of M-L and A-P axes in the two bones will be the same. To allow comparisons with other studies, however, the angle between the mediolateral axis as determined by the plateau centers and as determined by the talar surface orientation (the traditional osteometric tibial "torsion" angle) was carefully measured in each specimen, as described below (see Figure 2, angle 1).

Reference axes for the femur are somewhat more difficult to determine precisely and consistently from specimen to specimen. This is due to the complex shape of the proximal femur, which has no easily definable "center" of articulation, and the curvature of the femoral diaphysis in the sagittal plane. For the purposes of the present investigation, the femur was subdivided into two parts: the diaphysis, and the femoral head and neck. The longitudinal axis of the femoral head and neck--the cervical axis (Backman, 1957), runs through the middle of the femoral neck and intersects the center of the femoral head. Because of the anterior curvature of the femoral shaft, two longitudinal axes of the diaphysis can be identified: the "diaphyseal" axis, a curved line running through the middle of the diaphysis, and the "ideal" axis, a straight line connecting the midpoint of the femoral condyles and the intersection of the diaphyseal axis and the cervical axis (ibid.). For present purposes, the "ideal" axis is considered to be the longitudinal (z) axis of the femoral diaphysis. In practice, the two endpoints of this longitudinal axis are somewhat difficult to identify precisely, particularly in the sagittal (A-P) plane. Backman (ibid.), who has carried out perhaps the most complete and detailed study of the various femoral axes, planes, and angles, used lateral radiographs of the femur to help identify the intersection of the "diaphyseal" and cervical axes for the proximal point, and the intersection of the diaphyseal axis and "a normal to the underlying surface between the resting points of the condyles" for the distal point (ibid.: 25). Even with radiographs, however, Backman had to eliminate some specimens because of inability to estimate point positions accurately. Backman also noted that when resting on the femoral condyles and the greater trochanter (i.e., the normal position when laid dorsal side down on a flat surface), the "ideal" axis of the femur is not parallel to but rather forms an angle with the underlying surface plane, due to the relatively lower height of the proximal femur above the plane when in this position.

Having experimented with several different methods, I decided upon the following procedure as producing a longitudinal axis most closely approximating Backman's description of the femoral "ideal" axis, feasible without radiographs, and affording the most consistent and precise determination of the position of this axis from specimen to specimen (see Figure 3). As stated before, the femur was laid on a measuring board, dorsal side down. A point was located on the distal femoral diaphysis in the center, or lowest point, of the concavity present on the anterior surface just proximal to the lateral condyle. The height of this point above the measuring board surface was determined with the help of the mounted probe instrument described above, and the A-P breadth of the femur at this site measured using sliding calipers. The height of the anterior surface minus half the A-P breadth gave the height of the midpoint of the femoral shaft at this location. A similar procedure was carried out for the proximal diaphysis at a site immediately distal to the lesser trochanter. The proximal femur was then raised (usually about 1-2 cm) by inserting a cube of modeling clay under the greater trochanter until the two A-P midpoints were at equal heights above the measuring board surface.

This method of "leveling" the femur is very similar to that used by Stewart (1962) and Gilbert (1976) in their studies of anterior femoral curvature, except that these authors used only the height of the anterior surface to "level" (not the A-P breadth midpoint), and located their proximal point somewhat more proximally, between the levels of the greater and lesser trochanters. (It is noteworthy that Stewart had a great deal of trouble in precisely locating this latter point--this is one reason why I used the more distal location.)

The mediolateral position of the longitudinal axis of the femur was determined using the following two reference points (see Figure 3): the midpoint of the M-L breadth of the diaphysis at the same location where the proximal A-P breadth midpoint was taken (just distal to the lesser trochanter), and at the distal end, the deepest point in the patellar groove between medial and lateral condyles just above (anterior to) the intercondylar notch. The orientation of the frontal plane of the femur about the longitudinal axis was established by the posterior edges of the femoral condyles--that is, when laid dorsal side down on the measuring board, the natural resting position of the femur on the condyles was allowed to set the frontal plane orientation. The sagittal plane was taken as a perpendicular to the frontal plane passing through the mediolateral diaphyseal midpoints described above. The actual z axis used in the later 3-dimensional analyses of the femur was taken as the line falling in the sagittal plane parallel to the measuring board surface and at a height above it equal to the average height of the distal-most projecting points on the two femoral condyles, considered to be the distal articulation centers of the femur (the method of determining these points is described below).[1]

---

[1]The mean height of this line above the measuring board surface over the total sample was later found to fall within 1 mm of the height of a point measured on the greater trochanter between the gluteus medius and minimus insertion centers. This lends tangible evidence to the intuitive feeling that a "true" longitudinal axis of the diaphysis, running through the two articular endpoints, had been determined (as in the tibia), since in a sense the articulation "center" of the proximal femoral shaft, in the sagittal plane, is perhaps best approximated functionally by the location of the center of applied loadings of the thigh abductors.

The "leveling" of the femur in the sagittal plane is particularly important in the present study because of the structural analysis of transverse cross sections of the diaphysis. Allowing the femur to lie flat without raising the proximal end would result in very oblique cross sections being taken in the downward-curving proximal half of the diaphysis. Actually, any anterior curvature of the shaft will produce some obliquity (with respect to the curved "diaphyseal" axis) in transverse cross sections, if a straight "ideal" longitudinal axis is employed. (This is a point rarely if ever addressed in studies of this kind--i.e., how much the cross-sectional geometry of femoral transverse sections is affected, or "distorted" by femoral curvature). Some obliquity in femoral cross sections taken perpendicular to a straight line axis is therefore inevitable, however, raising the level of the proximal femur reduces the amount of obliquity to a minimum, and results in a more interpretable and perhaps less "biased" representation of proximal diaphyseal cross-sectional geometry.

As defined here, the frontal plane of the femur and tibia should be very similar when placed in true anatomical (<u>in vivo</u>) relationship to each other. However, the longitudinal (z) axis of the two bones will differ because the femoral diaphysis lies at an angle to the "whole body" sagittal plane as traditionally defined. Macquet (1976: 110) states that the longitudinal "mechanical" axis of the femur, running from the distal end through the femoral head center, coincides with the longitudinal axis of the tibia. Therefore, the angle between the femoral "mechanical" axis and the "ideal" axis of the diaphysis was measured in each specimen, to allow an estimate of the divergence of the longitudinal axes of the two bones (for technique of measurement, see below).

Location of cross sections.

Investigators of cortical bone geometry in the femur and tibia have used various methods to identify cross section sites. Because very few easily recognizable and precisely locatable anatomical landmarks exist on the tibial and femoral diaphyses (the tibial nutrient foramen being an exception), investigators have usually relied on percentages of bone length to identify locations for cross sections. Also, because the endosteal boundary between compact cortical and trabecular bone becomes increasingly difficult to distinguish as one approaches the bone ends (and trabecular bone becomes a significant factor in load bearing, and thus cannot be ignored), studies of cortical bone geometry have usually concentrated on about the middle 60% of the tibia and femur, that is, from about 20% of bone length to 80% of bone length. The number of cross sections taken from this area has varied to as high as 15 (Miller and Purkey,1980), although most studies have examined between 4 and 7 cross sections (Amtmann, 1971; Minns et al., 1975; Carlson et al., 1976; Lovejoy et al., 1976; Martin and Atkinson, 1977). From the graph curves connecting structural properties at different cross section locations reported by these authors, it is apparent that taking a larger number of cross sections would not add significantly to the discrimination of important geometrical changes along the femoral and tibial shafts (e.g., Minns et al., 1975: Figs. 1-4; Lovejoy et al., 1976: Fig. 10; Martin and Atkinson, 1977: Fig. 6).

A total of 5 cross sections from femoral and tibial diaphyses was considered sufficient to describe the structural variation present at different levels. These cross section locations correspond to 20%, 35%, 50%, 65%, and 80% of bone length, measured proximally-distally (see Figures 2 and 3). The measurement of bone length used for the identification of these cross section sites was not total bone length, either "maximum" or in the case of the femur, "bicondylar" (Bass, 1971). These measurements are either "biased" (in the sense discussed earlier in this chapter), or do not measure the true distance between the articulation center endpoints of the bones (or "beams"). In the tibia, the "biomechanical" length of the bone, used to calculate cross section positions, was taken as the average distance parallel to the z axis from the level of the talar center point (origin of the z axis) to the two tibial plateau center points (the mounted probe instrument was useful in making these determinations). Femoral "biomechanical" length was measured as the distance along the z axis from the average distal-most projections of the femoral condylar surfaces to the intersection of the z axis with the superior surface of the femoral neck. This latter point usually fell at about the junction of the femoral neck with the greater trochanter, just medial to the insertion area for obturator internus. All endpoints of the z axes were clearly marked with ink for later use in orienting the bones before sectioning.

Once the bones had been positioned correctly on the measuring board surface as illustrated in Figure 4, and "biomechanical" length determined, it was relatively easy to calculate and locate the longitudinal position of the diaphyseal cross sections. In addition to

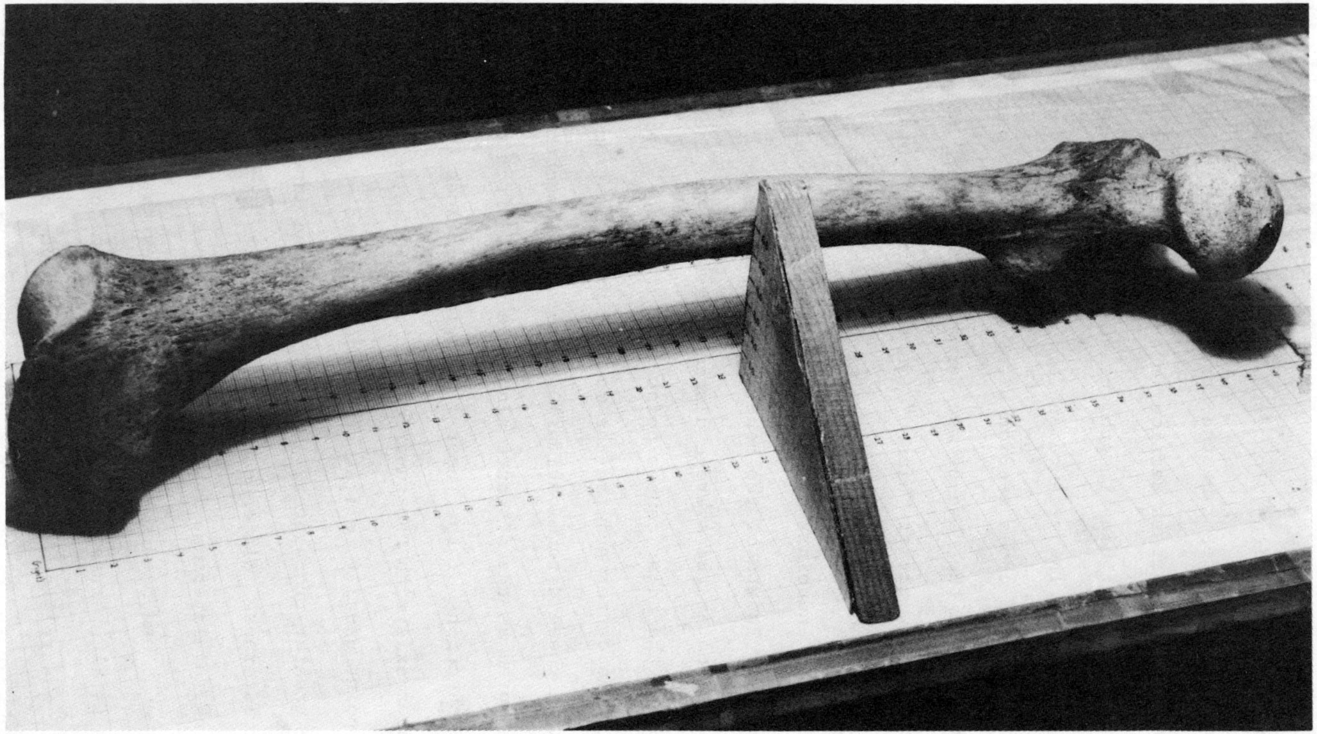

Fig. 4. Method of measuring 3-dimensional point coordinates on measuring board using triangular wedge.

these 10 sections (5 on each bone), an eleventh cross section through the femoral neck was located and marked. This femoral neck section was taken in a plane perpendicular to the cervical axis of the head and neck, at the location of minimum superior-inferior breadth (i.e., minimum breadth when viewed from above the frontal plane of the head and neck with the bone on the measuring board), usually falling about midway between the junction of the neck with the head and greater trochanter (see Figure 3).

The small wedge covered with graph paper shown in Figure 4 was used to locate each cross section in 3-dimensional space with respect to reference axes. With one perpendicular edge against the measuring board surface, aligned in the transverse plane, the wedge was moved until the vertical perpendicular edge made contact with the bone surface. X and z coordinates of bone surface points could then be measured using the intersection of the vertical perpendicular edge with the measuring board surface, while y coordinates were read as distances above this surface using the graph paper on the wedge. (This method was found to be more accurate and reproducible for taking these measurements than using the mounted probe instrument described above.) Two points on the outer surface of each cross section were measured and marked in this way: one on the medial, and one on the lateral side of the diaphysis (on the femoral neck, superior and inferior surfaces). Coordinates were read to the nearest .1 mm, however, repeated measurements of the same points indicated a true accuracy in point location of about .5 mm. Each point was clearly marked as the intersection of a horizontal and vertical line drawn with indelible ink on the bone surface; the vertical line defined the location along the longitudinal axis where the bone was to be sectioned, the horizontal line, bisected during sectioning, was used to define the location of the x, y reference points on the cut bone surface, as detailed in a later section of this chapter. Using this technique, each of the 11 cross sections could be located and oriented in 3-dimensional space relative to all other cross sections and certain other reference points, described below, even after sectioning.

Other whole bone measurements.

For subsequent 3-dimensional analyses and future stress analyses of the femur and tibia under simulated loadings, x, y, z coordinates of several points other than those at the cross section locations were measured and recorded. These points included first the coordinates of all of the centers of articulation already identified in the establishment of reference axes--the talar, two tibial plateau, and two femoral condyle centers. These last two points are difficult to locate precisely, due to the smooth convex curve of the condylar articulation surfaces and lack of any clear landmarks indicating the exact center of articulation with the tibia (this center of articulation is itself not constant, since the femoral condyles roll forward slightly on the tibia during extension of the leg--Morrison, 1970: 52). For lack of a better landmark, the distal-most projecting points on the condyles, with the femur oriented in the standardized frontal plane as described above were used as the articulation centers. These points were identified using the triangular wedge shown in Figure 4, holding the wedge distal to the bone in the sagittal plane and moving it proximally until the vertical perpendicular edge intersected the condylar surface (the ridges outlining the outer margin of the condylar articulation surface, which sometimes project further distally, were excluded from consideration). The two condyle points thus identified were usually fairly close in distance above the measuring board, although the lateral condyle "center" averaged higher than the medial "center" over the entire sample. The average between the two heights was used in setting the position in the sagittal plane of the femoral diaphyseal longitudinal reference axis, as detailed above. The y coordinates of all of these "center of articulation" points on the tibia and femur were later subtracted from the y point coordinates of the cross sections, which had been measured as distances above the measuring board surface, to arrive at the true positions of these cross sections in the sagittal plane relative to the reference axes.

X, y, z coordinates of two other reference points on the femur, important for possible future analyses, were also measured at this time. These were 1) the center of articulation of the femoral head, and 2) the center of the insertion area for the gluteal abductors of the thigh on the greater trochanter.

Seven angles of the tibia and femur were also measured, using a clear plastic protractor with a free swinging arm, a second protractor with free arm attached to a vertical perpendicular support, and another plastic vertical support. Some of these angles were taken because they provide important structural information (e.g., angles formed by the femoral neck and the shaft), and some principally to provide information allowing comparisons with other studies which used different sets of reference axes (e.g., the tibial "torsion" angle). All measured angles were taken to the nearest degree. They are illustrated in Figures 2 and 3 and summarized in Table 3.

It should be mentioned that these measurements, particularly the x, y, z coordinates of the two points on each cross section, constituted the most time-consuming and tedious phase of the entire project. All of the information gathered is vital for both 2- and 3-dimensional structural analyses of the bones. However, the author plans to develop in future studies of this kind, alternate, more rapid techniques for locating and measuring these points and angles, using either stereophotography or radiography with lead markers. Computerized tomography, complete with suitable fixed axes markers in the scans, would provide the most efficient means of conducting 3-dimensional structural analyses, and, if financially feasible for large samples, could eliminate much of the manual labor involved in the initial stages of a study of this kind.

Photography and radiography.

All bones used in the study were photographed and radiographed in both A-P and M-L planes before sectioning[2]. All photographs included scales oriented along standardized reference axes. Separate radiographs of the proximal femur in the antetorsion plane of the femoral neck, after sectioning of the diaphysis, were also made.

---

[2] I would like to thank Dr. Marjorie LeMay, Chief of Radiology, Harvard University Health Services, for providing free of charge the equipment and film necessary for radiographing the sample.

Table 3

Summary of Measurements Made on Whole Bones Before Sectioning (also see Figures 2 and 3)

I. 29 x, y, z point coordinates

   A. 22 points on the medial and lateral (femoral neck--inferior and superior) surfaces of 11 cross section locations

   B. 7 articulation center points

      1. tibia: talar and 2 plateau surfaces

      2. femur: 2 femoral condyle surfaces, femoral head surface, insertion area for gluteal abductors

II. 7 angles

   A. Tibia

      1. tibial "torsion" angle (talar M-L axis/ tibial plateau M-L axis)

      2. plateaus/M-L axis

      3. plateaus/A-P axis

   B. Femur

      1. cervico-diaphyseal (longitudinal axis neck/frontal plane shaft)

      2. antetorsion (longitudinal axis neck/sagittal plane shaft)

      3. "mechanical"/"ideal" longitudinal axes

      4. condyle articulation centers/M-L axis

III. 4 bone lengths

   Maximum and "biomechanical" lengths of the femur and tibia

This part of the study was carried out for two reasons: first, to preserve as complete as possible a visual representation of the intact bones before sectioning, and second, to secure additional structural data while it was still available, information which, while not used in the present study, might serve as a source of data for future studies. For example, only a few of the relevant physiological loading points--articulation centers and the thigh abductor insertaion center--were specifically located and measured before sectioning; the photographs taken are sufficiently detailed and complete to allow 3-dimensional placement of other major muscle attachments (e.g., the tibial tuberosity) or any other points of structural interest, which would be useful in future stress analyses of the tibia and femur, among other studies. Radiographs can provide further information of interest, for example, the presence of transverse lines of growth arrest (Harris Lines), and trabecular patterns, particularly in the proximal femur (this was one reason why the separate additional radiographs of the proximal femur oriented in the antetorsion plane were made--to better show trabecular patterns).

Perhaps most importantly, given this material and the measurements included in the "whole bone" analysis described in the preceding sections, very little information was lost by sectioning the bones obtained from the Pecos study sample. Alternate techniques of morphometric analysis of these bones are still possible, even after sectioning, using this material. It should also be remembered that a paired bone for each specimen included in the study sample was left intact and available for future study.

Bone Mineral Measurements

As discussed earlier in chapter 2, photon absorptiometry (Cameron and Sorenson, 1963) has been used extensively to measure bone mineral content (BMC) among both living and past populations. BMC determinations for all individuals included in the present study sample were carried out before sectioning using the Norland-Cameron ^{125}Bone Mineral Analyzer housed in the Department of Biobehavioral Sciences, University of Connecticut (courtesy of Dr. William S. Laughlin and Dr. Al Harper).

At first it was planned to take BMC scans of all 11 cross sections in all 120 individuals in the study sample. However, due to time constraints it was found to be impossible to scan all 1320 cross sections, therefore the following compromise was arrived at. In all 120 individuals, BMC determinations for at least two cross section locations were made: the femoral midshaft cross section, and the femoral neck cross section. These sites were chosen as representative of areas of chiefly compact cortical bone and a combination of compact and trabecular bone, respectively. Also, the midshaft femur has frequently been used as a standard site measurement of skeletal changes with aging (e.g., Smith and Walker, 1964), and the BMC of the femoral neck could be of critical importance in determining incidence of femoral neck fractures. In 40 individuals, all 11 cross section locations were scanned. This completely scanned sample was chosen to include equal numbers of males and females, right and left sides, and individuals under and over 40 years of age (within each of these age groups, subjects were equally distributed throughout the five or ten year original age groupings). Thus, comparisons of sex, age, and side differences in BMC at sites other than the femoral midshaft and neck can still be made, while keeping treatment cell sizes at a minimum of 10 individuals.

The ^{125}I Bone Mineral Analyzer was used with a 3.0 mm collimated scan beam, at a scanner speed of 2 mm/sec. Whole bones were submerged in 6 cm of water in a plexiglass box and scanned twice, with the average of the two scans recorded (if the first two scans differed by more than .10 g/cm in BMC--i.e., by more than about 3-4%--another scan was made; if this scan fell outside the range of the first two, a fourth was made and the average taken). Bone width (BW), an additional reading made by the Bone Mineral Analyzer, was also recorded. Both BMC and BW were recorded to three significant figures.

In order to ensure that the absorptiometric scan path coincided precisely with the standardized transverse plane of the tibia and femur as determined earlier, a piece of graph paper was taped to the bottom of the plexiglass holding box and marked with a longitudinal axis. This axis was then lined up by eye with the markings at the end of each bone indicating the position of the reference z axis. Dorsal side down in this position (clay supports in place), absorptiometric scans were taken across the mediolateral breadth of the bone perpendicular to the longitudinal axis at each diaphyseal cross section location.

Because of the different orientation of the femoral neck cross section, it was difficult to realign intact femora within the original plexiglass holding box for scanning of this section. Therefore, the femoral neck scans were taken later, after sectioning the diaphysis (but before sectioning the femoral neck). A smaller holding box was used, with the cervical axis lined up with a longitudinal axis, across the superior-inferior breadth of the neck at the marked location (the proximal femur was held in the antetorsion plane while making these scans, i.e., head and neck flat against the bottom of the holding box).

As part of the normal operating procedure in using the Bone Mineral Analyzer, the machine was recalibrated against known standards before each scanning session. In addition, several bone cross sections were scanned at more than one session to determine replicability of results. Differences in BMC and BW readings at different sessions were small, averaging under 5%. (There are, however, several technical difficulties which may present problems in the interpretation of BMC data that will be discussed elsewhere.)

Fortunately, another, independent method of determining bone mass in these specimens was available through the work of David Thompson (Department of Biobehavioral Science, University of Connecticut). As part of an ongoing study of histological changes in bone with aging, Thompson removed (with the Peabody Museum's permission) bone cores 4mm in diameter from a location just proximal to my femoral 50% location (approximately midshaft, as traditionally defined), in the anterior quadrant. These are currently being analyzed by him for osteon remodeling differences and other histological features of interest. As part of his analysis, Thompson is calculating (wet) bone density by directly measuring core colume and weighing cores to obtain bone mass. At present, 39 specimens have been completed, and he has graciously supplied the bone density data for these individuals. These directly measured density determinations are presented in chapter 5 in conjunction with BMC and BW data.

Bone Sectioning

After all whole bone measurements, radiography, ^{125}I scanning, and whole bone photography had been completed, bones were sectioned using a Felker diamond-bladed circular saw (see Figure 5). A special rig, illustrated in Figure 5, was built to hold each bone rigidly in the proper orientation to the saw blade for each diaphyseal cut. The saw was originally equipped with a metal platform which could be rolled on tracks parallel to and under the saw blade. To this platform was bolted a board approximately 20 X 60 cm, its long axis perpendicular to the saw blade. Graph paper marked with a longitudinal axis, covered by a sheet of plexiglass for protection, was attached to this surface. Two wooden end pieces projecting above the board surfaces and overhanging the near edge (away from the saw blade and towards the operator), held a metal rod 1.2 cm in diameter and 90 cm in length, oriented parallel to the marked longitudinal axis, about 15 cm to the near side of the axis and 5 cm above the board surface (see Figure 5). The metal rod could be slid through its supporting holes in the end pieces, thus moving parallel to and along the longitudinal axis.

Three standard laboratory 3-pronged clamps were clamped to this rod using small C-clamps. At the start of sectioning a new specimen, the bone was lined up on the plexiglass sheet using the marked longitudinal axis as a guide, and the three clamps attached, one to each end section and one to the section immediately proximal to the first (most

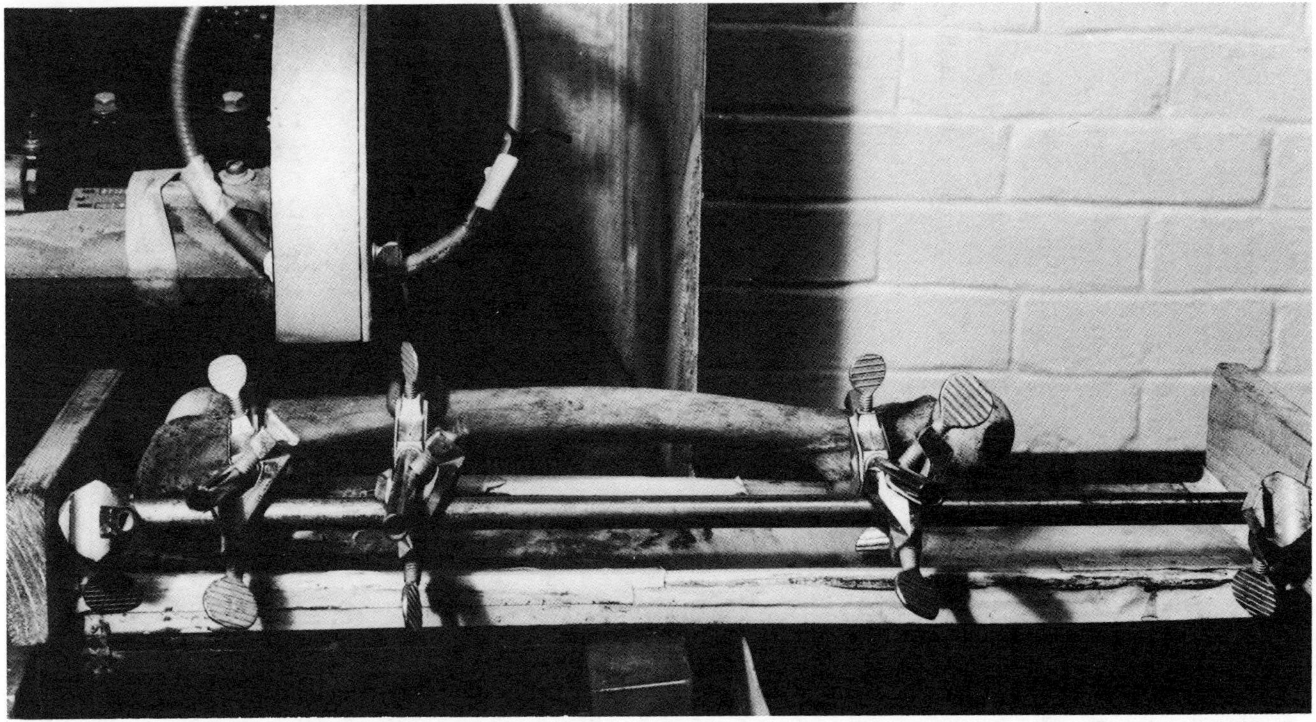

Fig. 5. Sectioning apparatus. Femur is positioned for first (20%) cross section cut.

distal) cross section to be cut. Small wooden blocks propped under the bone diaphysis were also frequently found to be useful in stabilizing the bone during sectioning. The first cross section was lined up with the plane of the saw blade using the previously made markings on the bone surface. The whole rig was then rolled on the movable platform through the saw blade (positioned just above the board surface), and the bone sectioned. After making this first cut, the distal bone section was removed and the 3-pronged clamp which had been holding it "leap-frogged" over the second clamp to a position just proximal to the <u>second</u> cross section location, and fastened. The metal rod was then slid down (distally) until the second cross section was lined up with the saw blade (two C-clamps kept the rod from moving during the cutting). This procedure was then repeated for each cut until reaching the proximal end.

Thus, each bone was held rigidly to the metal rod by at least two fastened clamps at all times, and always maintained in its original position relative to standardized reference axes, even after removing bone segments, by sliding this rod along parallel to the longitudinal axis for each new cross section cut (enough "slack", or overhang of the rod past the proximal wooden end support, had to be allowed for when first positioning the bone to permit the necessary distal movement of the rod while still holding it within the two supports). The bone cross sections obtained through this procedure are true transverse sections of the diaphysis, each taken precisely parallel to all other diaphyseal cross sections, and perpendicular to the longitudinal axis.

After completing all diaphyseal sectioning, the femoral neck was cut. Due to the complex shape of the proximal femur and the difficulty in clasping it in the correct orientation using the 3-pronged clamps, the femoral neck section was taken holding the proximal femur manually. Wooden blocks were used to help secure a stable position of the head and neck during the cut. Care was taken that this section was made perpendicular to the cervical axis (and the antetorsion plane), and intersected the two points on the neck marked earlier indicating the minimum superior-inferior breadth.

At first it was thought that coating the bones with a suitable material, such as PVA (polyvinyl acetate), would be necessary to avoid chipping or spalling of the outer surface during sectioning. However, after some experimentation, it was decided that a coating was unnecessary, since very little damage to the bone cross section surface occurred while sectioning untreated bones. This was probably due in part to the precision and smoothness of the action of the diamon blade (which used a water jet for cooling of the blade, which incidentally helped carry away bone dust from the cut during sectioning), and to the rigid stabilizing of the bones during sectioning. Any minor chipping which did occur was easily repaired using clay pressed into the bone surface and shaped using a scalpel blade.

Cross Section Photography

After sectioning, cross section surfaces were prepared for photographing and digital analysis. In each case, the proximal end of the bone segment distal to the cross section cut was used as the cross section surface to be analyzed. All geometrical analyses in the study depend upon accurate location of each cross section with respect to x, y, z reference axes. This was accomplished using the two reference points on the outer (subperiosteal) surface of each cross section which had been measured earlier, before sectioning. In order to make these reference points visible on the cross section photographs, lines from the reference points were extended inward from the medial and lateral outer edges onto the cut bone surface. (As noted earlier, cross section refeence points had been marked as the intersection of two perpendicular lines--one of these lines was obliterated during sectioning, the other bisected line served to identify the position of the reference point after sectioning.) The intersection of these lines drawn on the cross section cut surface with the subperiosteal boundary, clearly visible in the cross section photographs (see Figure 6), located the positions of the original two reference points and allowed each cross section to be correctly oriented, independently of other cross sections and of any other bone landmarks.

This method of orienting bone cross sections after sectioning should be just as accurate as embedding whole bones in plaster or epoxy jackets marked with reference axes before sectioning (Kimura, 1974; Lovejoy et al., 1976; Piziali et al., 1976; Miller and Purkey, 1980). Some authors do not indicate how their reference axes were identified after sectioning (Amtmann, 1971; Martin and Atkinson, 1977), and at least one group simply states that A-P and M-L planes "were clearly marked with indelible ink down the length of the shaft", without indicating how these planes were arrived at (Minns et al., 1975:

253). Even using plaster jackets can result in some inaccuracy in orientation of cross section axes unless the "jackets" are actually rectangular bins (cubes), with each face of the bin parallel to a reference plane (only Piziali et al., 1976; and Miller and Purkey, 1980 state specifically that this was the procedure followed). Given an accuracy of ±.5 mm in the measuring of x and y reference point coordinates, the maximum error in orientation of cross sections around the longitudinal axis using the method employed in the present study is 2-3°. Avoiding embedding of bones also has several advantages, especially in a large study of this kind, in terms of extra expense, time required waiting for 240 embedded bones to "set", difficulty of precisely identifying longitudinal positions of cross sections after embedding, awkwardness in storage, and perhaps most importantly, preservation of the bones for future study (all bone segments were returned to the Harvard Peabody Museum after photographing).

After marking the positions of the reference points on the cross section surface, each cross section was photographed as shown in Figure 6. A round hole, large enough to accomodate any bone cross section, was cut in a piece of stiff cardboard which was marked with anterior and medial directions (in Figure 6--the "M" designating "medial" could be moved to the right or left of each cross section, depending upon whether the bone was from the left or right limb, respectively). Other interchangable labels to the right of this "viewing hole" identified number of the individual, bone (tibia or femur), and cross section location.

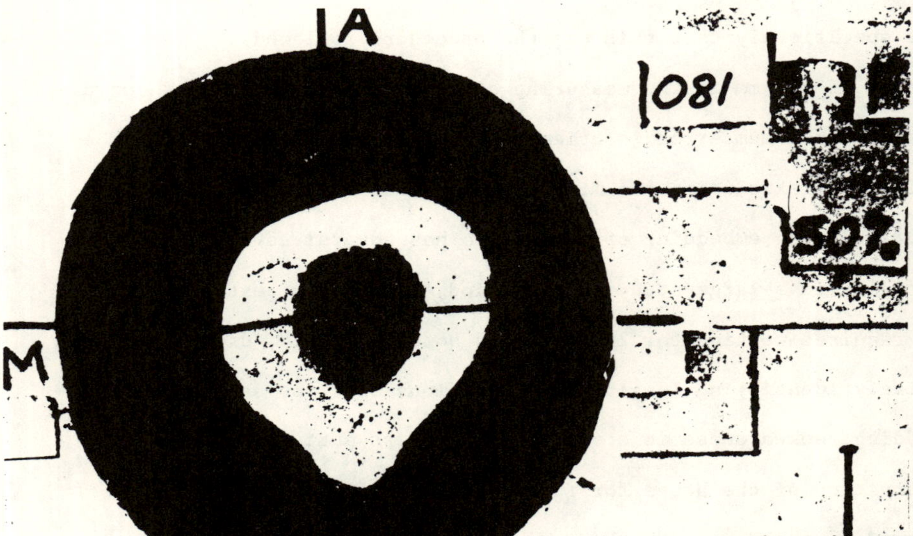

Fig. 6. Bone cross section (midshaft right femur) as photographed for digitizing (shown here 2X life size-- slide negatives used in actual digitizing enlarged to 8X life size). See Figure 8 for a computer representation of this cross section.

To avoid any distortion of the cross section surface geometry, it was vital that photographs be taken in a plane exactly parallel to the (transverse) plane of the cross surface. To ensure that this was the case and that all photographs were taken using the same standardized procedure, a small wooden brace was built which held the camera lense and cardboard display surface parallel to and at a constant distance from each other. Bone segments were held with clay on a wooden mount behind the cardboard surface, and adjusted using a square until the cross section surface was parallel to the display and camera lense surfaces. Using two magnifying lenses, the camera lense could be brought to within 12 cm of the cross section surface. Photographs were taken using Kodak Panatomic-X, and were commercially developed into slide negatives. Sharpness and contrast in the slides were generally excellent, and when enlarged for digital analysis presented a very detailed picture of bone cross-sectional morphology.

Analysis of Geometric Properties--Program SLICE

Computer program SLICE was originally developed by Mark Nagurka and Wilson C. Hayes, Department of Orthopaedic Surgical Research, University of Pennsylvania (Nagurka and Hayes, 1980)[3]. Given input of cross section perimeter boundary point coordinates, SLICE automatically calculates all important geometrical properties of the cross section: cross-sectional area, coordinates of the centroid, second

---

[3]Present address (Hayes): Department of Orthopaedic Surgery, Beth Israel Hospital, Boston.

moments of area (moments of inertia) and product of inertia about x and y axes translated to the centroid, the angle between translated x and y axes and the principal axes, and second moments of area about principal axes (the biomechanical significance of these various properties will be discussed during presentation of results in chapter 5). Cross sections of any irregular shape may be analyzed, and any number of interior sections, or "holes", can be accomodated (of course, only one "hole"--the medullary cavity--was entered in the present study).

SLICE works by dividing a cross-sectional area into a series of small trapezoids (or rectangles), and then adding or subtracting the properties of these elemental areas to determine the composite properties of the total area. The method replaces integration with the summation of finite elements, and assumes linear segments between consecutive perimeter coordinates. The algorithms used to calculate these properties, as well as a general outline of the program, are documented in Nagurka and Hayes (1980).

Input of perimeter boundary point coordinates was formerly accomplished manually, by tracing cross section outlines on graph paper, reading x and y coordinates of selected points around this outline, and entering these coordinates into the SLICE program (Ruff, n.d.). Including the initial tracing, and depending upon the number of perimeter points taken, this process could take up to one-half hour or more for a single cross section (_ibid_.). The procedure has since been speeded up tremendously through the addition of an electronic digitizer for input of point coordinates, and now averages less than 5 minutes per cross section. Use of the digitizer also eliminates several possible sources of observer error in manually reading and transcribing data, and helps to standardize the procedure (i.e., increase replicability, especially important in the present study, since over one thousand cross sections were analyzed over a six month period).

The SLICE program was modified slightly by the present author during the course of the investigation. First, two new properties were calculated from the input data: the maximum distances along the major and minor principal axes of the cross section from the area centroid to the outer perimeter boundary, properties used in stress analyses (commonly labeled "c"--e.g., Shigley, 1976). Second, SLICE was modified to save all geometrical properties of the total area within the outer (subperiosteal) perimeter, _not_ subtracting the inner "hole" (medullary cavity). These properties had originally been calculated by SLICE during an intermediate step of the program, but had not previously been saved. These data were particularly useful in documenting subperiosteal expansion, deriving per cent cortical area (cortical area/subperiosteal area X 100)--a commonly used index in studies of osteoporosis, and in analyzing all geometric properties of the femoral neck. The femoral neck was treated as a special case in the present study, due among other things to the relatively high proportion of trabecular bone at this site (which probably makes a significant contribution to femoral neck strength), and the difficulty in many specimens of identifying a clear endosteal border, making calculations of geometric properties of the cortical bone shell less accurate and meaningful. For these reasons, geometric properties of the total subperiosteal area of the femoral neck were always examined along with properties of the cortical shell alone.

SLICE has been shown to be accurate to within 7 significant figures when tested on simple geometric figures of known properties with manual input of boundary point coordinates (Nagurka and Hayes, 1980) (accuracy with input from the digitizer is discussed below). The method , especially as presently available with the digitizer, represents a significant advance in terms of time and labor involved over other commonly employed techniques which usually require placing a grid over a cross section outline and counting grid intersections or blocks falling over areas of bone (Amtmann, 1971; Kimura, 1974; Minns et al., 1975; Carlson et al., 1976; Lovejoy et al., 1976; Martin and Atkinson, 1977). As greater accuracy is desired, finer and finer grids, and thus more and more manual counting of intersections or blocks is necessary when using these other techniques. Using the present method, a high degree of accuracy can be obtained merely by specifying the particular intervals from which perimeter boundary points are to be taken. With the digitizer, the method is just as fast as a recent "radial grid" method decribed by Lovejoy and Barton (1980), and is more accurate (they reported inaccuracies of up to 8% in section properties when compared with planimeter tracings).

Use of the Digitizer

A 14" X 14" Active Surface Area RP614 electronic digitizer (Talos Systems, Inc.) was used for inputting cross section perimeter boundary coordinates to SLICE (see Figure 7). Slide images of the cross sections

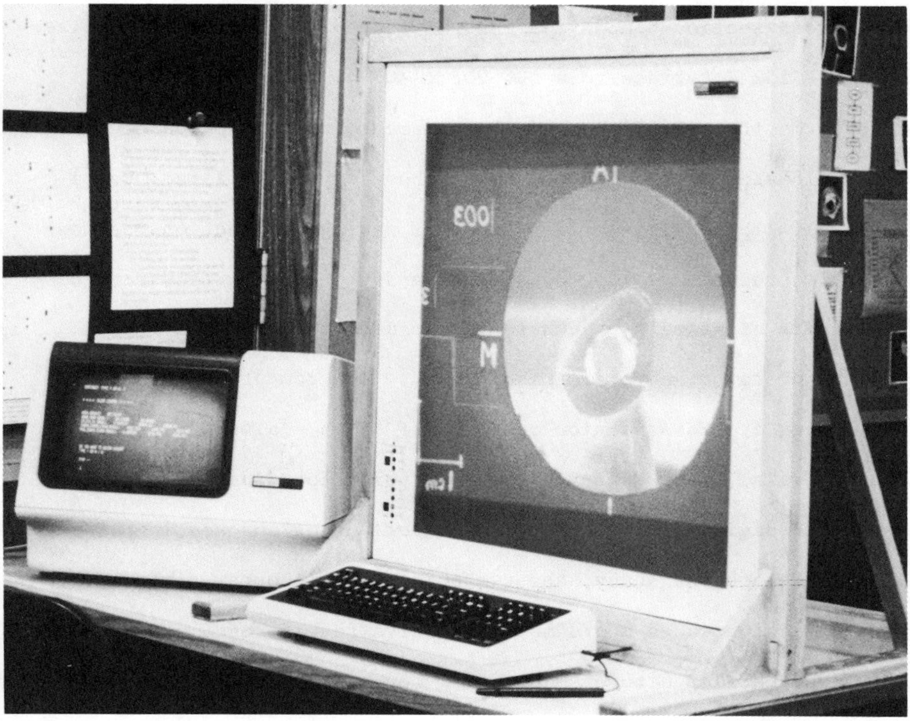

Fig. 7. Digitizer screen and DEC PDP 11/03 terminal. Cross section image is projected from behind the screen (out of picture to the right). Stylus used for tracing cross section outlines shown in lower foreground.

were rear-projected from a distance of about 2.5 meters onto the digitizer surface, magnifying cross section images (Figure 6) approximately 8 times. A stylus, visible in the bottom of Figure 7, is used to trace cross section boundaries manually, sending electronic impulses through an RS232 interface to a DEC PDP 11/03 minicomputer, where SLICE performs its calculations (all equipment for this phase of the project is located in the Biomechanics Laboratory, Beth Israel Hospital, Boston). The digitizer is capable of making distinctions between x, y point coordinates 1/1000 of an inch apart on the surface screen.

The entire digitizing procedure is carried out interactively between the operator and the minicomputer.[4] The computer first asks for the number of cross section deletions ("holes"--always 1 in the present analysis), then asks for the x, y point coordinates of two points on the digitizer screen for use in orienting and scaling of the image. In the present study, the two points on the outer surface of each cross section whose coordinates relative to reference axes had been measured earlier before sectioning, were used for this purpose. As described above, the intersection of lines marked on the cut cross section surface with the medial and lateral outer (subperiosteal) edges marked the exact position of these points in the cross section photographs. The x, y coordinates for these two points were entered into the program, and the stylus touched to the corresponding points on the digitizer screen. Use of this system eliminates the need for any

---

[4]David Black, working in the Biomechanics Laboratory, wrote and implemented the interactive program necessary for using SLICE with the digitizer.

elaborate and time-consuming positioning or enlargement to a particular size of the cross section image on the digitizer screen, since the orientation relative to reference axes and scale are reset each time a new cross section is analyzed. Theoretically, cross sections could be "twisted" almost 180 degrees, and still be "straightened out" by the reference point data entered into the program. Also, effects on image size of refocusing or slight forward or backward movement of the slide projector between tracings are inconsequential in terms of the analysis, since no matter what the screen size, each cross section is automatically reduced to its original true size. The only precaution which has to be taken is that the slide projector beam be perfectly perpendicular to the digitizer surface--otherwise the cross section image may be distorted during projection (as when taking cross section photographs--see above). Accordingly, positions of the digitizer surface and slide projector lense were rechecked for correct orientation using a tape measure and levels at the beginning of each digitizing session.

Before tracing cross section outlines, the digitizer must be set to "track" mode and the desired increment between boundary points whose coordinates are to be taken entered. In "track" mode, the digitizer reads x, y coordinates of points touched by the stylus point on the digitizer screen (contrasting with "point" mode, used in entering the two reference point positions, where just one point is taken each time the stylus makes contact with the screen). The increment between points to be read by the digitizer is entered by the operator in the same scale used in entering the two reference points (that is, if an increment of 1 mm between points is specified, this will be interpreted by the program as 1 mm in the original unmagnified scale of the cross section). Increments of 1 through 5 mm were tested to see what effect different intervals between boundary points would have on accuracy in determining geometrical properties. Although no difference in accuracy using 1, 2, or 3 mm intervals between points could be detected, a 1 mm interval was used throughout the present study, since the time involved in tracing the outline is the same no matter what increment is chosen, and there are few limitations on number of boundary perimeter points which can be processed and stored by the program.

After entering these specifications, the subperiosteal outline of the cross secion is traced, counterclockwise, with the digitizer automatically reading x, y point coordinates at intervals of 1 mm (in the original unmagnified scale). The perimeter boundary is automatically closed by the program when coming back to within a certain tolerance distance of the first point. The second (endosteal) perimeter is then traced in the same way. After tracing is completed, the program then asks if a representation of the cross section just completed is desired, and if answered in the affirmative, will plot a picture of the cross section on the minicomputer terminal screen. This was used to check for errors in tracing (and entering of point coordinates for scaling and orientation). If errors are detected, the operator may jump out of the program and begin again. If asked to continue, the program displays a list of the geometric properties calculated by SLICE, and asks whether the cross section is to be stored.

X, y point coordinates for the two perimeters are stored on floppy disks processed in the minicomputer. Using the 1 mm boundary point interval, coordinates for approximately 80-100 points on the subperiosteal perimeter and 60-80 points on the endosteal perimeter were stored for each cross section (about 100 cross sections, or close to 10 individuals, could be stored on one floppy disk).

Hard-copy printouts of cross section representations with associated geometrical properties may be produced for any cross section using a Tektronix 4010 graphics terminal linked with the PDP 11/03. These printouts can be used for demonstration purposes or for visually comparing different cross sections. An example of such a printout for the bone cross section depicted in the photograph in Figure 6 is shown in Figure 8.

Tests of accuracy and replicability in calculating properties using input from the digitizer were carried out on both simple geometric shapes of known characteristics and bone cross section photographs. Geometrical properties of the simple geometric shapes varied by an average of less than 2% from their known values; no values outside 5% of their known values were obtained. Replicability tests of bone cross sections were run on both repeated trials during the same digitizing session, and on sections traced and retraced at intervals of several months apart. Replicability was generally excellent, averaging under 1% difference between trials, even after long periods of separation between tracings. The principal difficulty in tracing bone cross sections is in identifying the endosteal border, particularly in sections with significant porportions of trabecular bone such as the tibia

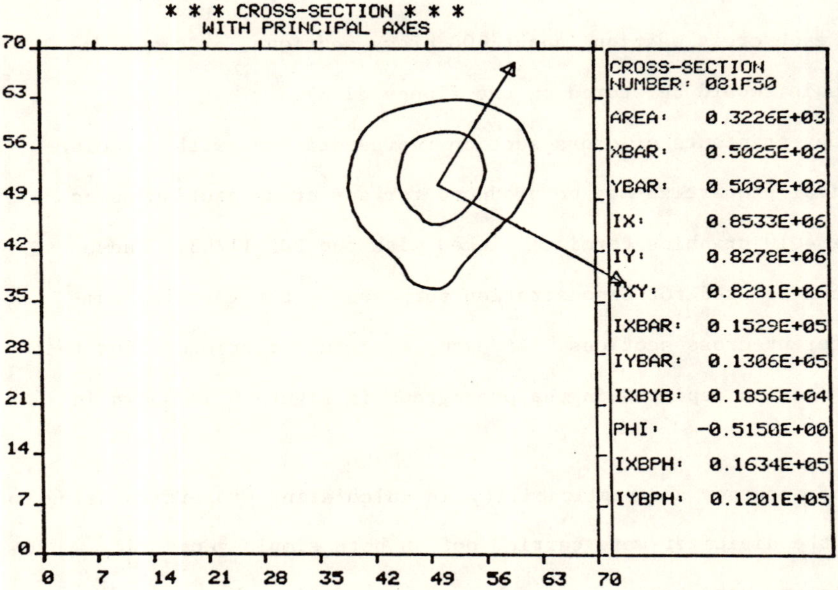

Fig. 8. Computer printout of femoral midshaft cross section illustrated earlier in Figure 6 (p. 120). Arrows show direction of principal axes; lengths indicate relative magnitudes of principal second moments of area about axes. Section properties listed at right (see chapter 5 for description of properties).

80% (most proximal) and femur 20% (most distal) sections (and of course the femoral neck). In general I included as part of the medullary cavity any areas filled with less than 50% bone, the only exceptions being obvious transverse sections of nutrient foramina embedded in compact cortical bone. Since this was done by eye, and the gradation between "compact" and "trabecular" (or cancellous) bone is continuous, this distinction was by necessity somewhat arbitrary. However, the good replicability of results indicates that the same distinctions were being made consistently throughout digitizing. The standardized conditions under which the cross sections were prepared, photographed, and projected onto the digitizer, as well as the general quality and detail of the photographs, also helped in maintaining as consistent and accurate a tracing of each cross section as possible. To further aid in determining cross section boundaries, the actual cut bone segments themselves were brought into the laboratory while digitizing, and were always available (and frequently used) for confirming the position of bone perimeters in the cross section photographs.

After all 1320 cross sections had been completed, a special program was written to recalculate and store all geometrical properties of the cross sections (including the ones added by the present author) on a separate disk (only point coordinates had previously been stored). These data were then transferred via phone line to the VAX 11/780 computer at the MIT Joint Computing Facility, where they were put on magnetic tape. Other structural information collected, e.g., BMC data and "whole bone" measurements, were later added to this tape. All statistical analyses of the data were carried out at the Boston University

Computing Center on an IBM 370 computer.

As stated earlier (chapter 1), studies of structural variation in long bone cross sections have tended to use quite small sample sizes, usually under 10 and never (previously) more than 40 individuals. It is safe to say that without SLICE and the electronic digitizer input of data, the present study would have been limited to a similar sample size range, due to the enormous time involved in past methods of calculating cross section properties necessary for structural analyses. With a time limitation of less than 5 minutes per cross section, and easy means of storage and retrieval of data, it was possible to conduct a large-scale investigation with adequate sampling of inter-individual variation along several lines (sex, age, side of body). The results of this investigation are presented in the following chapter.

Summary

1. A set of x, y, z reference axes was defined for each bone based primarily on the positions of the centers of articulation at the bone ends. Eleven cross section sites were located, five on the tibial and femoral diaphyses corresponding to 20%, 35%, 50%, 65%, and 80% of "biomechanical length", and one through the femoral neck. With the aid of a measuring board and several simple measuring tools, x, y, z coordinates of two points on the subperiosteal surface at each cross section location were measured and marked. These were later used in orienting cross sections in 3-dimensional space after sectioning. Several other measurements were also taken of each bone before sectioning, summarized in Table 3.

2. Bone mineral content (BMC) of cross sections was determined using an ^{125}I Norland-Cameron Bone Mineral Analyzer. BMC was measured at two sites on all individuals--the femoral midshaft and femoral neck, and at all 11 sites in 40 individuals chosen to represent sex, age, and side distribution of the sample. All bones were also radiographed and photographed before sectioning in standardized A-P and M-L planes.
3. Bones were sectioned using a diamond-bladed circular saw. A special rig attached to the saw allowed precise transverse cuts at each cross section site. The proximal surface of the bone segment distal to each cross section cut was photographed and made into a slide negative for analysis.
4. Analysis of cross section geometric properties was accomplished using computer program SLICE (Nagurka and Hayes, 1980). SLICE requires as input the x, y coordinates of perimeter boundary points. These were supplied through use of an electronic digitizer, projecting cross section images onto the digitizer screen and tracing subperiosteal and endosteal boundaries manually, with point coordinates taken automatically at 1 mm intervals. The procedure takes less than 5 minutes per cross section, and provides results which are accurate and replicable to within 2%.

Chapter 5

Results

The structural features of the femur and tibia that were investigated in this study fall into three general categories, based primarily on the method of measurement involved in each: (1) measurements of the "whole bone" taken before sectioning, including bone lengths and angular measurements of articular surface planes and the femoral neck; (2) geometrical properties of 11 cross sections, including bone areas and other properties of biomechanical interest, determined through the use of SLICE: and (3) bone mineral content data at these same 11 sites, obtained using photon absorptiometry. Analysis of these three sets of features forms the three major subdivisions of this chapter. While this format is somewhat arbitrary, in that structural properties do not vary independently of each other in accordance with measurement technique employed, it is a convenient means of organizing the initial presentation of a very large set of data; interactions between the different types of features in terms of overall functional adaptations are discussed in chapter 6. (Analysis of the 3-dimensional position of section area centroids relative to each other, while strictly a "whole bone" type of feature--i.e., involving the macro-geometry of the entire bone, not just a single cross section--is considered in the section on cross-sectional geometry, since these properties are derived through SLICE and are closely related to other cross section properties.)

Three factors are explored for their effects on variation between individuals with respect to these structural features: sex, age, and side of body. Although individuals were assigned to six different age groups in five or ten-year divisions (see chapter 3), for many of the following analyses comparisons between two larger age groupings--20-39 years ("younger" adults) and 40+ years ("older" adults)--are sufficient for summarizing general age differences (the larger n's in each age grouping also reduce within-group sampling error). In addition, in some analyses where the sample is categorized by more than two factors (i.e., by sex, age, and side), or in BMC analyses of cross sections other than the femoral midshaft or neck, where only 40 individuals were measured, only two age groupings could be included in order to maintain each treatment cell at a minimum of 10 individuals. For more detailed consideration of trends in structural properties throughout the adult life span, the six age group divisions were used, and analyses of variance employed to test the statistical significance of these trends.

It should be noted once again (see also chapter 3) that one individual was eliminated from the initial study sample of 120 because of gross pathological distortion, leaving a final sample size of 119 (this individual was a male, 35-39 years, left side). The computer programs used in this study for the analysis of variance--BMDP7D and BMDP2V (Dixon and BRown, 1979)--do not assume the equality of cell means, and so are unaffected by this loss of one individual. Other programs from the BMDP package employed in statistical analyses of the study data include BMDP2D, 3D, 6D, 6R. As stated earlier, all statistical analyses were carried out at the Boston University Computer Center on an IBM 370 computer.

Sex Differences

Table 4 summarizes whole bone lengths and angles of male and female tibiae and femora of the study sample. Two significant sex-related differences in these dimensions are apparent: 1) a general size difference, reflected in differences in bone length, and 2) a difference in orientation of the femur in the frontal plane resulting in a relatively more medial position of the femoral head in males, reflected in certain angular measurements.

The observed sexual dimorphism in bone lengths in the sample--approximately 9-10%--is about average for long bones in human populations (Krogman, 1962: 143). The tibia is slightly more sexually dimorphic in length than is the femur, and "biomechanical" and maximum lengths for each bone show almost identical degrees of dimorphism. The average values for maximum bone lengths obtained here are very close to those reported by Hooton in his earlier study of the Pecos sample, in no case deviating by more than 2.5 mm from his values (Hooton, 1930: 134, Table V-1; 144, Table V-16; right and left sides combined). This excellent agreement between the two studies is good evidence that a random sampling (with respect to body size) of the Pecos Collection is represented in the present study sample, or at least that both Hooton's larger sample and the present sample are sampling approximately the same range of size variation present in the total collection (standard deviations are also very similar in the two studies).

Table 4

Sex Differences in "Whole Bone" Dimensions

Dimension [1]	Male Mean	Male SE	Female Mean	Female SE	Male - Female	Male/Female X 100
Tibia						
Lengths [2]						
Maximum	358.8	2.0	325.1	2.2	33.7***	110.4
Biomech.	339.4	2.0	308.0	2.2	31.4***	110.2
Angles [3]						
Torsion	19.6	1.0	16.9	1.2	2.7†	---
Plat/M-L	3.9	.4	2.9	.4	1.0*	---
Plat/A-P	12.2	.5	12.1	.4	.1	---
Femur						
Lengths [2]						
Maximum	429.1	2.2	392.9	2.4	36.2***	109.2
Biomech.	405.2	2.0	371.4	2.3	33.8***	109.1
Angles [3]						
Cerv-Diaph	128.8	.6	130.0	.6	-1.2	---
Antetor.	23.0	.8	22.4	1.2	.6	---
Mech/Ideal	5.6	.1	5.3	.1	.3**	---
Cond/M-L	9.0	.2	9.1	.3	-.1	---

[1] See Table 3.

[2] mm

[3] degrees

† p<.10 * p<.05 ** p<.01 *** p<.001

The significantly larger tibial plateau/M-L axis angle in males (i.e., in anatomical position--tibial longitudinal axis vertical--tibial plateaus slant downwards more from lateral to medial edge in males), without a corresponding sex-related difference in the femoral condyles/M-L axis angle, indicates that the longitudinal axis of the femoral diaphysis is closer to the vertical in males. This, and the greater angle between the "mechanical" and "ideal" axes of the femur in males, places the femoral head in a relatively more medial position in males with respect to the center of the knee joint, the tibial longitudinal axis, and (if tibiae are equally spaced in males and females relative to body size) the mediolateral midline of the body. This finding is consistent with general morphological differences between male and female pelves, with female pelves relatively broad mediolaterally, with acetabula consequently placed relatively further apart than in the male, farther from the midline and center of gravity of the body (Lovejoy et al., 1973; Burr et al., 1977). The biomechanical consequences of this sexual distinction in terms of loading of the lower limb during weight-bearing and femoral and tibial cross-sectional morphology, will be explored further in succeeding parts of this chapter and chapter 6.

The other angles measured on the tibiae and femora of the study sample are not significantly different between males and females, although the significance level for the greater degree of tibial "torsion" (lateral twisting of the distal end) in males is .09. The values for the cervico-diaphyseal angle of the femur--128.8° in males and 130.0° in females--are very close to those of a large English sample reported by Pearson and Bell (1919), who apparently used a measurement method much like that in the present study (although their description does not make this completely clear). I measured the cervico-diaphyseal angle as a projection into the frontal plane of the femur (Figure 3). Most other authors (e.g., Parsons, 1914; von Lanx and Wachsmuth, 1938; Billing, 1954) give slightly lower angles, averaging 125-126° in the adult. The method of measurement of this angle, whether as a projection into the frontal plane (as done here) or as a direct measurement in the antetorsion plane (plane of the head and neck), will alter the values obtained, the magnitude of the alteration depending primarily upon the size of the antetorsion angle. In fact, I determined that the mean cervico-diaphyseal angle in the present study sample falls to 126.4° when measured in the antetorsion plane--a very similar reading to that obtained by other authors using a "non-projection" antetorsion plane measurement technique. Most investigators do not report a significant sex difference in this dimension, although Langer (1865, quoted in Backman, 1957) found a 10° average difference between males and females (females smaller).

Most other angles measured in the present study, e.g., the "bi-condylar" angle (condyles/M-L axis) and "mechanical"/"ideal" axes of the femur, and the tibial "torsion" angle, also agree well with the general range of values reported for these angles by other authors (Mikulicz, 1878, quoted in Elftman, 1945; Le Damany, 1901; Parsons, 1914; Pearson and Bell, 1919; Martin, 1928; Elftman, 1945; Hutton and Scott, 1949; Steindler, 1955). However, one angle--the femoral antetorsion angle--is significantly greater on average in the present study sample than average values reported by other investigators. Many authors have studied this angle (always as measured here--as a projection in the transverse plane of the femoral diaphysis), because of its clinical importance in diagnosing and treating various orthopedic disorders, particularly in children. Values for the antetorsion angle in a large series of studies, mainly of modern Europeans (both excised bones and _in vivo_), range from 8° to 18°, with the average about 12-15° (Parsons, 1914; Pearson and Bell, 1919; Elftman, 1945; Kingsley and Olmsted, 1948; von Lanx, 1949; Billing 1954; Backman, 1957; Shands and Steele, 1958; Roberts, 1962). This contrasts with 22-23° mean figures obtained for this angle in the present study (Table 4). Greater "torsion" of the proximal femur has been noted as a "racial" feature of American Indians. Stewart (1962) measured the degree of "torsion" through an indirect method (taking heights of several points above a measuring board surface) in American Black, White, and Indian samples, and found the Indian sample to average significantly greater degrees of proximal femoral torsion than the other two samples. He also found a relationship between the antetorsion angle and the degree of femoral bowing, or anterior curvature of the diaphysis, which was again more marked in the Indian sample. Hooton observed "medium" or "pronounced" femoral torsion and anterior curvature in the majority of his sample from Pecos, and also noted the bowing as an "Indian" trait (1930: 138). Some reports in the literature also indicate a higher antetorsion able in certain other archaeological samples, e.g., a Neolithic sample from France which averaged 29° (quoted in Pearson, 1919).

The factors responsible for these apparent differences among human populations are not clear, and are outside the scope of the present investigation. The relationships between the femoral antetorsion angle, other "whole bone" angles, femoral anterior curvature, and individual cross section properties in the present study sample will be discussed further in future sections of this study.

Age Differences

Age differences in whole bone dimensions, by sex, are presented in Tables 5a and 5b. To simplify this presentation, means for only the two broader age groupings (20-39 and 40+ years) are given here (analysis of the six finer age divisions demonstrates the same general age-related patterns of change).

It should also be noted that in this section, and in future sections, differences in cross section properties between age groups are presented as "age changes" or "age trends", although the data here are derived from a cross-sectional sample and do not actually document longitudinal "growth" changes in individuals directly. As discussed earlier in chapter 3, there are several possible explanations for age group differences in cross-sectional samples, although true ontogenetic (growth) change is probably the most important factor in the present

Table 5a

Age Differences in "Whole Bone" Dimensions: Males

Dimension [1]	20-39 yrs Mean	20-39 yrs SE	40+ yrs Mean	40+ yrs SE	40+ yrs - 20-39 yrs	%dif	F [2]
Tibia							
Lengths [3]							
Maximum	356.8	2.4	362.8	3.7	6.0	1.7	2.96*
Biomech.	337.5	2.5	343.1	3.6	5.6	1.6	2.89*
Angles [4]							
Torsion	19.9	1.3	19.0	1.7	-.9	---	.92
Plat/M-L	4.1	2.9	3.4	.6	-.7	---	.98
Plat/A-P	12.0	.4	12.8	1.0	-.8	---	.96
Femur							
Lengths [3]							
Maximum	425.8	2.6	435.7	3.5	9.9*	2.3	2.61*
Biomech.	402.1	2.6	411.2	3.0	9.1*	2.3	2.16†
Angles [4]							
Cerv-Diaph	129.2	.7	127.9	.9	-1.3	---	1.11
Antetor.	23.6	1.2	22.0	1.0	-1.6	---	.28
Mech/Ideal	5.6	.1	5.7	.1	.1	---	.33
Cond/M-L	8.8	.2	9.2	.4	.4	---	.34

[1] See Table 3.

[2] Results of 6 age group ANOVA--see text.

[3] mm

[4] degrees

† $p<.10$ * $p<.05$

Table 5b

Age Differences in "Whole Bone" Dimensions: Females

Dimension [1]	20-39 yrs Mean	20-39 yrs SE	40+ yrs Mean	40+ yrs SE	40+ yrs - 20-39 yrs	%dif	F [2]
Tibia							
Lengths [3]							
Maximum	324.3	2.9	326.7	3.4	2.4	.7	.78
Biomech.	307.8	2.9	308.5	3.3	.7	.2	.59
Angles [4]							
Plat/M-L	17.2	1.5	16.3	.7	-.9	---	.41
Plat/A-P	3.0	.4	2.7	.7	-.3	---	.48
Femur							
Lengths [3]							
Maximum	392.9	3.1	393.0	3.8	.1	0	.72
Biomech.	370.4	3.0	373.4	3.7	3.0	.8	1.01
Angles [4]							
Cerv-Diaph	131.1	.6	127.9	1.1	-3.7**	---	1.70
Antetor.	24.6	1.3	17.8	2.1	-6.8**	---	1.83
Mech/Ideal	5.2	.1	5.7	.2	.5**	---	3.85**
Cond/M-L	9.2	.3	8.8	.6	-.4	---	1.93†

1 See Table 3.

2 Results of 6 age group ANOVA--see text.

3 mm

4 degrees

† p<.10 ** p<.01

sample. For convenience, then, in the following discussions age differences are presented as representative of longitudinal age changes, but it should be borne in mind when interpreting results that alternative explanations are possible (these points are discussed again below in reference to bone length age differences).

The most interesting age changes apparent from Tables 5a, b are those in the femoral angles, particularly in females. Both cervico-diaphyseal and antetorsion angles of the femur show significant declines with age in females. Male declines in these angles are much less marked and therefore are nonsignificant, but show the same pattern of change with age. Females also show a significant <u>increase</u> in the "mechanical"/"ideal" femoral axes angle with age, while males remain essentially unchanged in this angle. As will be shown later, the decrease in the cervico-diaphyseal angle as measured here (projected into the frontal plane) is brought about mainly by the decline in antetorsion angle with age. A decreased antetorsion angle will also produce an increase in the "mechanical"/"ideal" axes angle of the femur, since this angle was also measured as a frontal plane projection, and a lowered antetorsion angle (without any corresponding change in the length of the femoral neck) will bring the femoral head into a relatively more medial position, viewed in the frontal plane. Thus, changes in the antetorsion angle with age account for essentially all observed age-related changes in other proximal femoral angles. Interestingly, these changes bring about a reduction in sexual dimorphism in these other angles, although the antetorsion angle itself becomes more sexually dimorphic with age.

The age-related decline in the antetorsion angle of the femur in adults of the study sample appears to be a continuation of a general trend established during childhood and adolescence in many population samples. Several investigators have documented a decrease of 20° or more in the antetorsion angle from infancy through adolecence (using cross-sectional samples) (Kingsley and Olmsted, 1948; Lanz and Mayet, 1953, quoted in Roberts, 1952; Billing, 1954; Shands and Steele, 1958; Crane, 1959). Torsion of the proximal femur progressively develops during the prenatal period, increasing from no torsion or even negative torsion (backward incliniation of the head and neck) in the early fetal period to a maximum of 35-40° at birth (Roberts, 1962). From birth onwards it declines, with the sharpest drops at about 1-2 years and again during early adolescence (Billing, 1954; Shands and Steele, 1958). Proposed explanations for this decline usually involve "postural mechanisms and alterations concerned with growth and weight transmission" (Frazer, 1965: 120). Pathologies such as congenital dislocation of the hip, achondroplasia, cerebral palsy, and abnormal mechanical stresses (such as unusual habitual sitting or sleeping postures) have been shown to affect the magnitude of the antetorsion angle in children (Shands and Steele, 1958; Crane, 1959; Roberts, 1962).

The cervico-diaphyseal angle has also been shown to decrease during childhood and adolescence, from an average of 145° at one year of age to 130° or less in adulthood (von Lanz, 1940; Shands and Steele, 1958). Again, the sharpest declines in the angle occur from one to two years of age and again during early adolescence. It should be noted, however, that there are difficulties in obtaining an accurate measurement of this angle from *in vivo* radiographs (e.g., Billing, 1954), and that most investigators apparently take the angle as a projection in the frontal plane, therefore, the great decline in the antetorsion angle during the pre-adult years may partially account for this observed decline in the cervico-diaphyseal angle. At least one study suggested that the cervico-diaphyseal angle might continue to decrease throughout adulthood, "and in advanced age could be 120 degrees without necessarily being regarded as pathologic" (Backman, 1957: 13, quoting study by von Lanz and Wachsmuth, 1938). The results of the present study do not support this contention, especially when changes in the antetorsion angle and consequent changes in perspective are accounted for.

Before leaving the subject of age differences in whole bone angles, mention should be made of reported pre-adult age changes in the degree of distal tibial torsion. Unlike proximal femoral torsion, distal tibial torsion has been shown to *increase* during childhood, from very slight or no torsion at birth to close to adult values by the age of 5 or 6 years, the most rapid increase occuring during the first year of life (data quoted in Elftman, 1945; Hutter and Scott, 1949). Like the femoral antetorsion angle, tibial torsion can be affected by various pathologies and abnormal mechanical stresses applied by various pathologies and abnormal mechanical stresses applied during early childhood (*ibid.*) In the present study sample, with the exception of one individual (female) with a tibial torsion angle of -1°, no individuals were found with "negative" (counter-clockwise as viewed from above) torsion of either the tibia or femur, the range for the tibial torsion angle extending from -1° to 36°, and the femoral antetorsion angle from 4° to 43°. No significant change with age in tibial torsion in either sex was observed.

Because proximal femoral and distal tibial torsion angles are similar in direction (clockwise viewed from above) and magnitude in the present study (averaging 23° and 18°, respectively), it might be hypothesized that some direct functional relationship between the two angles exists. However, the correlation between the angles over the entire sample is only .230 (males .190, females .250), which, while statistically significant, indicates only a very weak association between femoral and tibial torsion on an individual basis. Interestingly, Elftman (1945), who carried out a similar analysis on 35 male cadavers, found a correlation between the two angles of .225, very close to that of the present study. It should also be noted that the antetorsion angle of the femur declines significantly with age in females in the present study sample, while tibial torsion is nonsignificantly different in the two groups. Thus, the direct relationship between the two angles appears to be weak, although a more complex structural-functional relationship, involving other angles and other structural features, may exist (also see the discussion of bilateral asymmetry below, where the two angles do appear to vary in concordance).

The other significant age-related difference in whole bone dimensions in the study sample apparent in Tables 5a, b is an average increase in bone length in males. The difference is statistically significant only in femora, averaging 2.3%, while tibiae average a 1.7% gain in maximum length ("biomechanical" lengths show almost identical age differences). Females show no significant age changes in these dimensions, averaging less than a 1% change for either bone.

Since long bones do not increase appreciably in length after epiphyseal union occurs, the age differences in bone length observed in the adult males of the study sample cannot be a result of true ontogenetic (growth) changes with age. The observed differences could be caused either by 1) selective preservation of the material, i.e., only the largest, most robust of the thinner and weaker long bones of older individuals withstanding burial at Pecos; 2) selective survival, with larger males in the Pecos population more likely to survive to old age; or 3) some other nonrandom sampling of the site with respect to older and younger adult males (also see the discussion of these issues in chapter 3). Tibial and femoral maximum bone lengths for males and females are plotted by the narrower six age group divisions in Figure 9. Figure 9 shows that there is substantial fluctuation in bone lengths from age group to age group, particularly in males, and that the "trend" to longer bones in older males indicated in Table 5a is in fact neither regular nor consistent. In fact, closer examination of the data indicates that a few individuals may be inordinately influencing the magnitude of the observed age differences. For example, if a single very large individual (more than 1 cm longer than any other in the entire study sample) is eliminated from the 50+ year male age group, the mean maximum femoral bone length for this group is reduced by more than 4 mm, and the difference between 20-39 and 40+ year age groups becomes nonsignificant. Similarly, if one very small male (shortest by more than 2 cm of all males) is eliminated from

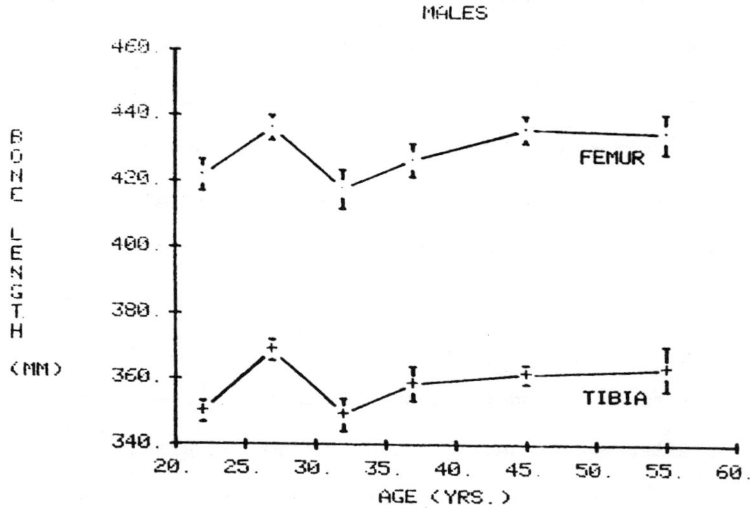

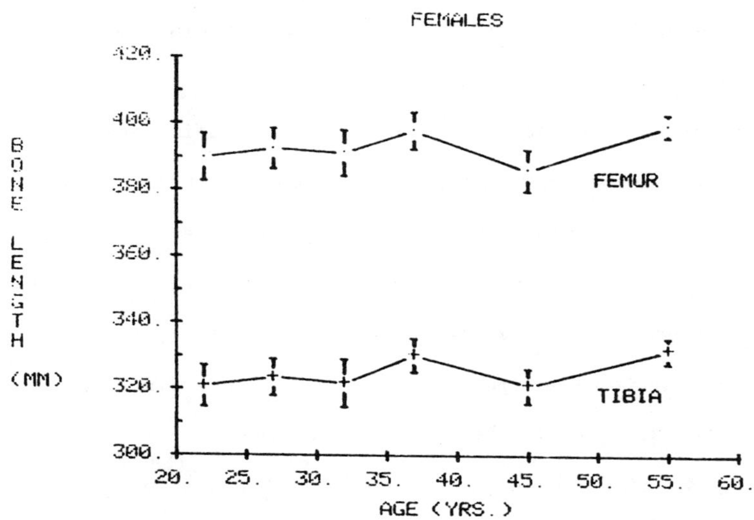

Fig. 9. Age changes in maximum femoral and tibial lengths.

the 30-34 age group, the mean femoral length for this group is raised by almost 5 mm, substantially reducing the age-related fluctuation in this dimension (see Figure 9) and further reducing the average difference between the two larger age groupings from 2.3% to 1.5%. Thus, random sampling error may very well play an important part in creating apparent "age trends" in bone length in males, although it is still possible that some real age-related changes, for example, those which would be brought about by selective survival in favor of larger individuals, are also operating. (In this case, one would have to postulate that this mechanism operated only in males, since females do not show a similar age trend. The lack of age differences in bone length in females also appears to disfavor the archaeological preservation hypothesis, since females lose more bone than males with aging and thus would be expected to show even more disproportionate preservation of larger, more robust individuals.)

In terms of the present study results, the most important consequence of fluctuations in bone lengths is its effect on other bone dimensions, e.g., the geometric areal properties of the bone cross sections. Partitioning out the effects of "size" variation between individuals in a biomechanically meaningful way is a very difficult task, and the entire issue of "size standardization" needs much more investigation before it is resolved.

In reviewing the differences in bone length between male age groups, it should also be remembered that, although statistically significant, the differences between the two larger age groupings are in reality relatively small, averaging only about 2% of bone length.

Side Differences

Differences between right and left whole bone dimensions, by sex, are shown in Table 6 (note that, as explained in the previous chapter, these are not true "bilateral" differences, since right and left sides were obtained from different individuals). The most conspicuous side-related differences in the study sample involve the tibial "torsion" and femoral antetorsion angles, both of which are smaller on the left side. Although significant in both sexes, this difference is particularly marked in females, where the difference reaches more than 10° in the tibia, and 8° in the femur. In effect, at both proximal and distal ends, the left lower limb bones appear to be oriented in a direction closer to the anatomical A-P and M-L planes of the body, while the right limb bone ends present a more "twisted" orientation relative to the standard frontal plane defined by the knee axes (it is also conceivable, although seemingly less likeley, that the knee joint is more medially "twisted" in right lower limbs). Several other investigators have noted a similar bilateral difference in tibiae (Le Damany, 1909; Hutter and Scott, 1949) and femora (Parsons, 1914--females only; Kingsley and Olmsted, 1948). Kingsley and Olmsted reported the bilateral difference in the femoral antetorsion angle--only about 1.5° in their sample--to develop only from childhood on, with infants under 2 years of age exhibiting no side difference. One study (Pearson and Bell, 1919) has reported greater proximal femoral torsion on the left side, contrary to most other studies, including the present study.

Although, as stated earlier, there is only a very weak correlation between the torsional angles of the proximal femur and distal tibia on

Table 6

Side Differences in "Whole Bone" Dimensions

Dimension [1]	Males			Females		
	Right	Left	%dif [2]	Right	Left	%dif [2]
Tibia						
Lengths [3]						
Maximum	358.4	359.3	-.2	326.9	323.4	1.1
Biomech.	338.7	340.1	.4	309.4	306.6	.9
Angles [4]						
Torsion	22.0	17.1	4.9* [5]	22.0	11.7	10.3*** [5]
Plat/M-L	4.3	3.5	.8 [5]	2.0	3.7	-1.7* [5]
Plat/A-P	11.6	12.9	-1.3 [5]	12.7	11.5	1.2 [5]
Femur						
Lengths [3]						
Maximum	428.6	429.7	-.3	393.7	392.2	.4
Biomech.	405.6	404.8	.2	372.8	370.1	.7
Angles [4]						
Cerv-Diaph	129.0	128.6	.4 [5]	129.7	130.4	-.7 [5]
Antetor.	25.1	21.0	4.1** [5]	26.4	18.4	8.0*** [5]
Mech/Ideal	5.6	5.7	-.1 [5]	5.4	5.3	.1 [5]
Cond/M-L	9.2	8.7	.5 [5]	9.0	9.2	-.2 [5]

[1] See Table 3.

[2] [(Rt-Lt)/Rt] X 100

[3] mm

[4] degrees

[5] Rt-Lt

* p<.05   ** p<.01   *** p<.001

an individual basis, the concordance of the bilateral differences shown by the two angles, even to the point of similar sex/side differences (see Table 6), seems strongly suggestive of some kind of functional relationship, a relationship which may be partially obscured in a simple correlation statistic by the large degree of individual variation shown by the two angles. Whether the side-related differences in these angles observed here indicate a difference in use of right and left limbs during locomotion or other activities, or a difference in how the lower limbs are held during characteristic postures such as sleeping or sitting positions, is difficult to determine from these data. Relationships between the two angles and side differences in cross-sectional properties are discussed in a future section of this chapter.

The other significant side difference shown in Table 6 is in the tibial plateau/M-L axis angle in females, where the left side is greater. In males exactly the opposite (but nonsignificant) difference exists. These differences do not appear to appreciably affect the "mechanical"/"ideal" axes angle of the femur in either sex, which may be a result of compensating differences in the femoral condyles/M-L axis angle of the femur (larger on left in females, on right in males). No side differences in bone lengths were observed for either bone. Several authors have reported slightly longer left limb bones in paired samples (Munter, 1936; Schultz, 1937; Trotter and Gleser, 1952; Latimer and Lowrance, 1965). The lack of any consistent side-related differences in bone length in the present sample is probably due to the nature of the sampling procedure employed (bones not paired from the same individual).

CROSS SECTION GEOMETRY

Before presenting the results of the analysis of cross-sectional properties in the study sample, several issues regarding the interpretation of these properties will be discussed. First, the general biomechanical significance of the various geometric properties examined will be reviewed. While some of these properties, such as cross-sectional bone areas, are well known and often used in studies outside the field of biomechanics, others, such as second moments of area (area moments of inertia) are less familiar. In addition, all geometric properties, including simple areas, take on somewhat different meanings when interpreted in a functional (biomechanical) sense.

The second issue explored is that of "size" standardization, and in particular, the reasons why such standardizations are not used in the present study. Finally, some technical details regarding the method of measuring and reporting the angle between principal axes and x, y reference axes (phi) will be discussed. As will be seen, it was found necessary to perform certain transformations of the data in order to obtain meaningful estimates of the range of variation and central tendencies exhibited by these angles.

Biomechanical Significance of Cross Section Geometric Properties

Bones *in vivo* are subjected to a complex array of different forces brought about by gravity, muscle action, and resistance to bone movement through tendonous and ligamentous connections and the shapes of bone joints and articulation surfaces. In order to apply the findings of engineering mechanics to the musculoskeletal system, certain simplifications regarding the forces applied to bones, or bone loadings, must be made. In the present study, the tibia and femur are modeled as beams, supported at each end and subjected to axial, bending, and torsional loadings. Both bones undergo axial and bending stresses; in addition, the tibia is subjected to significant torsional stress during normal locomotion (Lanyon, 1975; Carter, 1978).

In this discussion, bone is assumed to behave as a homogeneous, isotropic material (Reilly and Burstein, 1974) (further discussion of material properties is presented later in connection with the interpretation of BMC data). The relationships between bone cross-sectional geometry and stress described here (e.g., the "flexure formula"--see below) are strictly valid only up to the elastic limit of bone, i.e., within the straight line portion of the stress-strain curve before the occurence of plastic deformation (_ibid_.; Currey, 1970). While this factor may affect estimates of the magnitude of ultimate stress in bone (_ibid_.: 220-222), the purpose of the present investigation is not to arrive at specific predictions for the maximum stress at failure, but rather to measure _relative_ degrees of stress under loadings given particular cross-sectional geometric configurations. Therefore, assuming that equivalent plastic deformation would occur in all specimens analyzed, this factor may be ignored in the following discussion.[2]

---

[2]Also see the discussion of future stress analyses in chapter 6.

It should also be noted that in some of the following discussions, bone "strength" and "rigidity" have been used almost as if interchangeable. In fact, "strength", as defined mechanically, refers to the failure characteristics of a specimen, usually taken as the ultimate stress, or stress at which failure (fracture) occurs under bending or some other loading configuration (e.g., Klenerman et al., 1967; Currey, 1970; Reilly and Burstein, 1974). "Rigidity" is defined in terms of specimen deflection, or strain, under loading _prior_ to failure, with more rigid specimens showing less deflection (_ibid._). Both rigidity and strength are affected by material properties (modulus of elasticity, or stiffness) and geometric properties of the specimen (see below). Strength is a direct measure of the likelihood of failure, and thus is the most important characteristic in evaluating fracture probability. However, bone normally undergoes remodeling under _in vivo_ loadings which do not cause fractures but do set up local strains. In other words, while the ultimate result of remodeling may be to produce stronger bones, the actual process by which this is brought about may be controlled more by localized stresses, strains, and rigidity characteristics. Therefore, while the emphasis here is on the optimization of strength in bones, it should be realized that other parameters, e.g., rigidity, may be more important in the actual remodeling process.

Axial loading.

Axial loading of a cross section results from forces applied to beam (or bone) ends which run along the longitudinal axis of the beam through the section centroid. For the beam to be in static equilibrium, the forces applied to the two ends must be equal in magnitude and opposite in sense. Axial loading produces normal stress (internal stress parallel to the longitudinal axis)--compressive if both forces are directed toward the beam, tensile if both forces are directed away from the beam ends. Since stress is defined as force/area over which force is applied, the magnitude of tensile or compressive stress under a particular axial loading will be dependent upon the cross-sectional area of the beam (or bone) (Shigley, 1976). Thus, cortical bone area in a long bone diaphyseal cross section is an appropriate dimension for evaluating relative bone strength under pure axial loading.

Bending loading.

However, skeletal elements *in vivo* are rarely, if ever, subjected to pure axial loading. In the first place, many forces (and almost all muscular forces) acting upon bones have lines of application which are not coincident with the longitudinal axis running through the center of the bone (more precisely, the neutral axis of the bone which contains section centroids). For example, during normal stance or bipedal locomotion, the center of gravity of the body, producing a downward-directed force, is medial to the longitudinal axes of both lower limbs. Such a force produces a mediolateral bending moment, the magnitude of the moment determined by the magnitude of the applied force (i.e., body weight) and the perpendicular distance from the line of application of

---

[3]Actual physiological forces may be distributed over a relatively large surface area, however, for analysis of stresses distant from the point of application of the force (e.g., a long bone diaphysis relative to an articular end), these forces may be idealized as a single equivalent force vector applied to a single point (Frankel and Burstein, 1970).

the force to the longitudinal axis of the bone (section centroid)- the greater the distance, the greater the moment. Secondly, long bones themselves are never perfectly straight, for example, the tibia and femur both show marked curvature in the anteroposterior plane relative to articular endpoints (and some mediolateral curvature as well, as will be shown later). Such deviations from a straight longitudinal axis (e.g., the "ideal" axis of the femoral diaphysis) will also produce bending moments within the bone, varying according to the degree of curvature and the eccentricity of the loading (perpendicular distance from the straight longitudinal axis to the area centroid of the cross section). Thus, even axial loadings applied to the articulation end centerpoints of bones can produce bending moments if the bones are curved.

Bending moments produce both compressive and tensile normal stresses, compression occurring on the concave surface under bending, tension on the convex surface. (Shear stress, or stress perpendicular to the longitudinal axis, also occurs under bending, but except in very short beams is relatively small (Piziali et al., 1976), and for present purposes can be ignored.) Because bone is stronger in compression than tension (Currey, 1970), and because axial stress in the lower limb is almost exclusively compressive, loadings which produce bending, and thus tensile stress, are very important in determining failure of the lower limb bones under normal physiological conditions (torsional loading is considered below). Thus, the cross-sectional geometrical properties which determine the magnitude of stress under a bending load are of considerable importance in evaluating bone rigidity, bone strength and probability of failure (fracture).

Between the concave (compression) and convex (tension) sides of a beam under a bending load lies a plane in which strain, or bone displacement, is zero. This is known as the neutral plane, and within a single cross section, as the neutral axis (for more detailed discussion of these properties and those which follow, see standard engineering texts, such as Timoshenko and Gere, 1972; Shigley, 1976). The magnitude of compressive or tensile normal atress in any unit area of a beam cross section under bending can be shown to be a direct function of the distance of the area from the neutral axis, and a reciprocal function of a geometrical property known as the second moment of area, or area moment of inertia. This relationship is expressed in the "flexure formula", given below:

$$\sigma = \frac{M\,y}{I} \tag{1}$$

where σ = normal stress (compressive or tensile in in the unit area

M = externally applied bending moment

y = distance from neutral axis to unit area

I = second moment of area of cross section

Maximum stress occurs in the outermost fibers of the cross section in the plane of bending, where $y = y_{max}$; y_{max} is often denoted as "c", and substituted into formula (1) to obtain :

$$\sigma_{max} = \frac{M\,c}{I} \tag{2}$$

$$\text{or } \sigma_{max} = \frac{M}{I/c}$$

where $I/c = Z$ = section modulus

The second moment of area, I, is calculated as the sum of squared distances from the neutral axis of elemental areas times these elemental areas, taken over the entire cross section ($I = \int y^2\, dA$). Thus, both total area and distribution of area around the neutral axis ("shape") are important in determining the magnitude of this property (also see Figure 10 below). I is expressed in units to the fourth power, e.g., mm^4.

Whereas the second moment of area is a geometrical property of an entire cross section, Z, the section modulus, indicates the stress at the point of maximum stress (or points, if the section if symmetrical), given a specified bending moment. Both properties will be examined in the present study. However, the emphasis will be on second moments of area, since these are better indicators of total bone cross-sectional geometry the principal focus of this study. (There are other problems involved in interpreting the section modulus, since the location(s) of maximum stress in a cross section will actually vary in a complex way according to the geometry of the section and the direction of the applied bending moment--for further discussions of these points, see Lovejoy et al., 1976; Jungers and Minns, 1979; Lovejoy and Barton, 1980).

Second moments of area of cross sections may be calculated for any plane of bending, that is, around any neutral axis. In beams of regular, symmetric cross-sectional geometry, such as rectangles or I-beams, second moments of area, denoted I_x and I_y, can be calculated about x and y axes of symmetry, respectively, using relatively simple, widely available formulas (e.g., Timoshenko and Gere, 1972). For irregular, asymmetric cross sections, the calculation of these properties is much more difficult. As stated earlier (chapter 4), SLICE uses a summation of small elemental area properties to calculate second moments of area, with the size of the area elements determined by the digitizer increment between boundary point coordinates. SLICE calculates I_x and I_y around both original x and y reference axes (always chosen to fall completely outside the section area--see Nagurka and Hayes, 1980), and x and y axes translated to the section centroid. In the present study, x and y axes correspond to mediolateral and anteposterior directions, respectively, as described in chapter 4.

However, in long bone cross sections, M-L and A-P directions rarely correspond exactly to the axes about which second area moments reach a maximum and a minimum. These axes--the principal axes of the cross section--are of particular interest since they indicate the directions of maximum and minimum bending rigidity and strength in the cross section. The <u>major axis</u> runs perpendicular to the axis about which the second moment of area is greatest--it thus indicates the direction of maximum bending rigidity. The <u>minor axis</u> is perpendicular to the major axis and indicates the direction of minimum bending rigidity. The angle between x and y reference axes and the principal axes is denoted phi.[4] In the present study, phi is measured as the angle between the x (M-L) axis and the major axis of the section, expressed in degrees. Maximum and minimum second moments of area, referred to by various notations in other reports, are here simply denoted as I_{max} and I_{min}, respectively. These properties are illustrated in Figure 10.[5]

Another property found to be useful in describing variation in cross-sectional geometry is the ratio I_{max}/I_{min}. This is both an index of relative bending strength in two planes, and an index of the relative distribution of cross-sectional area in two dimensions. As such, it functions in effect as a biomechanical interpretation of bone cross-sectional "shape", and may be compared, for example, with the traditional osteometric platycnemic or platymeric indices based on outer dimensional breadths (e.g., Jungers and Minns, 1980: Fig. 3).

---

[4]Algorithms used by SLICE for calculating all of these section properties are listed in Nagurka and Hayes, 1980.

[5]Figure 8, p. 130, illustrates the computer graphics capability for representation of bone cross-sectional properties. In this figure, abbreviations for section properties listed in the right column are as follows: XBAR, YBAR: centroid coordinates relative to original x and y axes (plotted); IX, IY: second moments of area about original x and y axes; IXBAR, IYBAR: second moments of area about x and y axes translated to the section contriod; IXBPH, IYBPH: second moments of area about principal (rotated) axes (IXBPH is I_{max}). Phi is expressed in this figure as the angle between the x axis and minor axis, in radians (see also footnote 8, p. 181). IXY and IXBYB refer to products of inertia about original and translated x and y axes, respectively (not defined here--used in determining direction of principal axes). Lengths of plotted principal axes indicate the relative magnitude of second moments of area about these axes.

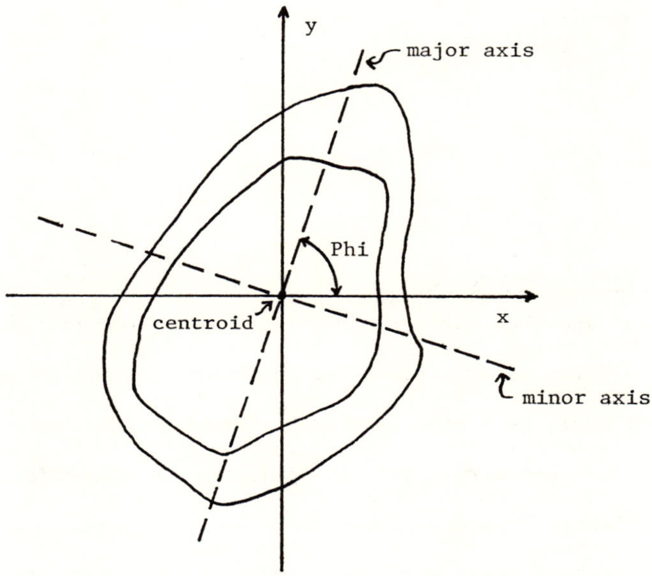

Cortical area = 216 mm^2

Subperiosteal area = 506 mm^2

I_{max} = 24290 mm^4

I_{min} = 7976 mm^4

Phi = 71.9°

Fig. 10. Tibial cross section (65% location) illustrating several geometric properties calculated in the study (see text).

Angle (Phi) Transformations

Phi, as described earlier, is defined in the present study as the angle between the major axis of the cross section and the mediolateral axis. This angle is a very important element in the description of cross-sectional geometry, since it indicates the direction of maximum bending rigidity. Several authors have reported values for phi for femoral and tibial cross sections (Kimura, 1974; Minns et al., 1975; Lovejoy et al., 1976; Piziali et al., 1976; Martin and Atkinson, 1977; Miller and Purkey, 1980). The specific findings of these investigators with regard to this property will be discussed in a later section in connection with the present study results; at this time, several points regarding the method of analysis of phi, in particular the problems arising where phi shows great interindividual variation, will be presented and discussed.

The angle between the principal axes of a cross section and standard anatomical axes has been presented in several ways by different investigators. In studies where only small samples are used, individual values for this property for all members of the sample may be presented and interpreted (e.g., Minns et al., 1975: Fig. 4). Where larger samples are involved, individual variation in phi increases, and interpretation becomes more difficult. This is particularly true in the femur, which because of its relative circularity in most diaphyseal regions shows greater variability in this property than the tibia, a more triangular-shaped bone with more strongly defined directions of maximum bending strength. As an example of the problem this can create, Martin and Atkinson (1977) in their study of femoral cross-sectional geometry, presented their findings with regard to phi only as frequency distribution graphs, with no attempt to calculate means or to examine sex or age-related differences in this property. Although they suggested some tendencies in these angles to approach certain values in a majority of individuals (which will be discussed later), the great variation betwen individuals in this property clearly made any attempts at defining "central tendencies" very difficult (_ibid._: Fig. 12). Amtmann (1971: 52-53) also commented on "the enormous variation in cross-sectional shape and size" of his femoral cross sections. Again, a "tendency" for cross-sectional area to be distributed in certain directions is observed, as well as possible age changes in these characteristics, but he notes that "a significance for this cannot be proved". No statistical treatment of phi is presented, and one has only tracings of cross section outlines (_ibid._: Fig. 31) upon which to form a subjective judgment regarding the validity of his observations.

Possible reasons for the observed variability in shape of the femoral diaphysis, particularly in its middle region, will be discussed later; of concern here is how to present and interpret this variation in such a way as to make meaningful generalizations about phi, including comparisons between differenct sex and age groups. The difficulty in determining central tendencies for this angle in past studies (particularly of the femur) is partly due to the lack of sufficient numbers of individuals in the samples analyzed to date (maximum N = 37), and also to the method of reporting these data. In the present study, a large enough sample was available so that central tendencies in phi could be clearly distinguished for each cross section. The availability of more complete frequency distrubutions also made it possible to distinguish certain modifications in how these angles are measured which are necessary in order to carry out meaningful statistical comparisons between groups.

First, it should be noted that principal axes run through a cross section from one outer edge to the other, passing through the section centroid, which is also the origin of the translated x and y axes (see Figure 10). Thus, no matter how measured, the phi's for a particular cross section can vary by a maximum of 180°--angles outside this range are equivalent to phi - 180° (e.g., an angle of 190° between the major axis and M-L axis is the same as an angle of 10° between these axes). A decision must be made where to take this 180° range (e.g., 0° to 180°, or -90° to +90°) when devising the calculations performed to determine phi.[8] The problem arises when phi "overlaps" the chosen

---

[8] SLICE's algorithms (as developed by Nagurka and Hayes--see chapter 4) calculate the angle between the x (M-L) axis and the minor axis, and use a range of -90° to +90°, reported in radians (i.e., $-\pi/2$ to $+\pi/2$)--for example, an angle of 95° is recorded as one of -85°. For the pur-

upper or lower limit of this range. For example, the range of phi might be chosen as 0° to 180°, with 0° (and 180°) representing the mediolateral (x) axis. If the major axis of a particular bone cross section were to approximate, on average, the mediolateral axis (as does the 20% femoral cross section, as shown below), phi's would be measured under this system as varying from, say, 0° to 30°, and from 150° to 180°. This creates the absurdity of measuring angles of 1° and 179° as "178 degrees" apart, when in fact they refer to almost the same angle, and of deriving a "mean" phi of about 90° when the true mean is actually close to 0° (or 180°). This particular problem could be solved by shifting the range of possible phi's to -90° to +90°, thus measuring angles of 150° to 180° as -30° to 0°, and deriving a mean phi of close to 0°, with a range of -30° to +30°. However, for similar reasons, this new range will not work for phi's which vary around 90° or -90°, and the original (0° to 180°) range must be reverted to for measuring this angle.

It might be thought that these two ranges--0° to 180° and -90° to +90°--would be sufficient for measuring all angles of this kind. However, this still leaves the problem of what to do when an angle varies by more than 90°, and "overlaps" both the x (0°, 180°) and y (-90°, +90°) axes. For example, phi for a particular cross section might have a true central tendency of 45°, but vary from -20° to 110°. Using either of the two ranges above, some individuals in the sample will be

---

poses of the present study, radians were converted to degrees, and 90° added to measure the angle between the major axis and the M-L axis, thus, all original "untransformed" phi's presented here (see text and Figures 11 and 12) vary from 0° to 180°

misrepresented, i.e., either those of -20° to -1° (which using the 0° to 180° range will be recorded as 160° to 179°), or those of 91° to 110° (using the -90° to +90° range recorded as -70° to -89°). In this case a different 180 degree range, perhaps from -45° to +135°, would be appropriate.

Thus, the method of measuring and recording phi will depend on the particular cross section in question, and must be reevaluated for each new cross section on the basis of observed frequency distributions. Graphs of phi distributions for the 11 cross section locations included in the present study are presented in Figures 11 and 12a and 12b. (The abscissa (degrees) in these graphs is reversed in direction, increasing in magnitude from right to left, in order to make comparisons with future illustrations of phi easier, e.g., Figures 20a, b. Approach these graphs as if looking down at the proximal surface of a cut cross section from a right femur or tibia which has been laid out with the distal end away from you.)

Figure 11 illustrates the frequency distributions for phi in the five tibial cross sections. With no exceptions, all 119 individuals in the study sample fall between 0° and 90° in all tibial sections (i.e., the major axis is directed anterolaterally in all sections). Thus, no adjustments to the original range of phi chosen (0° to 180°--see footnote 8) was necessary. In fact, the narrow range of variability of the tibial cross sections in this property, with the exception of the distal-most section, is remarkable, varying by no more than 60° degrees, and in some cases by less than 40° degrees. Even the 20% section frequency distribution is relatively narrow in range compared

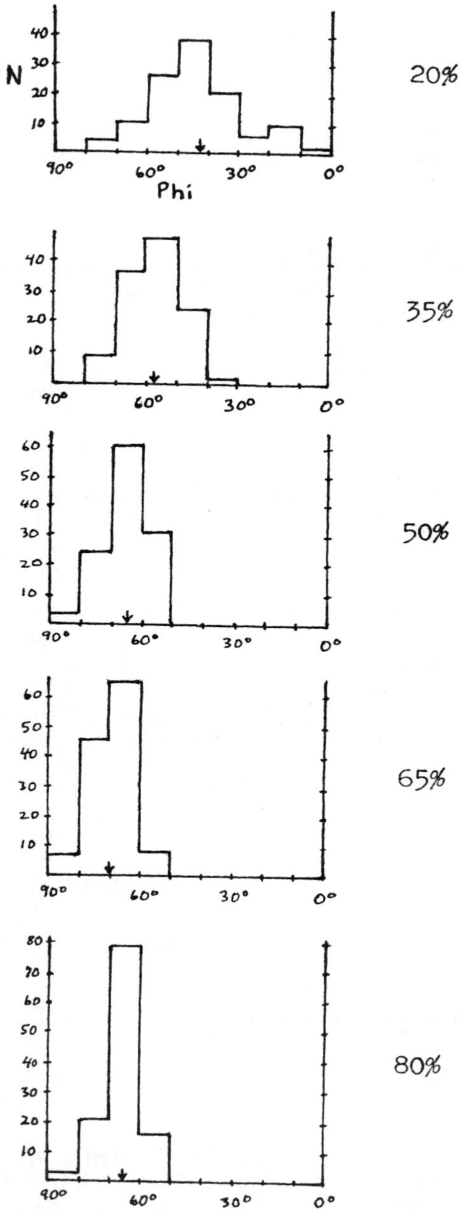

Fig. 11. Frequency distributions of Phi for tibial cross sections (sexes combined). Arrow indicates mean.

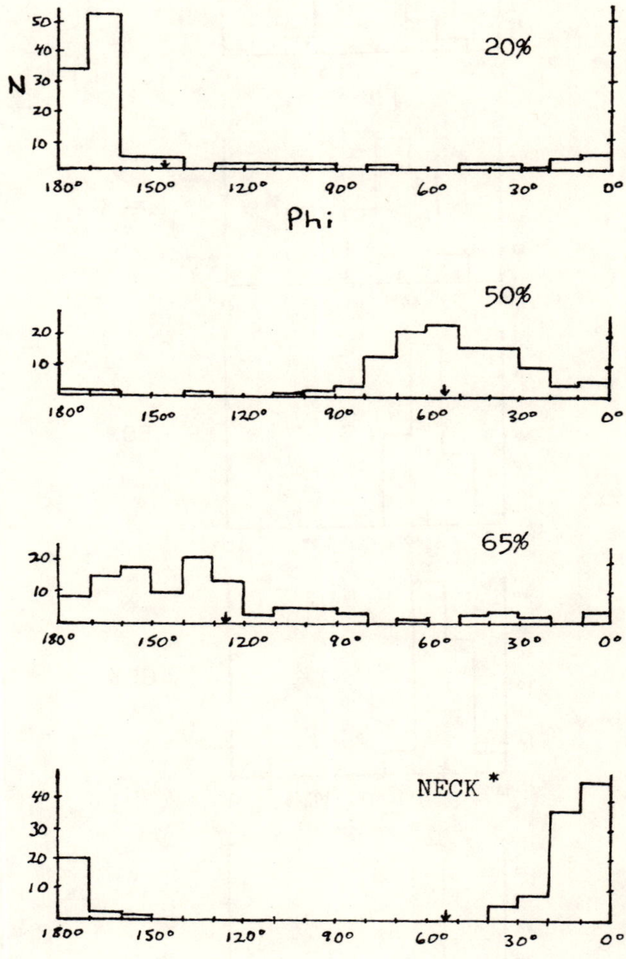

Fig. 12a. Frequency distributions of Phi for original untransformed femoral cross sections, later transformed (see Fig. 12b and text). Arrow indicates mean.

* Phi calculated from total subperiosteal area of neck.

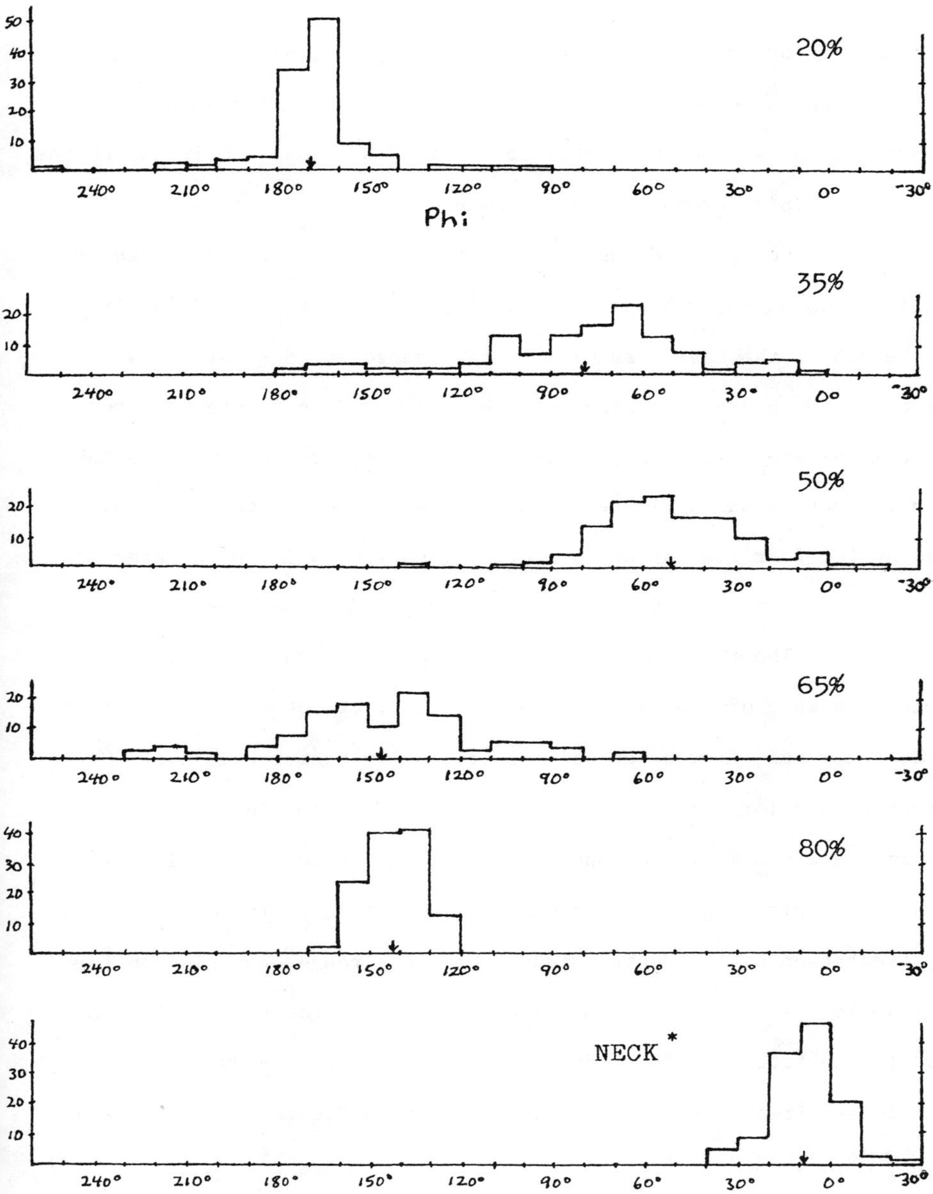

Fig. 12b. Frequency distributions of Phi for femoral cross sections--20%, 50%, 65% and neck transformed (see text). Arrow indicates mean.

* Phi calculated from total subperiosteal area of neck.

to femoral cross sections, most of which vary by close to 180 degrees. Another interesting observation is that the range in tibial phi's decreases steadily from the distal to proximal sections (fully two-thirds of the 80% tibial cross sections examined are characterized by phi's between 60° and 70°). Means for each cross section are indicated by small arrows on the graphs. The variation in these mean phi's along the length of the tibia, as well as the change in range of variation present at each cross section, will be discussed in future sections.

Femoral cross sections, as indicated above, are much more variable with respect to phi. Figure 12a shows the original frequency distributions for four femoral cross section locations which were later "transformed" by shifting the possible ranges for phi (see Figure 12b and below). The femoral neck section[9] provides perhaps the most obvious example of the necessity of shifting, or rotating a phi range. Here, the x axis from which phi was originally measured is the inferior-superior axis of the femoral neck; as shown in Figure 12a, phi for this section varies around the x or inferior-superior axis. The "mean" for this original distribution is 43.1°--clearly not an accurate representation of the central tendency of the range of variation for this angle. In this case, the range for phi was rotated to -90° to +90° (by subtracting 180° from all angles greater than 90°). The transformed frequency distribution is shown in Figure 12b, and has a mean of 8.3°.

---

[9]The phi ahown here is for total subperiosteal area of the femoral neck--the phi distribution for compact cortical area alone is very similar.

The problem with the original phi range in the femoral 20% section is similar to that of the femoral neck, again involving variation around 0° (or 180°), although in this case there is substantial variation through a total range of 180° degrees (note that rotating phi ranges does not reduce the range of variation of this property, only shifts the range of variation to one side or the other). For this cross section, the phi range was rotated to 90° to 270° (by adding 180° to all angles less than 90°); results are shown in Figure 12b.

The other two femoral cross sections which were "transformed"--50% and 65%--required somewhat more judgment in determining the necessity of and then the degree to which the range of phi was to be rotated. Two guidelines were used in deciding these questions: 1) there must be significant skewing of the frequency distribution with obvious "overrun" of the data at either the 0° or 180° end of the distribution, and 2) there must be a distinct break, or gap in the distribution at some point providing a "natural" location for beginning a transformation (i.e., adding or subtracting 180°). In the 65% cross section, this "break" was determined to fall at about 60°, with angles originally measured as less than 60° transformed by adding 180° (thus, the new phi range for this cross secion is 60° to 240°). In the 50% cross section, the breakpoint was taken at 150°, with two individuals in the 160° to 180° range considered to belong to the (right) tail end of the frequency distribution for this angle (see Figure 12a). These two individuals were transformed by subtracting 180° (thus, the new phi range for this cross section is -30° to

+150°). One phi of 138° could have been included at either end of the 50% section distribution following the criteria noted above; an examination of the photographic slide made of this particular cross section indicated that the original untransformed angle value was more appropriate (also, this reduces the skew in the final frequency distribution--see Figure 12b). Both new frequency distributions for the 50% and 65% cross sections are shown in Figure 12b.

The frequency distribution for the femoral 35% cross section, shown in Figure 12b, is somewhat skewed to the left, and it was considered whether to "transform" the few individuals at this end (say, 150° to 180°) by subracting 180°, thus shifting the phi range to the right. However, it was decided not to do this, first, because no obvious "overrun", or cutting off of the distribution at either 0° or 180° is present (contrast this with the original 50% and 65% section distributions), and second, because no breaks in the distributions are apparent. Thus, the decision where to make the cutoff point for transformation would have had to have been completely arbitrary and depend upon what appeared to reduce skewness in the distribution the most. In fact, a slightly skewed distribution may actually be more "natural" for this cross section than an artificially produced distribution more closely approximating the normal distribution (note, for example, the large number of individuals--13--with 35% phi's between 100° and 110°, well to the left of the mean of 79°.

It must be kept in mind, however, for this and other cross sections, both transformed and untransformed, that the ranges of variation and means obtained for phi are only approximate, and represent simply the best estimates which could be made regarding the "true" frequency distributions of these angles. Also, it should be noted that mean values for phi sometimes represent only relatively weak "central tendencies", e.g., in the femoral 35% and 65% cross sections (Figure 12b). Thus, the conclusions drawn from comparisons of these means should be approached as broad generalizations, with a wide range of individual variation possible around these general trends.

3-Dimensional Position of Area Centroids

As described earlier, one of the outputs of program SLICE is the area centroid, or center of cortical bone area of each cross section analyzed. Because all cross sections were located precisely with respect to standartized x, y, and z axes for each bone prior to sectioning (see chapter 4), the coordinates of these centroids may be used to provide a description of the 3-dimensional geometry of the femoral and tibial diaphysis included in the study. For example, anterior femoral curvature is reflected in an increase in values for the y (A-P) coordinate of mid-diaphyseal femoral cross section area centroids, while medial or lateral displacements of the shaft produce changes in the x coordinates of area centroids. Using this system, more precise descriptions of 3-dimensional features of traditional interest to anthropologists (e.g., anterior femoral curvature) are possible. Perhaps of more importance in the present study, as discussed earlier in this chapter, eccentricity of cross sections (deviations from straight longitudinal axis drawn between bone articulation end-points) will produce bending moments in the bone, even under purely axially applied loadings. Thus, the 3-dimensional positions of area centroids relative to those of other cross sections and to bone loading points are critical in determining the relative magnitude and direction of bending moments, and thus stresses developed within bone cross sections under *in vivo* loadings.

Two factors must be borne in mind when interpreting these data, however. First , in some instances, particularly in the proximal femur, due to lack of better landmarks diaphyseal breadths were used to help locate reference axes (see Figure 3, chapter 4). Thus, variation in the 3-dimensional position of cross section centroids is not always strictly independent of the reference axes used to determine these positions, i.e., some circularity of reasoning may be involved in the measurement of shaft curvatures. However, the effects of these slight "biases" in the orientation of reference axes should be minimal when compared to overall diaphyseal morphology. Also, it should be noted that the 3-dimensional position of all cross section centroids *relative* to each other, and to articulation endpoints, will remain the same no matter how reference axes are originally set up. Theoretically, any arbitrary set of axes could be chosen and as long as all features of interest were related accurately to these axes, the complete 3-dimensional configuration of the bone could be reconstructed. In the present study, reference axes were chosen in an attempt to reflect loadings patterns (i.e., articulation endpoints) and to approximate the references axes used in provious studies (e.g., Backman, 1957; Stewart, 1962). The various "whole bone" features measured prior to sectioning (see chapter 4) permit evaluation of several different types of loadings, for example, loadings through the femoral head, which will be discussed further in chapter 6.

A second consideration in interpreting the following data is that the 3-dimensional position of cross section centroids is dependent upon both larger--"whole bone"--variation in form, and site-specific variation in cross-sectional shape. For example, as will be shown below, a relatively more laterally placed tibial cross section centroid may be the result of either a relatively more laterally curved tibial diaphysis, or a more laterally "twisted" cross-sectional geometry (i.e., a more mediolaterally directed major axis, bringing the anterior tibial crest to a more lateral position), or to a combination of both factors. In reality, as mentioned at the beginning of this chapter, it is somewhat artificial to divide bone geometry into "whole bone" and individual cross section features, since the two form a continuum and are interdependent. Therefore, in the following sections, sex, age, and side-related variation in centroid positions will be discussed as reflections of both overall bone remodeling, and specific within-cross section remodeling, or "cortical drift" (Enlow, 1963).

In the following presentations, primarily in order to make direct comparisons between the sexes easier, centroid coordinates are divided by bone length ("biomechanical") before graphing. This effectively reduces all individuals to the same length and allows plotting of points for individuals of different length on the same scale. Implicit in this transformation is the concept of geometric similarity in form (that is, individuals with the same "shape" will exhibit the same ratios of linear dimensions). While this assumption is very questionable with regard to cross-sectional geometric properties such as areas and areal moments, for reasons discussed elsewhere,

dividing by bone length is probably justified in the present context because the positions of area centroids are themselves linear dimensions similar to bone length, not second or fourth powers of a linear dimension as are areas or moments. Also, as will be shown below, in most cases groups with average differences in bone length, e.g., males and females, show a remarkable similarity in position of area centroids after division by bone length. Thus, geometric similarity in 3-dimensional shape does appear to characterize these larger, "whole bone" dimensions, at least within the present sample.

Sex differences.

Comparisons of cross section area centroid positions in male and female femoral and tibial diaphyses are shown in Figures 13a, b. As noted above, all x (M-L) and y (A-P) centroid coordinates have been divided by bone length and are indicated in the figures as A-P and M-L "indices". The ordinate scale for these indices has been exaggerated 7.5 times to better show details of variation along the length of the shafts. Dotted lines indicate extrapolations of area centroid curves to the intersection of the z (longitudinal) axis with articulation endpoints (proximal and distal femoral y (A-P) and distal femoral x (M-L) endpoints in these figures represent only approximations of true articulation centers of the femur--see Figure 3, chapter 4).

At first glance, one of the most striking featrues of these graphs is the close similarity between males and females in y (A-P) and x (M-L) positions of centroids, relative to bone length. With the

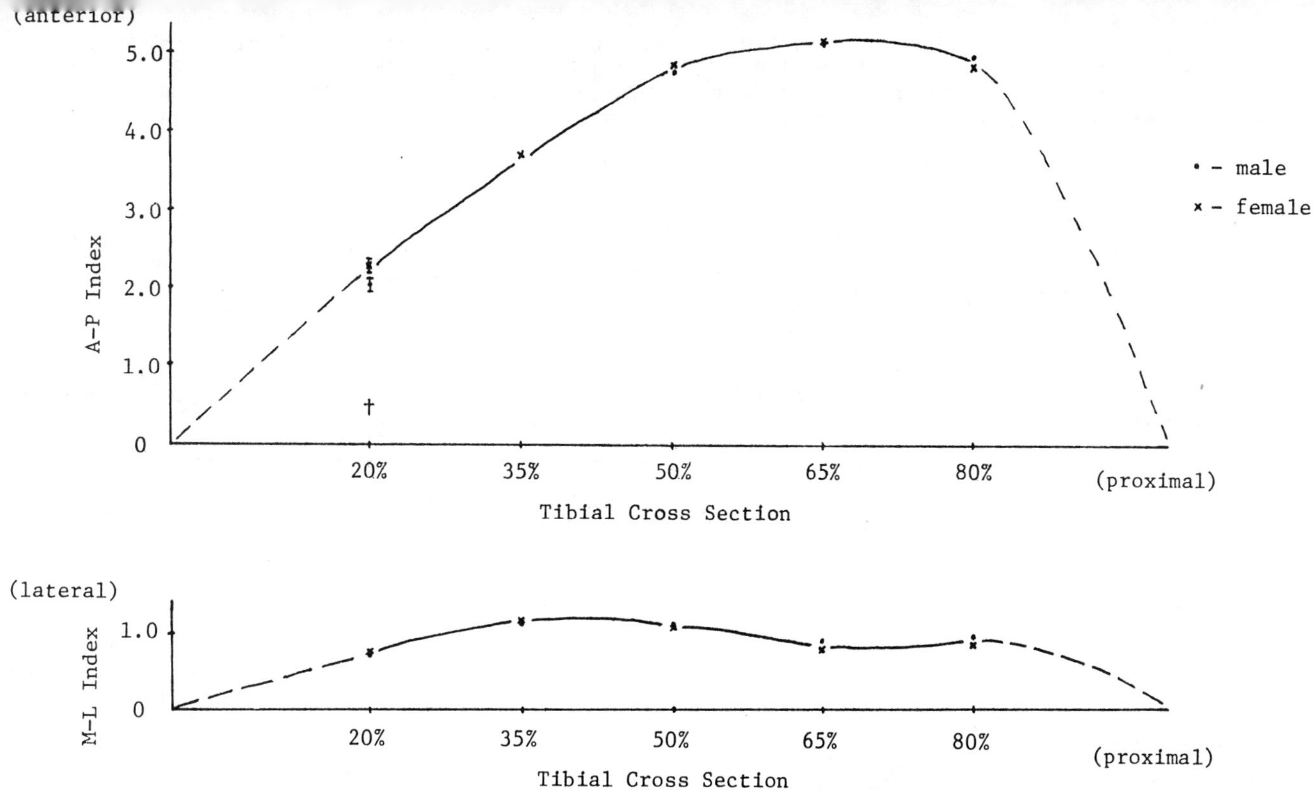

Figure 13a. 3-dimensional positions of tibial cross section centroids: anteroposterior displacement (top) and mediolateral displacement (bottom)--ordinate scales exaggerated 7.5 times (standard error bars shown for statistically significant and near-significant male-female differences only: † = $p<.10$).

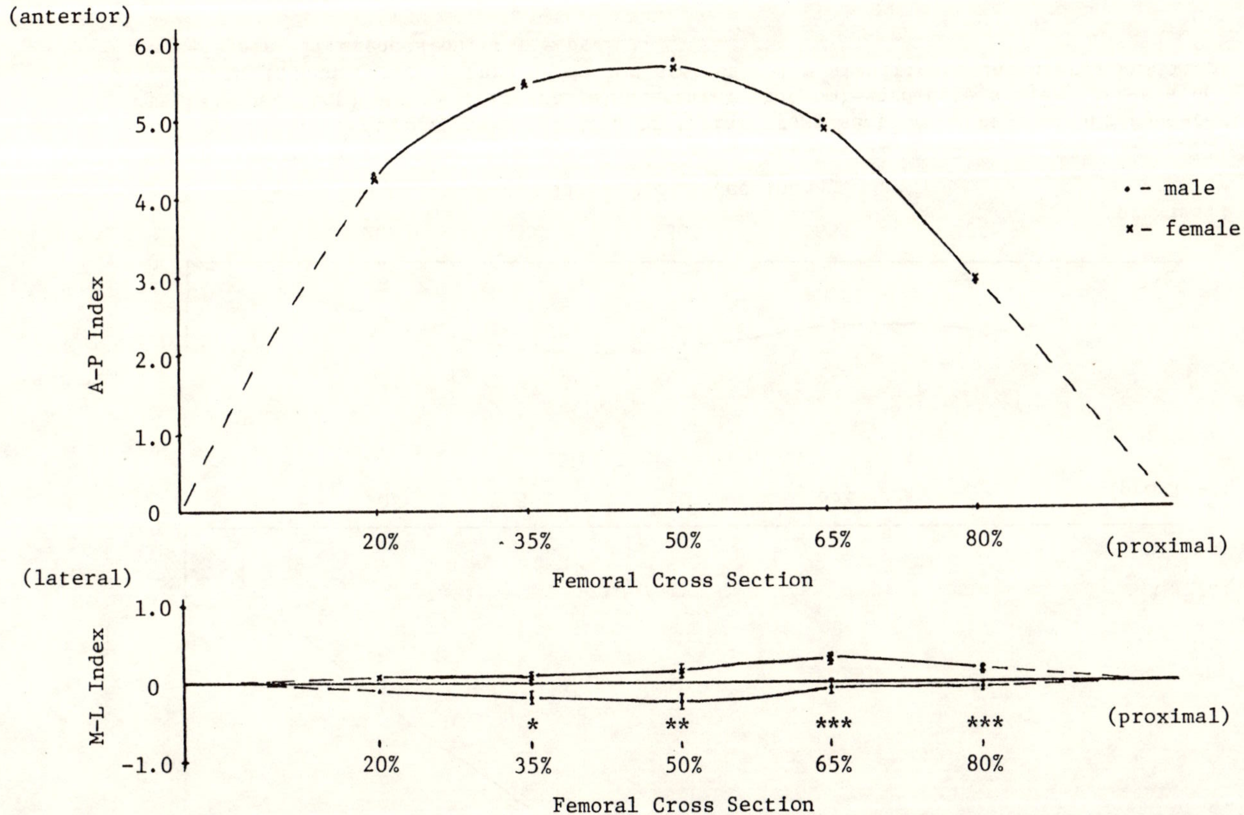

Figure 13b. 3-dimensional positions of femoral cross section centroids: anteroposterior displacement (top) and mediolateral displacement (bottom)--ordinate scales exaggerated 7.5 times (standard error bars shown for statistically significant and near-significant male-female differences only: * = p<.05, ** = p<.01, *** = p<.001).

exception of the M-L positions of femoral cross sections, no significant or near-significant ($p < .10$) differences in relative location of area centroids between the sexes is present. In other words, except for mediolateral displacement of the femoral shaft, males and females are remarkably similar in overall 3-dimensional geometry of the femoral and tibial diaphyses. Another observation pertinent here is the relatively narrow range of individual variation in position of area centroids, as indicated by the small standard error bars (shown for significant and near-significant differences only). In general, then, a particular pattern of 3-dimensional morphology appears to be characteristic of the great majority of individuals in this sample.

Figures 13a, b demonstrate the sensitivity of the technique used in this study to even minor variations in position of cross section centroids. While the displacement of centroids in the A-P plane is more marked, reflecting the general anterior curvature of both femoral and tibial shafts, medial and lateral displacements of section centroids are also clearly apparent. As noted above, these displacements reflect both larger, whole bone morphology and localized variation in cross-sectional shape. For example, the graph of tibial centroid locations in the M-L plane (Figure 13a, bottom) indicates a general lateral displacement of the tibial diaphysis relative to the longitudinal z axis. However, this graph also demonstrates a distinct pattern of variation in the degree of lateral displacement at different levels of the diaphysis, which follows a sigmoidal path from the 20% through 80% sections. Observation of an intact tibia shows that this pattern is closely related to the sigmoidal track of the anterior tibial crest along the length of the tibia, the result of a complex "twisting" of the shaft which is also reflected in changes in the orientation of the major axes of individual cross sections, or phi (see Figure 11 above). Figure 11 shows that phi is closest to 90° (i.e., anteriorly directed) in the 65% cross section and more mediolaterally directed in the 80%, 50%, and 35% sections. The corresponding changes in relative position of the anterior crest, which forms the thickest part of the tibial cortex, produce the characteristice curve of mediolateral centroid displacements shown in Figure 13a. This general relationship changes in the distal tibia (20% section), where the anterior crest is greatly reduced and the diaphysis assumes a more circular cross-sectional shape. Here the presence of a strong distal-medial crest produces a still more mediolaterally directed major axis, while the section centroid moves medially towards the distal articulation centerpoint.

Mediolateral displacements of cross section centroids are much smaller in the femur than in the tibia (Figure 13b). However, certain regularities in the pattern of these displacements are apparent, in particular a slight lateral curvature of the shaft, reaching its greatest extent at the 65% cross section, more marked in females. The apparent medial curvature of the shaft in the distal half, more marked in males, may be an artifact of the measuring procedure employed. The distal landmark for the longitudinal axis of the femoral diaphysis was taken at the deepest point of the intercondylar notch (Figure 3, chapter 4). However, because the medial condyle of the femur is larger than the lateral condyle, the actual midpoint between the two distal centers of articulation is slightly medial to the middle of the intercondylar notch--as calculated from other "whole bone" measurements taken (Table 3, chapter 4), an average of 3.3 mm medial to the longitudinal axis of the diaphysis. If this correction is made in Figure 13b, the "medial" curve of the distal half of the femoral diaphysis becomes a part of the general lateral curvature, perhaps even forming a separate lateral curve reaching its maximum at the 20% section. Thus, relative to a diaphyseal axis drawn through the true midpoint of the distal articulation, all femoral cross section centroids are laterally displaced, although still not to the same extent as tibial cross sections.

Perhaps of greater interest, however, is the observed sex difference in relative mediolateral position of femoral cross section centroids. From the 35% to 80% cross sections, female centroids are consistently and significantly more laterally placed than male centroids relative to the diaphyseal longitudinal axis (whether as drawn in Figure 13b or as "corrected" above). As mentioned earlier, this is the only feature of 3-dimensional bone morphology of this kind which showed a sex-related difference. This is of particular interest in that mediolateral bending, especially in the proximal femur, is one form of loading of the lower limb which shows a clear sex-related difference due to differences in pelvic morphology in males and females (see "whole bone measurements" above, also chapter 6). Later analyses in this chapter will show that females in the study sample have, on average, more A-P flattened (or M-L broadened) proximal femoral shafts, and thus, greater resistance to M-L bending. It is possible that a majority of this mediolateral broadening of the proximal shaft in females is a result of thickening of the lateral cortex, since this area would be expected to experience the greatest tensile stress (the most critical stress in determining bone failure) under M-L bending. This would tend to move the area centroid for these sections laterally in females, resulting in the pattern shown in Figure 13b. It is also possible that the method used to locate the proximal landmark for the femoral longitudinal axis--using the midpoint of the M-L diaphyseal breadth just distal to the lesser trochanter (see Figure 3 above)--in some way biases the positioning of the axis and thus artificially produces this sex-related difference in centroid location. However, this proximal landmark is in fact very close to the level of the 80% cross section location (Figure 3). The fact that there is still a highly significant sex difference in M-L placement of the 80% section centroid, even though both sexes were oriented using the midpoint of the M-L breadth at this location, indicates that some imbalance in cortical area, i.e., more cortical area on the lateral side in females, must be present. Direct measurement of cortical thicknesses at various sites around the cortex will be needed to conclusively prove or disprove this hypothesis.

The pattern of area centroid displacements in the A-P plane for both tibia and femur are very similar in males and females. Both bones exhibit anterior curvature, reaching a peak at the 65% cross section in tibiae and the 50% cross section in femora. It is interesting that although femora appear to show more curved diaphyses than tibiae when intact bone external contours are compared, the magnitude of actual anterior displacement of section centroids from their respective longitudinal axes, relative to bone length, is about the same--

a little over 5% of bone length. Also of interest are the "mirror image" A-P profiles in the two bones, with both increasing gradually in degree of anterior curvature towards the knee joint, then dipping sharply to the knee articulation center (Figures 13a, b). The functional role of anterior curvature in the femur and tibia will be discussed in more detail in chapter 6, however, it should be noted that this sudden posterior "dip" in the distal femur and proximal tibia brings the knee joint back under the center of gravity of the body in the sagittal plane (Steindler, 1955: 107).

Another way to measure relative displacement of section centroids is through measurement of angles formed by these centroids with the longitudinal axis. The closest previous approximation to this approach has been the study of Backman (1957: 24-25), who measured the angle between the "straight, proximal anterior contour of the shaft" and the "ideal" (longitudinal) anxis of the femoral diaphysis using M-L radiographic projections of the femur. His average for 97 cases (European ancestry, sexed mixed) was 8.0°, with a standard error of .2°. The angle available from the present study data which most closely approximates Backman's angle is that formed by a line drawn through the 65% and 80% femoral cross section centroids with the longitudinal axis of the femur. The mean over the entire study sample for this angle is 7.53 ± .18°, with males averaging slightly (nonsignificantly) greater values than females (males: 7.79 ± .30°, females: 7.28 ± .22°). Thus, the figures obtained for this angle in the two studies are very similar. However, it should be noted that the angle measured in the present study is probably an underestimate relative to Backman's, since the anterior surface of the proximal femur, used by Backman, is not straight at the level of the 65% cross section but has begun to slope downwards (with the bone held on the measuring board--see Figure 3 above). Also, use of area centroids rather than the anterior contour of the shaft probably decreases the angle measured, since the development of the linea aspera, present at the 65% level but not at the 80% level, would tend to move the centroid of the 65% section posteriorly while not significantly affecting the anterior contour. Thus, the Pecos sample may actually exhibit angles of femoral anterior curvature, measured according to Backman, several degrees greater than 7.53°, possibly significantly greater than his figure of 8.0°.

Several authors have studied differences among human populations in anterior femoral curvature, using the anterior diaphyseal contour of the bone to measure this characteristic (Stewart, 1962; Walensky, 1965; Gilbert, 1976). Although the magnitude of anterior femoral curvature measured in this way shows variation among local populations (e.g., Gilbert, 1976: Table 1), in general Amerindians as a group have been found to exhibit greater curvature than European ancestry Whites, who in turn show more curvature on average than Blacks. No sex-related differences have been reported. The population differences have still not been adequately explained, although several factors, including relative body weight (Gilbert, 1976), activity levels and genetic predisposition (Walensky, 1965) have been suggested as possible population-specific influences on anterior curvature.

While Stewart's (1962) study of anterior femoral curvature is based on a different measuring technique, employing the anterior contour of the diaphysis rather than centroid locations, it is interesting to compare some of this results with those of the present study. For example, the point along the femoral shaft where anterior curvature reached a maximum among his Amerindian sample (18th century Dakota) averaged 47% of bone length from the distal edge of the lateral condyle to the proximal edge of the greater trochanter. American Whites and Blacks both exhibited more proximal locations of maximum curvature, averaging 54% and 55% of bone length, measured from the distal end, respectively.

The present study results (Figure 13b) indicate an area of maximum femoral displacement in the sagittal plane at the 50% cross section, although an extrapolation of this curve would probably place the point a little more distally, perhaps between 45% and 50% of bone length. Thus, Stewart's figure and that of the present study of this landmark are quite similar. Stewart also measured an "indicator of torsion" of the femur by taking the A-P distance of the highest point on the femoral head above his femoral leveling points, and dividing by bone length. He found that greater "torsion" showed some association with reduced femoral diaphyseal curvature, although at least part of this association appeared to be due to the measurement technique used (specifically, the influence of "torsion" on his proximal leveling point). In the present study, several methods were used to test the association between femoral antetorsion and anterior diaphyseal curvature. The method most closely approximating that of Stewart was testing the correlation between my index of A-P displacement of the femoral 50% cross section (Figure 13b) with an index of femoral head A-P displacement, derived from the measurement of the femoral head articulation point taken as a "whole bone" measurement earlier (see chapter 4), divided by bone length. Contrary to Stewart's findings, a positive correlation between these two indices was obtained for males (r = .422) (it should be noted that Stewart used only males in his study). I also tested the association between the antetorsion angle of the femoral neck and diaphyseal A-P displacement indices, and between the antetorsion angle and the angle of anterior diaphyseal curvature, measured from centroid locations as described above. Neither of these tests showed any positive associations between antetorsion and anterior curvature of the shaft, in fact, females exhibited small negative correlations between the antetorsion angle and A-P displacement of the 65% and 80% sections (r = -.377 and -.343, respectively). This negative association may be due to the fact that as the proximal femoral shaft is moved posteriorly (decreasing the A-P displacement indices of these sections), the antetorsion angle of the neck must be increased to maintain the femoral head in the same position relative to the rest of the diaphysis (this is a similar problem to the one faced by Stewart in his study--see above). Thus, due to the nature of the measurements involved and the interrelationships between various proximal femoral morphological features, it is doubtful that any simple relationship between antetorsion and anterior curvature, as measured here or in other studies, will be found. A somewhat different approach to the interpretation of these features, including consideration of stress distributions in the diaphysis under loadings applied to the femoral head, will be presented in the next chapter.

Age differences.

For the analysis of age differences in cross section centroid positions, the larger, 2 age group division of the sample was used (20-39 and 40+ years) to facilitate graphic presentation of the data--Figures 14a, b and 15a, b. Analyses of variance using the 6 smaller age group divisions reveal essentially the same age trends as those indicated here, with the addition of some minor unsystematic variation between groups probably attributable to sampling error.

Several apparent age trends are present in Figures 14a b, and 15a, b, however, only the M-L displacement of femoral cross sections among females (Figure 15b, bottom) exhibits a consistent and statistically significant difference between age groups (this was also found in the 6 group ANOVA). As shown above, the relative M-L displacement of femoral cross sections is also the only characteristice of centroid position in either bone which exhibits a significant sex difference, with females relatively more laterally placed, particularly in the proximal femur. From the graph in Figure 15b, it is evident that this sex difference progressively increases throughout life, with older females showing greater lateral displacement than younger females, while males show no significant age changes in this characteristic. If the hypothesis presented above is correct--that relative lateral displacement of femoral centroids is brought about by thickening of the lateral cortex--then the age trends shown here may indicate a process of continuing remodeling of the femoral cortex throughout life among females, brought about by relatively higher M-L loadings in this sex.

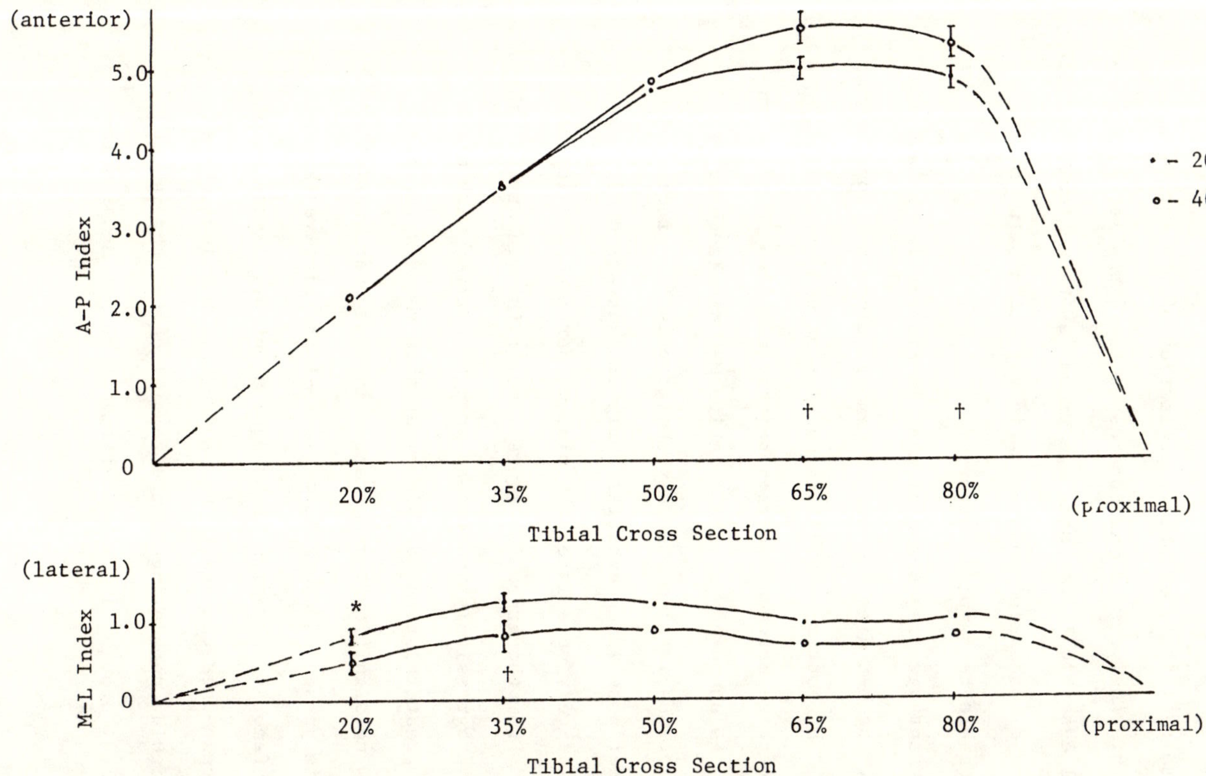

Figure 14a. Age differences in male tibial cross section centroid positions: A-P displacement (top) and M-L displacement (bottom)--ordinate scales exaggerated 7.5 times (standard error bars shown for statistically significant and near-significant differences between 20-39 year and 40+ year age groups only: † = $p<.10$, * $p<.05$).

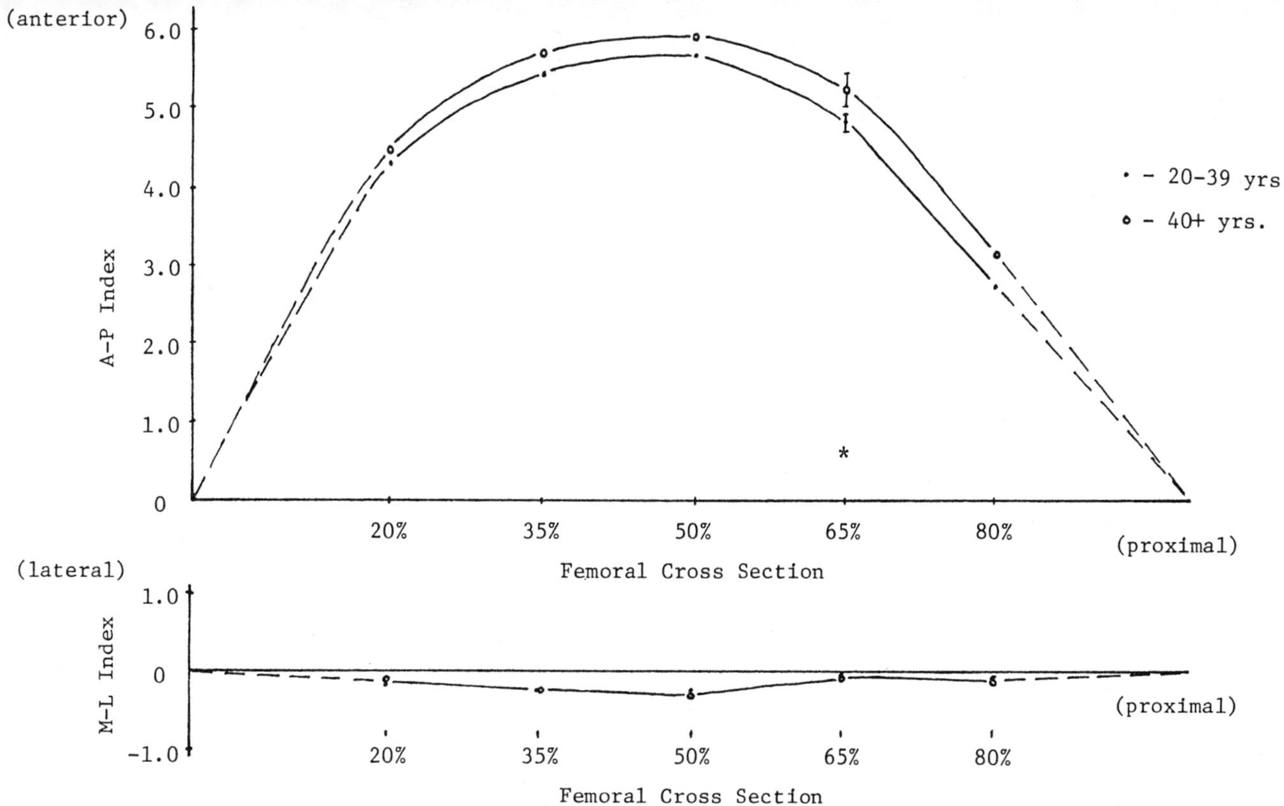

Figure 14b. Age differences in male femoral cross section centroid positions: A-P displacement (top) and M-L displacement (bottom)--ordinate scales exaggerated 7.5 times (standard error bars shown for statistically significant and near-significant differences between 20-39 year and 40+ year age groups only: * = $p<.05$).

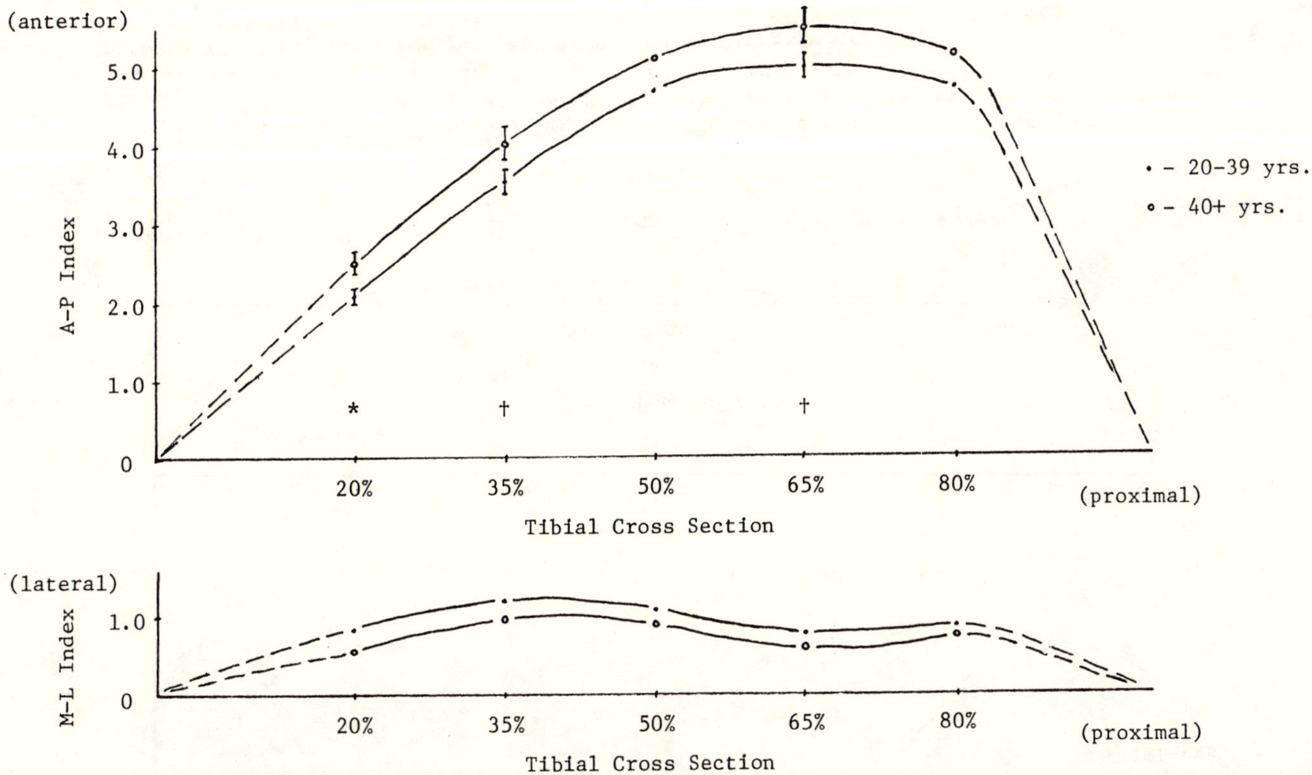

Figure 15a. Age differences in female tibial cross section centroid positions: A-P displacement (top) and M-L displacement (bottom)--ordinate scales exaggerated 7.5 times (standard error bars shown for statistically significant and near-significant differences between 20-39 year and 40+ year age groups only: † = $p<.10$, * = $p<.05$).

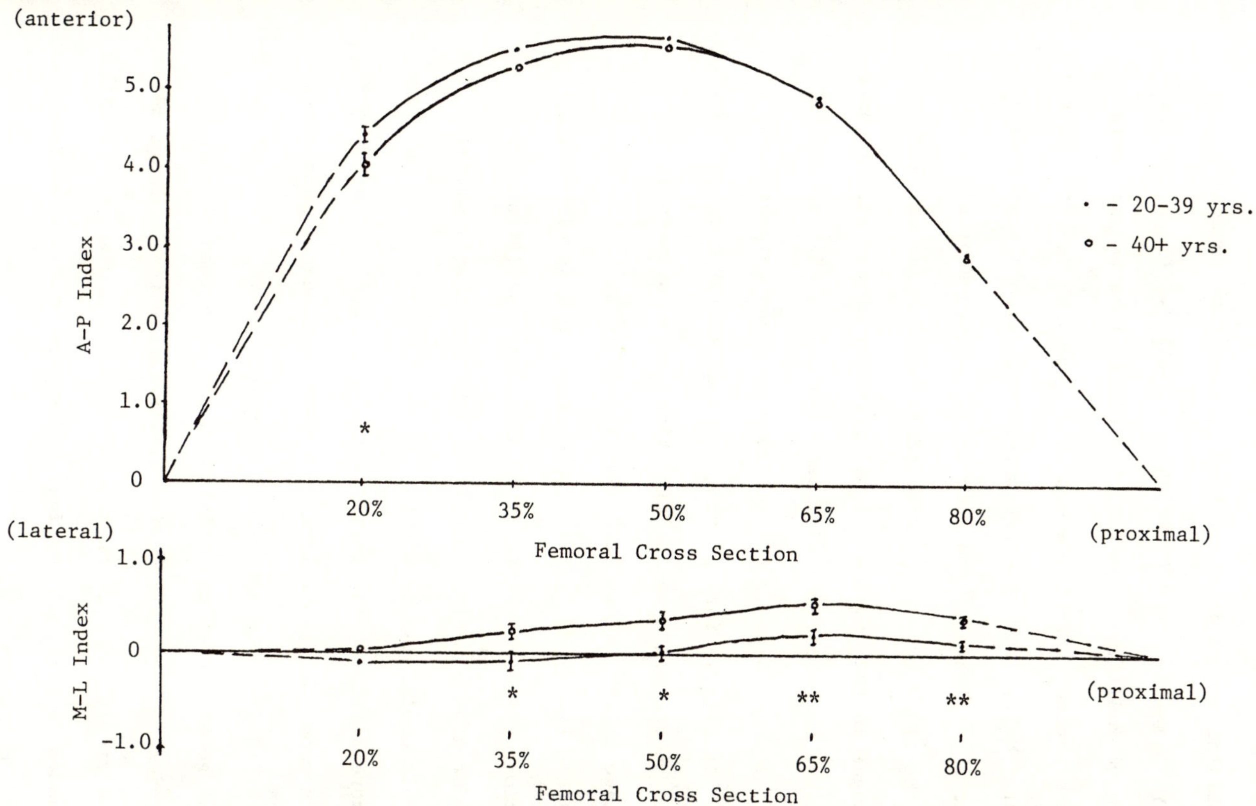

Figure 15b. Age differences in female femoral cross section centroid positions: A-P displacement (top) and M-L displacement (bottom)--ordinate scales exaggerated 7.5 times (standard error bars shown for statistically significant and near-significant differences between 20-39 year and 40+ year age groups only: * = p<.05, ** = p<.01).

Some change with age in the mediolateral position of cross section centroids also occurs in the tibia, although statistically significant at only one site. In both males and females, area centroids of the tibia move medially--closer to the tibial longitudinal axis--with aging, the change averaging about .2-.3 index units, or slightly less than 1 mm. Males show somewhat more change with age in this characteristic than females. The more medial position of tibial centroids in the older age group is probably best explained as a result of changes in the cross-sectional geometry of the tibial shaft, again more marked in males. As will be shown below, with aging the direction of the major axes of tibial cross sections becomes more A-P oriented. If this is interpreted as a result of "cortical drift" of the anterior cortex in a medial direction, this would explain both changes with age in phi and the relatively more medially placed section centroids in the tibia in the older group. Change with age in the A-P position of tibial centroids is also apparent (Figures 14a, 15a), with older adults having relatively more anteriorly placed centroids. This could also be brought about by drift of the anterior cortex to a more antero-medial position (assuming the posterior cortex remains in the same position relative to the longitudinal axis of the tibia). Selective preservation of the anterior cortex of the tibia with aging and general cortical thinning would also tend to move the section centroids in an anterior direction relative to the longitudinal axis. The significance of these age-related changes in the tibia will be discussed in terms of loadings and stress distributions in this bone in the following chapter.

Relative A-P displacement of femoral cross section centroids also changes with aging, although here the pattern of age-related changes differs in males and females. Males exhibit an increase in the degree of anterior displacement of the shaft (Figure 14b) which follows a fairly parallel path along the length of the diaphysis. This more anteriorly placed femoral diaphysis (relative to articular endpoints) is thus not due to any changes in the basic shape of the shaft as a whole (for example, the femoral anterior curvature angle, calculated as described above, shows no significant age change in males), but may be caused by remodeling within cross sections, or cortical drift with aging, which leads to a thicker anterior cortex. This age change would thus parallel the hypothesized age change in the tibia whereby the anterior cortex is selectively preserved. It is interesting, however, that only males show this pattern of age change in the femur, with females actually decreasing slightly in anterior displacement of the distal femoral diaphysis, and showing no change in A-P position of section centroids in the proximal femur (Figure 15b). The most significant change with age in femoral centroid positions in females, as noted above, is the relative lateral displacement of the shaft, possibly brought about by selective preservation (or increase in thickness) of the lateral cortex. This change is seen to be most marked in the proximal and mid-diaphysis (Figure 15b), while in the distal diaphysis a relative posterior displacement of centroids is evident. While it is difficult to interpret all of these age-related changes without actual cortical thickness measurements, it seems clear that male and female femora of this sample undergo fundamentally different types of geometrical remodeling with aging. This stands in direct contrast to the tibia, where quite similar patterns of age-related change in centroid positions were observed in the two sexes (Figures 14a, 15a). As will be shown below, this same difference in the aging pattern of the two bones--tibia more similar in males and females, femur showing a contrasting pattern in the two sexes--is also characteristic of other geometrical properties such as principal moments of bone cross sections. It has previously been mentioned at several points in this investigation that female femora are probably subjected to relatively higher M-L bending moments than male femora _in vivo_. Thus, if bone structural differences reflect _in vivo_ loading differences, one would expect more marked sexual dimorphism in the structure of the femur than the structure of the tibia. What is particularly significant in the present context is the fact that these differences in sexual dimorphism are also reflected in differences in the aging patterns of the two bones. This general theme will be developed further in succeeding segments of this study as additional geometric data are presented.

Side differences.

Differences between right and left tibiae and femora were also analyzed, including analysis of side differences in different sex/age groupings. Because of the complexity and number of graphs (or tables) necessary to present all of these data, a summary of these results is limited here to a brief description of the more important findings. More discussion of side differences is included in the presentation of cross section geometric properties below.

Three significant side-related differences in relative position of section centroids were noted. In the tibia, both sexes exhibit a more medial position of section centroids on the left side. The difference is greater in males, and increases with age, again more so in males (in 40+ year males, the right-left M-L difference for some tibial cross sections reaches more than 4 mm, with left tibial centroids falling very close to the longitudinal axis). As with the general pattern of variation in tibial M-L centroid positions, this side difference appears to be directly related to a more A-P oriented major axis in left tibial cross sections, a difference which coincides with the smaller angle of tibial "torsion" on the left side noted earlier (see Table 6). This lends further evidence to the hypothesis that left tibiae are more adapted for A-P bending stress, a side-related difference observed in both sexes but more markedly so in males.

Females also exhibit a more posteriorly placed centroid position in left tibiae, while males show no significant side-related difference in this characteristic. This side difference in females is difficult to explain, since one might expect exactly the opposite effect (left side more anteriorly placed) given the difference in phi on right and left sides and the factor of the anterior crest position discussed above. Other features of tibial cross-sectional geometry, such as relative development of _posterior_ crests, may be important here, or it may be that a general "whole bone" difference in position of the tibial shaft relative to articular end centerpoints exists in right and left female tibiae. This side difference decreases with age, which may be related to the increasing side difference in degree of tibial torsion (and M-L centroid position) noted above.

No significant diffences in A-P placement of section centroids in right and left femora were observed in either sex. No evidence was found to support the findings of Walensky (1965) that the point of maximum anterior curvature of the femoral shaft is more distally placed in right femora (he found this to be particularly ture in his Amerindian sample). Males of the present sample also showed no difference in the M-L placement of femoral centroids on right and left sides. Females, however, exhibited highly significant differences in this characteristic, with right femora relatively more laterally placed along the entire length of the diaphysis except at the 80% location. This side-related difference in females is difficult to interpret, since, following the line of reasoning established above, it would seem to suggest that right femora are more adapted for M-L bending stress among females. However, as will be shown below, cross-sectional properties such as bone areas and second moments of area all indicate relatively stronger left femora in females. It is conceivable that both medial and lateral cortices are thicker in female left femora (thus the greater cortical areas and principal moments on the left side), while right femora, with less total bone volume (because of less overall stress) still preserve some M-L bending strength by maintaining a relatively thicker lateral cortex. The same kind of explanation (left side overall more robust) could apply to the side-related difference in A-P placement of tibial centroids among females described in the preceding paragraph. Alternatively, these side differences in females could indicate sex/side-related differences in loadings of bone bones. Again, it will be difficult to arrive at any firm conclusions regarding these various hypotheses until direct measurements of cortical thickness are available.

Cross-Sectional Areas

Cross section geometric properties determined by SLICE and analyzed by sex, age, and side in this and subsequent sections, are listed in Table 7. Bone areas (CA, TA, and MA) are examined first, followed by orientation of principal axes (phi), principal second moments of area (I_{max} and I_{min}), and section moduli (Z_{max} and Z_{min}).

As discussed earlier in this chapter, the area of cortical bone in a cross section of a long bone diaphysis can be used as a measure of relative strength at the site under pure axial loading. Since *in vivo* loadings of bones usually produce bending moments rather than purely axial loads, the magnitudes of other cross section characteristics such as second moments of area are probably more crucial in determining the likelihood of bone failure. However, cortical bone area (CA) is an important feature of diaphyseal morphology to evaluate, first, because it is one dimension which has been measured (or extrapolated from breadths) in many previous investigations of skeletal aging, and second, because it reflects changes in total bone volume of the skeleton which are important in evaluating skeletal metabolism, e.g., alterations in the balance between bone resorption and formation. For similar reasons, total subperiosteal area (TA) and medullary area (MA) are important variables, in that they reflect changes in bone metabolism occurring at the two surfaces of the diaphyseal cortex, i.e., endosteal and subperiosteal

Table 7

Cross Section Geometric Properties

Property	Abbreviation
cortical area	CA
total subperiosteal area	TA
medullary area	MA
percent cortical area	%CA (CA/TA X 100)
orientation of major axis	phi
maximum principal second moment of area	I_{max}
minimum principal second moment of area	I_{min}
ratio of maximum to minimum principal second moments of area	I_{max}/I_{min}
section modulus, major axis	I_{max}/c_{maj}
section modulus, minor axis	I_{min}/c_{min}

resorption and apposition.

Sex differences.

Tables 8a, b summarize differences between males and females of the study sample in CA, MA, and TA of the 11 tibial and femoral cross sections. As expected, due to sexual dimorphism in size (see Table 4 above), males have significantly greater bone areas than females for all cross sections. What is more interesting is the variation in degree of sexual dimorphism in bone areas from section to section. This variation is made clear by taking the ratio of male/female values for each dimension, shown in the last columns of Tables 8a, b. Looking first at cortical areas in the tibia (Table 8a, top), we see that while both sexes increase in CA from the distal end of the tibia to a maximum at the 65% cross section, males show relatively larger gains, leading to an increase in the male/female CA ratio of more than 5 percentile points from the 20% to the 50% and 65% cross section. The male/female ratio for CA then falls again in the most proximal section. This pattern is graphically illustrated in Figure 16a, which clearly shows the increase in sexual dimorphism in the middle region of the tibial diaphysis. From the summaries of TA and MA in Table 8a it can be seen that the relatively larger male cortical areas in these sites is due to both relatively greater subperiosteal dimensions and smaller medullary cavity dimensions in the middle of the tibia in males. Males are relatively larger in TA at the proximal end (80% section) of the tibia, but because of an even larger increase in MA are relatively smaller in CA at this site.[10] Sex-related differences in cross-

---

[10]Of course, this does not take into account the presence of cancellous bone in the medullary cavity, which at this site may make a significant contribution to bone strength.

Table 8a

Sex Differences in Bone Areas: Tibia

CORTICAL AREA

Cross Section	Male Mean	Male SE	Female Mean	Female SE	Male - Female[1]	Male/Female X 100
20%	187.2[2]	3.5	140.4	3.4	46.8	133.3
35%	255.5	4.1	186.3	4.1	69.2	137.2
50%	287.4	5.2	207.4	4.8	80.0	138.5
65%	298.9	5.9	215.5	5.4	83.4	138.7
80%	271.6	4.9	204.3	4.7	67.3	133.0

TOTAL SUBPERIOSTEAL AREA

Cross Section	Male Mean	Male SE	Female Mean	Female SE	Male - Female[1]	Male/Female X 100
20%	450.6[2]	7.3	356.0	6.5	94.6	126.6
35%	402.3	6.7	314.7	6.9	87.6	127.9
50%	448.8	7.6	345.2	7.3	103.6	130.0
65%	530.3	8.0	405.9	7.7	124.4	130.7
80%	749.2	12.8	573.7	11.3	175.5	130.5

MEDULLARY AREA

Cross Section	Male Mean	Male SE	Female Mean	Female SE	Male - Female[1]	Male/Female X 100
20%	263.4[2]	7.2	215.6	6.6	47.8	122.2
35%	146.8	6.3	128.4	6.5	18.4	114.3
50%	161.4	6.8	137.8	7.4	23.6	117.1
65%	231.4	6.4	190.4	7.5	41.0	121.5
80%	477.5	12.4	369.5	11.4	108.0	129.2

[1] All male-female differences significant at .001 level, except medullary areas of 35% and 50% sections, significant at .05 level.

[2] mm^2

Table 8b

Sex Differences in Bone Areas: Femur

CORTICAL AREA

Cross Section	Male Mean	Male SE	Female Mean	Female SE	Male − Female[1]	Male/Female X 100
20%	261.6[2]	3.6	212.1	4.5	49.5	123.3
35%	318.3	4.2	252.0	4.8	66.3	126.3
50%	358.9	4.8	292.4	5.7	66.5	122.7
65%	366.8	5.6	300.6	5.7	66.2	122.0
80%	349.0	4.8	283.4	5.4	65.6	123.1
Neck	115.3	2.0	96.9	2.6	18.4	119.0

TOTAL SUBPERIOSTEAL AREA

Cross Section	Male Mean	Male SE	Female Mean	Female SE	Male − Female[1]	Male/Female X 100
20%	760.5[2]	12.9	613.3	11.1	147.2	124.1
35%	545.2	6.9	438.8	7.3	106.4	124.2
50%	499.3	5.9	414.1	6.8	85.2	120.6
65%	520.8	6.6	432.6	7.4	88.2	120.3
80%	574.7	6.6	470.1	8.5	104.6	122.2
Neck	579.8	9.1	462.4	8.0	117.4	125.3

MEDULLARY AREA

Cross Section	Male Mean	Male SE	Female Mean	Female SE	Male − Female[1]	Male/Female X 100
20%	498.9[2]	12.5	401.1	12.0	97.8	124.4
35%	226.8	6.7	186.7	7.1	40.1	121.5
50%	140.5	4.7	121.6	6.6	18.9	115.6
65%	154.0	4.7	131.9	7.2	22.1	116.8
80%	225.7	4.9	186.7	7.7	39.0	120.9
Neck	464.5	8.9	365.5	8.0	99.0	127.1

[1] All male-female differences significant at .001 level, except medullary areas, 50% and 65% sections, significant at .05 level.

[2] mm^2

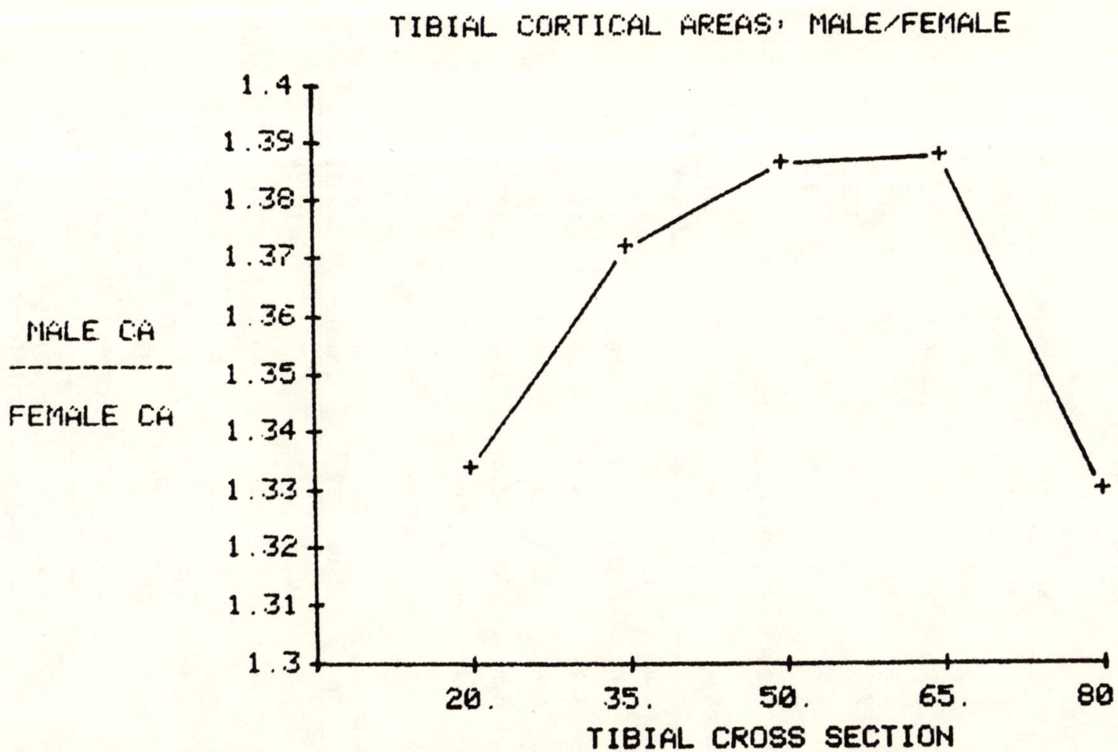
TIBIAL CORTICAL AREAS: MALE/FEMALE
MALE CA
FEMALE CA
1.4
1.39
1.38
1.37
1.36
1.35
1.34
1.33
1.32
1.31
1.3
20.
35.
50.
65.
80.
TIBIAL CROSS SECTION

Figure 16a

sectional shape of the tibial diaphysis coincident with these relative size differences are explored later in this chapter.

Bone areas of the femur show a different pattern of sexual dimorphism (Table 8b). Again, both sexes rise in CA from the distal end of the bone to a peak at the 65% section. However, unlike the tibia, the male/female ratio reaches its lowest point in this section, i.e., females exhibit their relatively greatest CA here. This is shown graphically in Figure 16b. In general, all bone areas--TA, MA and CA--show a greater degree of sexual dimorphism in the distal femur, and less in the proximal femur (i.e., males are relatively larger distally, females proximally). The drop in the male/female CA ratio in the 20% section shown in Figure 16b is the result of a relatively larger medullary area in that section in males, since males show their greatest TA relative to females at that location (Table 8b) (see also footnote 10 above). Otherwise reduction in the CA sex ratio in other sections of the femur are due primarily to reduced TA in males relative to females, with MA paralleling TA.

It is interesting that on the whole, sexual dimorphism in diaphyseal cortical bone area averages much less in the femur--23.5%--than in the tibia--36.1%. As will be shown in another section below, the same is true for second moments of area. Thus, based on geometrical properties at least, the female femur would appear to be relatively stronger than the tibia for withstanding both axial and bending forms of loading. Differences in sexual dimorphism of bone length do not account for this observation, since the average male/female ratio of bone lengths ("biomechanical") is about the same in the tibia and femur (110 and 109,

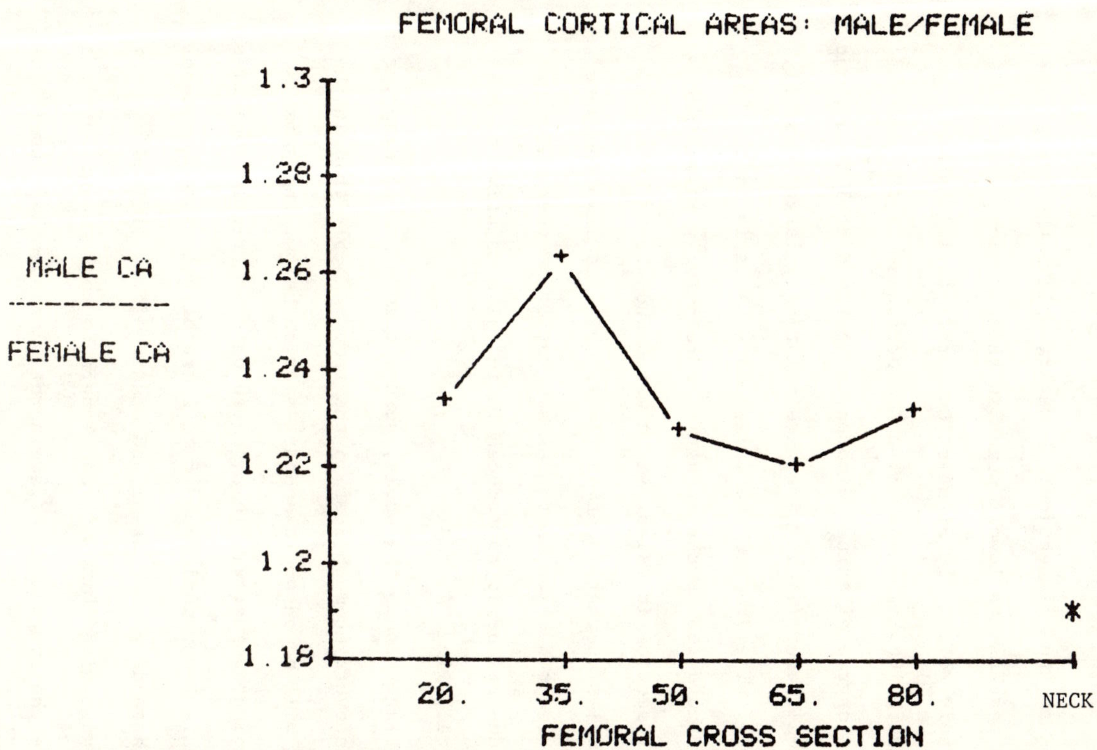

FEMORAL CORTICAL AREAS: MALE/FEMALE
MALE CA
FEMALE CA
1.3
1.28
1.26
1.24
1.22
1.2
1.18
20.
35.
50.
65.
80.
NECK
FEMORAL CROSS SECTION

Figure 16b

respectively). Patterns of bone loss with aging are partially responsible, since as will be shown below, females lose relatively more bone volume from the tibia than the femur with aging, while males lose about the same from each bone (thus, female averages over all age groups will be relatively lower for the tibia than for the femur). However, this only serves to emphasize the importance in females of maintaining a relatively greater volume of bone in the femur than in the tibia with aging. All of this suggests a sexual dimorphism in the loading of the two bones, with females subjected to relatively greater loadings of the femur, an observation made previously in this study in several different contexts. Alternatively, the present evidence could be used to argue for relatively greater stresses in male tibiae, although this seems less likely given that male tibiae and femora show about equal changes in bone volume with aging (i.e., neither is selectively "protected" against bone loss). The significance of the patterns of sex-related differences <u>within</u> bones illustrated in Figures 16a, b will discussed further below in the context of principal second moments of area, and chapter 6.

The general difference in degree of sexual dimorphism in CA in the femur and tibia is particularly apparent in the femoral neck cross section. Here the male/female ratio in CA is only 119.0, lower than any other cross section in either bone. This is especially significant in that this is the area of the lower limb which would be expected to most directly reflect differences in mediolateral bending moments applied through the femoral head. TA of the femoral neck section is relatively small (for femoral cross sections) in females (Table 8 a), so that the relatively larger values for CA are due entirely to a smaller medullary cavity, or area of trabecular bone in females. In other words, more of the total subperiosteal area of the femoral neck is filled with compact cortical bone in females, and less with trabecular bone. Since compact bone is stronger than trabecular bone, this should lead to relatively stronger cross section in females. A more precise analysis of sex-related differences in relative bending strengh of the femoral neck and other cross sections will be presented in conjunction with further cross section properties reviewed below.

Age differences.

Tables 9a, b (tibiae) and 10a, b (femora) summarize age changes in bone areas in the study sample. Concentrating first on differences between the two larger age groupings (20-39 years and 40+ years), on average over the 10 diaphyseal cross sections, we find essentially no change with age among males in cortical area (.5% loss), while females show a mean loss in CA of 10.0%. Females show a much greater average loss CA in tibia (14.2%) than in the femur (5.8%), while males show no bone-related differences in CA loss with age (0% and .9% for the tibia and femur, respectively). Both bones in both sexes display significant gains in TA at almost all cross section sites, averaging 7.5% overall in males and 11.2% in females. The lower figure for TA gain in males is primarily due to less gain in TA in the femur, averaging only 5.3% (to females' 11.5% gain). Gain in TA in the tibia on average is about the same in males (9.7%) and females (10.8%). Several other authors have previously noted the phenomenon of subperiosteal expansion with aging in long bones and other skeletal elements (see chapter 2). The

Table 9a

Age Differences in Tibial Bone Areas: Males

Cross Section	20-39 yrs Mean	20-39 yrs SE	40+ yrs Mean	40+ yrs SE	40+ yrs - 20-39 yrs	%dif	F[1]
CORTICAL AREA							
20%	187.9[2]	3.8	185.9	7.5	-2.0	-1.1	.59
35%	254.8	5.1	256.8	7.2	2.0	.8	1.51
50%	286.8	6.3	288.6	9.3	1.8	.2	.94
65%	296.5	7.0	303.4	11.1	6.9	2.3	1.16
80%	273.8	5.4	267.5	10.1	-6.3	-2.3	1.22
TOTAL SUBPERIOSTEAL AREA							
20%	439.7[2]	8.3	472.0	13.0	32.3*	7.3	3.50**
35%	385.7	5.9	434.5	13.7	48.8***	12.6	4.34**
50%	430.8	7.3	484.1	14.4	53.3***	12.4	4.00**
65%	511.2	9.6	567.5	10.7	56.3***	11.0	3.96**
80%	735.8	16.2	775.3	19.6	39.5	5.4	1.06
MEDULLARY AREA							
20%	251.8[2]	8.2	286.1	12.8	34.3*	13.6	3.63**
35%	131.0	5.1	177.6	13.3	46.6***	35.6	4.30***
50%	144.0	5.7	195.4	14.1	51.4***	35.7	3.92***
65%	214.7	6.5	264.1	10.7	49.4***	23.0	3.99**
80%	462.0	15.5	507.9	19.6	45.9†	9.9	1.81

Results of 6 age group ANOVA--see text.

mm^2

p<.10 * p<.05 ** p<.01 *** p<.001

Table 9b

Age Differences in Tibial Bone Areas: Females

CORTICAL AREA

Cross Section	20-39 yrs Mean	SE	40+ yrs Mean	SE	40+ yrs - 20-39 yrs	%dif	F [1]
20%	147.6[2]	3.6	126.0	6.0	-21.6**	-14.6	2.93*
35%	193.0	4.4	172.8	7.8	-20.2*	-10.5	1.36
50%	216.9	5.1	188.4	8.8	-28.5**	-13.1	1.82
65%	229.6	5.6	187.2	9.0	-42.4***	-18.5	4.37**
80%	214.5	4.7	183.8	9.0	-30.7***	-14.3	3.32**

TOTAL SUBPERIOSTEAL AREA

Cross Section	20-39 yrs Mean	SE	40+ yrs Mean	SE	40+ yrs - 20-39 yrs	%dif	F [1]
20%	345.5[2]	7.9	376.9	10.3	31.4*	9.1	1.56
35%	300.6	8.4	342.7	9.3	42.1**	14.0	1.80
50%	330.9	8.3	373.7	11.9	42.8**	12.9	1.79
65%	394.9	10.0	427.9	10.1	33.0*	8.4	1.20
80%	556.4	13.5	608.5	18.8	52.1*	9.4	1.60

MEDULLARY AREA

Cross Section	20-39 yrs Mean	SE	40+ yrs Mean	SE	40+ yrs - 20-39 yrs	%dif	F [1]
20%	197.9[2]	7.0	250.9	10.3	31.4*	9.1	1.56
35%	107.7	5.9	169.9	10.4	62.2***	57.8	6.04***
50%	114.0	6.4	185.2	12.8	71.2***	62.4	6.12***
65%	165.3	7.0	240.7	11.3	75.4***	45.6	7.12***
80%	341.9	12.3	424.7	18.9	82.8***	24.2	3.21**

1 Results of 6 age group ANOVA--see text.

2 mm^2

* p<.05   ** p<.01   *** p<.001

Table 10a

Age Differences in Femoral Bone Areas: Males

CORTICAL AREA

Cross Section	20-39 yrs Mean	SE	40+ yrs Mean	SE	40+ yrs - 20-39 yrs	%dif	F[1]
20%	266.4[2]	4.8	252.3	4.4	-14.1*	-5.3	1.03
35%	319.2	5.5	316.6	6.5	-2.6	-.8	.57
50%	359.0	6.4	358.7	6.7	-.3	.1	.73
65%	365.2	7.5	370.0	7.7	4.8	1.3	.82
80%	349.6	6.1	347.8	7.7	-1.8	-.5	.12
Neck	116.5	2.5	113.0	3.6	-3.5	-3.0	.31

TOTAL SUBPERIOSTEAL AREA

Cross Section	20-39 yrs Mean	SE	40+ yrs Mean	SE	40+ yrs - 20-39 yrs	%dif	F[1]
20%	751.4[2]	17.0	778.5	18.5	27.1	3.6	1.72
35%	536.1	9.1	563.0	9.3	26.9†	5.0	.84
50%	488.3	7.6	520.9	7.4	32.6**	6.7	2.36*
65%	509.5	8.5	542.8	8.4	33.3*	6.5	1.98
80%	565.7	8.0	592.3	10.6	26.6*	4.7	1.28
Neck	569.5	11.0	600.0	15.7	30.5	5.4	.95

MEDULLARY AREA

Cross Section	20-39 yrs Mean	SE	40+ yrs Mean	SE	40+ yrs - 20-39 yrs	%dif	F[1]
20%	484.9[2]	15.8	526.2	19.6	41.3	8.5	2.43*
35%	216.8	7.8	246.4	11.6	29.6*	13.6	1.48
50%	129.3	4.8	162.2	8.2	32.9***	25.4	3.17*
65%	144.3	5.0	172.8	8.7	28.5**	19.8	2.62*
80%	216.1	5.5	244.5	8.1	28.4**	13.1	3.06*
Neck	453.0	10.7	486.9	15.1	33.9†	7.5	1.06

[1] Results of 6 age group ANOVA--see text.

[2] mm^2

† p<.10 * p<.05 ** p<.01 *** p<.001

Table 10b

Age Differences in Femoral Bone Areas: Females

CORTICAL AREA

Cross Section	20-39 yrs Mean	SE	40+ yrs Mean	SE	40+ yrs - 20-39 yrs	%dif	F [1]
20%	221.3[2]	5.1	193.8	7.3	-27.5**	-12.4	2.42*
35%	255.8	5.4	244.6	9.4	-11.2	-4.4	.87
50%	299.0	6.5	279.4	10.6	-19.6†	-6.6	1.30
65%	303.7	6.4	294.5	11.4	-9.2	-3.0	.65
80%	285.7	5.7	278.9	11.8	-6.8	-2.4	.46
Neck	98.7	2.6	93.1	5.7	-5.6	-5.7	.41

TOTAL SUBPERIOSTEAL AREA

Cross Section	20-39 yrs Mean	SE	40+ yrs Mean	SE	40+ yrs - 20-39 yrs	%dif	F [1]
20%	601.0[2]	14.4	637.9	16.0	36.9	6.1	.97
35%	423.7	9.0	468.8	9.8	45.1**	10.6	3.04*
50%	398.5	7.6	445.2	11.0	46.7***	11.7	5.02***
65%	411.5	7.8	474.6	11.2	63.1***	15.3	6.71***
80%	449.7	9.0	510.7	14.3	61.0***	13.6	4.31**
Neck	454.8	10.8	477.5	11.3	22.7	5.0	.95

MEDULLARY AREA

Cross Section	20-39 yrs Mean	SE	40+ yrs Mean	SE	40+ yrs - 20-39 yrs	%dif	F [1]
20%	379.7[2]	14.8	444.0	17.4	64.3**	16.9	1.77
35%	167.9	6.8	224.2	12.7	56.3***	33.5	3.80**
50%	99.5	3.5	165.8	14.1	66.3***	66.6	8.20***
65%	107.8	3.7	180.2	15.8	72.4***	67.2	7.98***
80%	164.1	5.9	231.8	15.6	67.7***	41.2	5.85***
Neck	356.1	9.8	384.4	13.0	28.3†	7.9	1.20

[1] Results of 6 age group ANOVA--see text.

[2] mm^2

† $p < .10$ * $p<.05$ ** $p<.01$ *** $p<.001$

present data clearly show that the loss in CA in females with aging is due not to lack of subperiosteal expansion, but rather, greater expansion of the medullary cavity than in males (i.e., greater endosteal resorption of bone). In other words, both sexes gain in TA with aging (females more so than males on average), but the larger increase in MA in females eventually overtakes this gain and leads to a net loss in CA, while in males TA and MA increase about equally, leading to essentially no change in CA.

Analysis of bone area differences using the 6 smaller age groupings makes these contrasting patterns of age change in males and females even more apparent. Means and standard errors of all cross-sectional properties for all 6 age group divisions (20-24, 25-29, 30-34, 35-39, 40-49, and 50+ years) were also calculated. Results of the analysis of variance tests over these six age groups for bone areas are shown in the form of F's and their significance in the last column of Tables 1 a, b and 12a, b. More detailed presentation of this data here is limited to graphic representation of age changes in two typical cross sections--tibia 65% in Figures 17a, b and femur 50% in Figures 18a, b. Modifications of these "typical" patterns for individual cross sections are noted within the text.

First looking at general broad patterns of age changes in bone areas, we find confirmation for the sex-related differences in bone loss and gain above. In males (Figures 17a, 18a) TA and MA follow almost parallel courses over the 6 age groups (both increasing), leading to very little age-related variation in CA. In females (Figures 17b, 18b) increases in MA in the oldest age groups are greater than

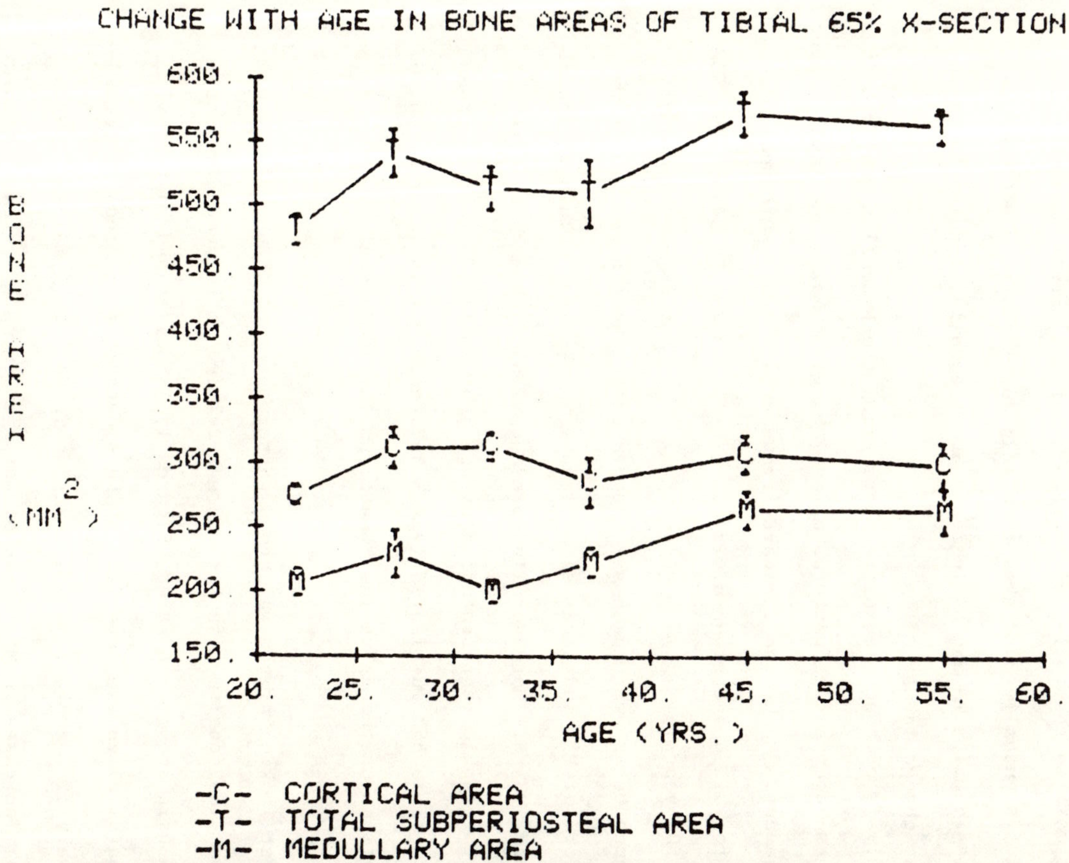
CHANGE WITH AGE IN BONE AREAS OF TIBIAL 65% X-SECTION: MALES
BONE AREA (MM²)
600.
550.
500.
450.
400.
350.
300.
250.
200.
150.
20.
25.
30.
35.
40.
45.
50.
55.
60.
AGE (YRS.)
-C- CORTICAL AREA
-T- TOTAL SUBPERIOSTEAL AREA
-M- MEDULLARY AREA

Figure 17a

CHANGE WITH AGE IN BONE AREAS OF TIBIAL 65% X-SECTION: FEMALES

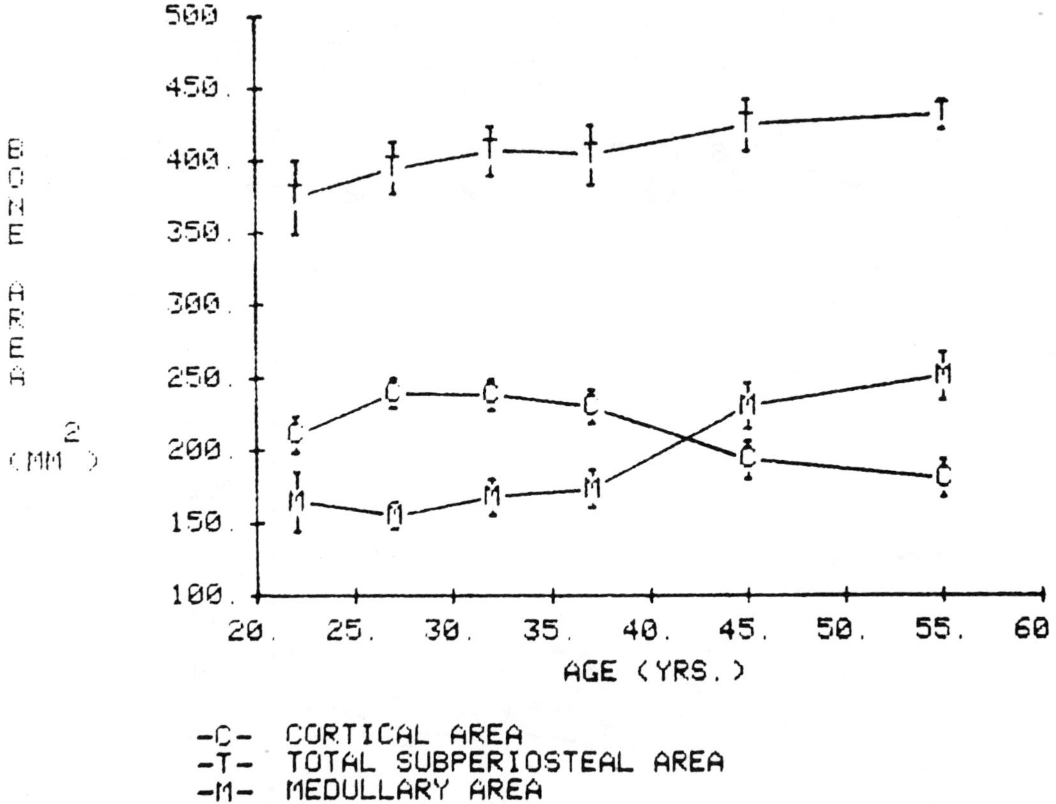

Figure 17b

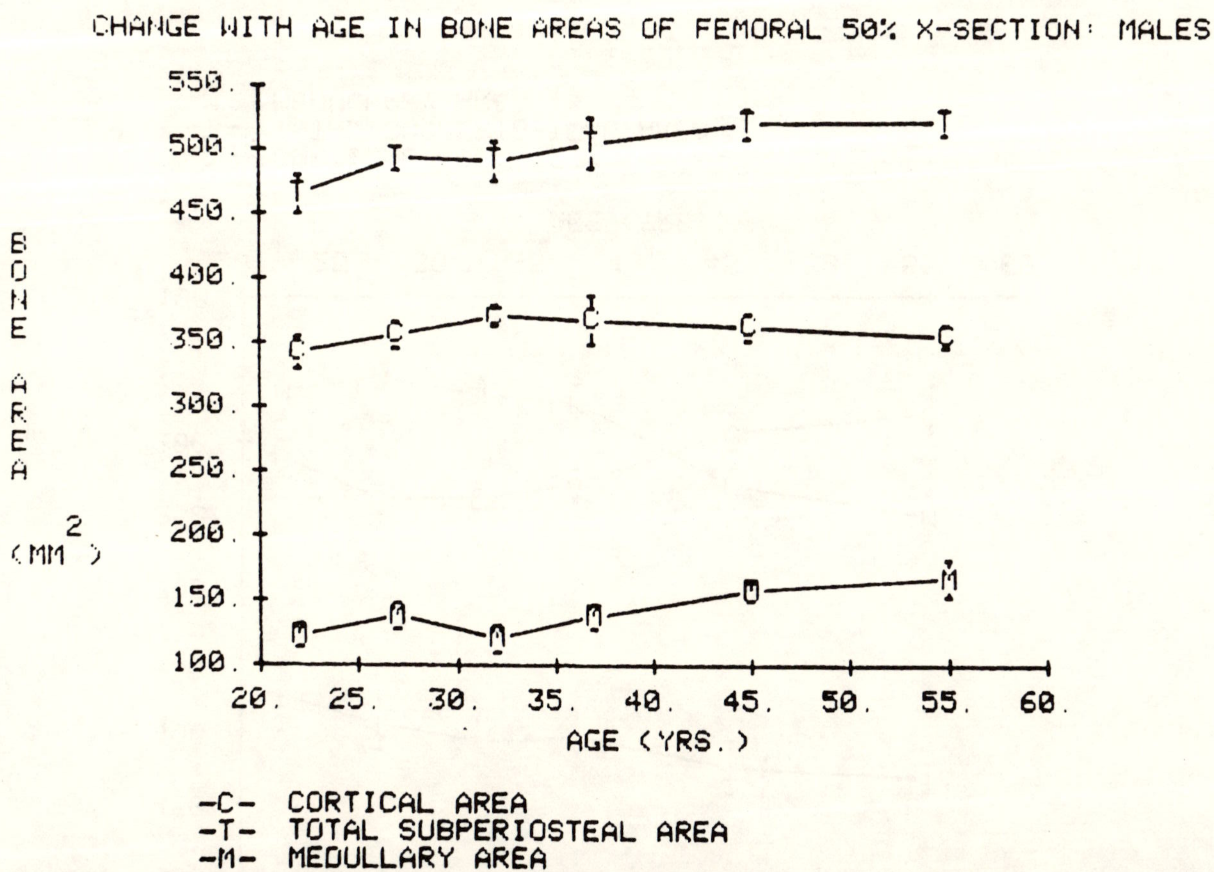
CHANGE WITH AGE IN BONE AREAS OF FEMORAL 50% X-SECTION: MALES
BONE AREA (MM²)
550. 500. 450. 400. 350. 300. 250. 200. 150. 100.
20. 25. 30. 35. 40. 45. 50. 55. 60.
AGE (YRS.)
-C- CORTICAL AREA
-T- TOTAL SUBPERIOSTEAL AREA
-M- MEDULLARY AREA

Figure 18a

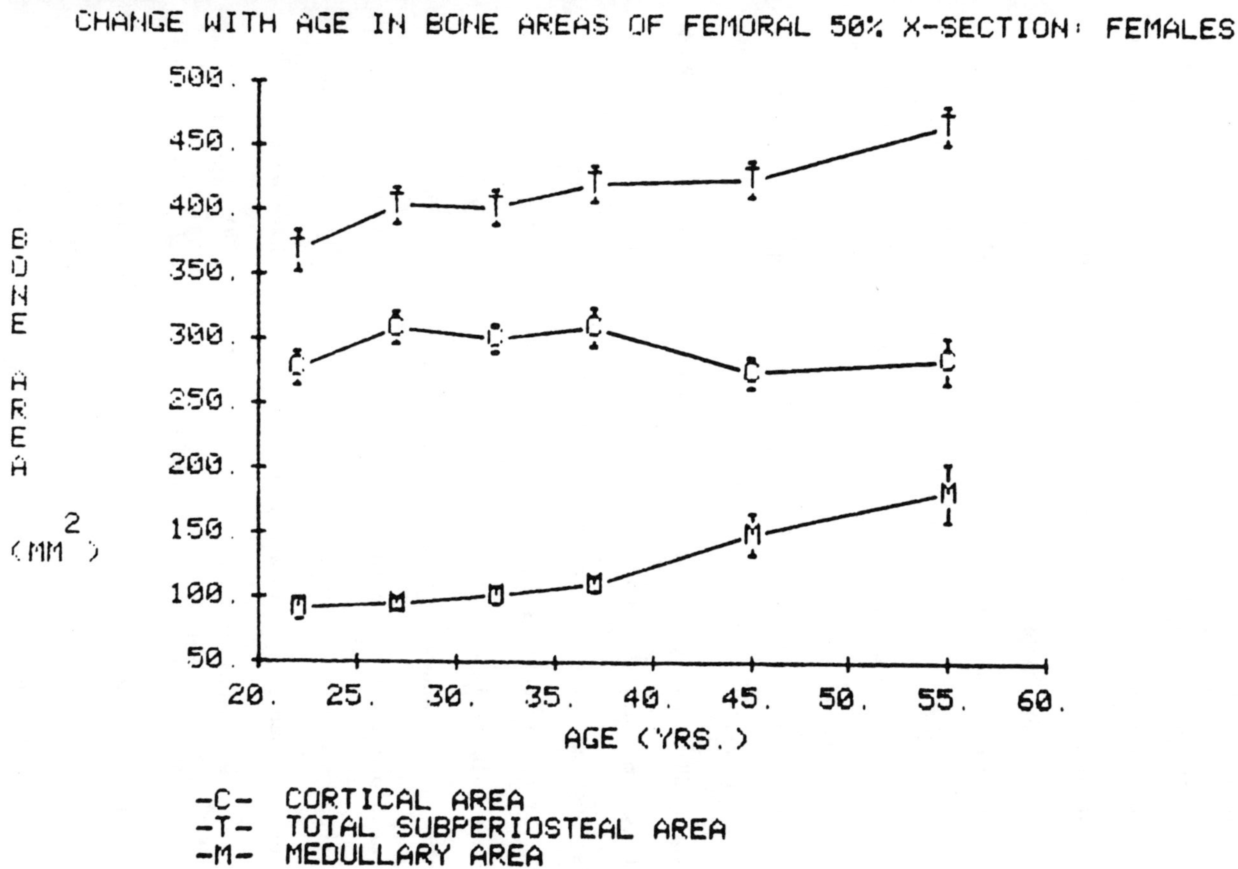
CHANGE WITH AGE IN BONE AREAS OF FEMORAL 50% X-SECTION: FEMALES
BONE AREA
(MM 2)
500.
450.
400.
350.
300.
250.
200.
150.
100.
50.
20.
25.
30.
35.
40.
45.
50.
55.
60.
AGE (YRS.)
-C- CORTICAL AREA
-T- TOTAL SUBPERIOSTEAL AREA
-M- MEDULLARY AREA

Figure 18b

increases in TA, leading to a loss in CA in these groups. Inspection of these figures also reveals further, nonlinear age trends not apparent in the 2 age group analyses presented above. Analysis of age trends in CA show that area of compact cortical bone reaches a maximum between 25 and 40 years, depending upon sex and cross section site. Thus, age changes in CA follow a curvilinear pattern: an initial increase in early adulthood, followed by a leveling off in males and a decrease in females in the fifth and sixth decades. The initial increase in CA in early adulthood is brought about by an increase in TA without a corresponding increase in MA--in fact, in most cross sections, MA actually decreases slightly in the 25-29 or 30-34 year age group, indicating some endosteal as well as subperiosteal bone appositional growth during this time period. Subperiosteal apposition continues throughout life; however, after 35 years endosteal resorption of bone has begun in all cross sections, leading to a net decline (over "peak" young adult values) in CA.

In the largest study of it kind to be carried out to date, Garn (1970) found precisely the same pattern of age change in second metacarpal breadths measured from radiographs. In his study, males peaked in cortical breadth at age 30 and females at age 40, which were also the ages when endosteal breadths reached a minimum. Doyle (1970), Dequeker (1975) and other have reported similar findings. The increase in CA in early adulthood due to both subperiosteal and endosteal bone apposition is apparently a continuation of a trend begun in middle adolescence in both sexes (Johnston and Watts, 1969; Frishancho et al., 1970).

A further aspect of age changes in bone areas of the lower limb is revealed by contrasting the patterns of age differences in different cross section locations. First, a difference in timing of the onset of bone loss (decline in CA) is apparent in female femora and tibiae. In all tibial cross sections, females reach a peak in CA in the 25-29 year age group, after which they begin to decline, more rapidly after age 35 (Figure 17b). In most femoral cross sections (the femoral 20% and femoral neck sections being the exceptions) females reach a peak in CA in the 35-39 year age group, after which they decline (Figure 18b). The increase in CA over a more extended period in time in female femora is due to a proportionately greater increase in TA in the third and fourth decades, rather than any endosteal appositional growth, since femoral medullary area actually increases steadily throughout this period, unlike the tibia. Thus, in females a net loss of cortical bone in the femur is delayed until after age 40 by increased subperiosteal apposition of bone, while in the tibia this net loss begins in the 30's. No such bone-specific difference in timing of bone loss or gain is apparent in males, where almost all cross sections peak in CA at age 30-34 (Figures 17a, 18a).

Thus, in terms of both total bone loss (Tables 9a, b; 10a, b) and timing of bone loss (Figures 17a, b; 18a, b), the female femur appears to be more "protected" than the tibia from reductions in cortical bone volume with aging. Biomechanically, this could be explained on the basis of relatively greater loadings of the femur in females, which would stimulate increased bone appositional growth to reduce stress. This hypothesis gains support from the observation that reduced loss in CA in female femora is brought about by greater apposition of bone on the subperiosteal surface, the region of beam cross sections subjected to the highest stresses under bending. Further discussion of these points will be carried out in connection with the presentation of age differences in principal moments of the femur and tibia.

Finally, some differences in patterns of age change in bone areas within the same bone were noted. In both sexes, percentage reduction in CA with age tended to be greater in areas with significant amounts of trabecular bone--the distal-most femoral cross section and the most proximal and distal tibial cross sections (Tables 9a, b; 10a, b). The one exception to this general pattern is the tibia 65% in females, which shows an even greater percentage loss in CA than the tibia 20% and 80% sections. The femoral neck shows a relatively large loss in CA in males (3.0%) and an average loss in CA (for the femur) in females (5.7%). The greatest percentage <u>gain</u> in CA in males occurs at the tibia 65% site, while in females the smallest loss occurs in the proximal femoral diaphysis.

Two factors, possibly operating in combination, may explain these site-specific differences in bone loss. First, the presence of trabecular bone may enhance endosteal resorption of the cortical shell by increasing access to compact cortical bone surfaces. Several authors have noted more rapid loss of trabecular than compact bone with aging (e.g., Nordin et al., 1965; Doyle, 1970), due presumably to the easier mobolization of calcium from trabecular bone (<u>op. cit.</u>). Since in sites near the ends of long bone diaphyses trabecular bone interdigitates with compact bone along the endosteal surface, natural grooves and furrows in the cortical shell are created which provide more surface area for osteoblasts and clasts to work. This could lead to a higher bone turnover rate at these sites and a greater loss in older adults, where endosteal resorption exceeds formation. Also, since some judgment is exercised by the researcher when determining the exact position of the endosteal border during tracing of bone outlines (see chapter 4), it is possible that some increased porosity along the endosteal surface could be interpreted as an increase in size of the medullary cavity (due to the unevenness of the endosteal boundary in some cross sections, it was frequently very difficult to arrive at a precise location for this bone "surface", particularly in older adults). Thus, the presence of trabecular bone could produce both real and apparent more-than-average increases in the rate of medullary cavity expansion with age.

The second factor which could lead to greater losses of cortical bone in some areas and smaller losses in others is differential stress along the length of the diaphysis. This is based on the theory that greater mechanical stresses will stimulate the production of bone, and smaller stresses the removal of bone ("Wolff's Law"). If this is true, then the areas of greatest bone loss with age should be those under least mechanical stress. In general, bending stresses in curved beams brought about by axial loadings applied to the bone ends will be greatest at the point of maximum curvature, or displacement from the straight longitudinal axis between beam ends. This occurs in the femur at about midshaft and in the tibia in the proximal sections (see Figures 13a, b above). In the femur, an additional M-L bending moment is present, created by the medial position of the femoral head, which will be greatest in the proximal part of the femoral shaft. Thus, regions of lowest bending stress in the two bones should be in the distal-most sections of the tibia and femur.[11] With the exception of female tibiae, these are in fact the cross sections which show the most loss in CA with aging (Tables 9a, b; 10a, b). Female tibiae show a somewhat different pattern, losing bone in the 35% section, but this too could be a result of differential stress under *in vivo* loadings. As stated earlier, the tibia is subjected to relatively high torsional loadings during gait. The most common site for torsional fractures of the tibia is about one-third the distance from the distal end (Bohler, 1958: 1695), or about at the 35% cross section level. If, as will be elaborated further in the following chapter, torsional loadings of the tibia were relatively greater in females of the Pecos sample (and A-P loading greater in males), this could explain why less bone is lost from the 35% section in females.

---

[11]This assumes that bending moments of the tibia and femur are produced only by axial forces (e.g., body weight) applied to the bone ends, and ignores the effects of muscular action. A fuller description of bending stresses in the lower limb will be presented in chapter 6.

The femoral neck cross section, as mentioned above, exhibits an average femoral loss in CA in females, and an above average loss in males. These results may be explained on the basis of a combination of the two factors described in the preceding paragraphs. First, cross sections with significant proportions of trabecular bone, including the femoral neck, lose (or appear to lose) more compact cortical bone due to enhanced mechanisms of endosteal resorption. Using the femoral 20% cross section as a model, this factor alone would lead to a prediction of at least a 5% loss in CA in males and a 12% loss in CA in females from the femoral neck (Tables 10a, b). Second, the femoral neck is subjected to unusually high bending (and possibly torsional) stress _in vivo_, particulary in females. This would lead to a _reduction_ in the loss of CA with aging, due to high stresses, more marked in females. Thus, the 3% and 6% average femoral neck losses of CA for males and females, respectively, seem reasonable given a combination of the two factors. Of course, the contribution of trabecular bone to bone strength at this site is probably significant (Paul, 1971), thus, any consideration of stresses and bone remodeling of the femoral neck should take into account changes in trabecular bone density (see later section of this chapter on BMC measurements) and other features of trabecular architecture (e.g., Arnold et al., 1966).

Another index of bone volume which has been used by some investigators to quantify skeletal age changes is percent cortical area, or

%CA = (CA/TA) X 100 (Garn, 1970; Himes et al., 1975). Although this index has no inherent biomechanical or even metabolic significance, since both CA and TA (and principal moments) may increase and still lead to a decrease in %CA, figures for this index are presented here for comparative purposes with other studies. Figures 19a, b illustrate changes in %CA with aging in the study sample at the tibia 65% and femur 50% sites. The two cross sections show remarkably similar patterns of age change within sex groups. In both, males reach a peak in %CA in the 30-34 year age group, and females in the 25-29 year age group, after which both sexes decline, females much more rapidly after age 40. If comparisons are made between the first four age groups and the last two age groups combined (i.e., similar to those in Tables 9a, b; 10a, b), the average reduction with aging in %CA over all diaphyseal sections in males is 4.1%, in females 10.8%. Similar calculations for the femoral neck average a 1.7% reduction in males and a 2.3% reduction in females. The tibia averages a somewhat greater decrease in %CA than the femur in both sexes, although the difference is not as marked as in the case of CA in females. Midshaft regions of both bones tend to show the greatest reductions in %CA, except in male femora, where %CA decreases are about equal at all levels of the diaphysis (this points up the fallacy of using %CA as a guide to bone loss, since in both sexes CA (and second moments of area) tend to show the greatest <u>increase</u> (or least loss) in the midshaft region).

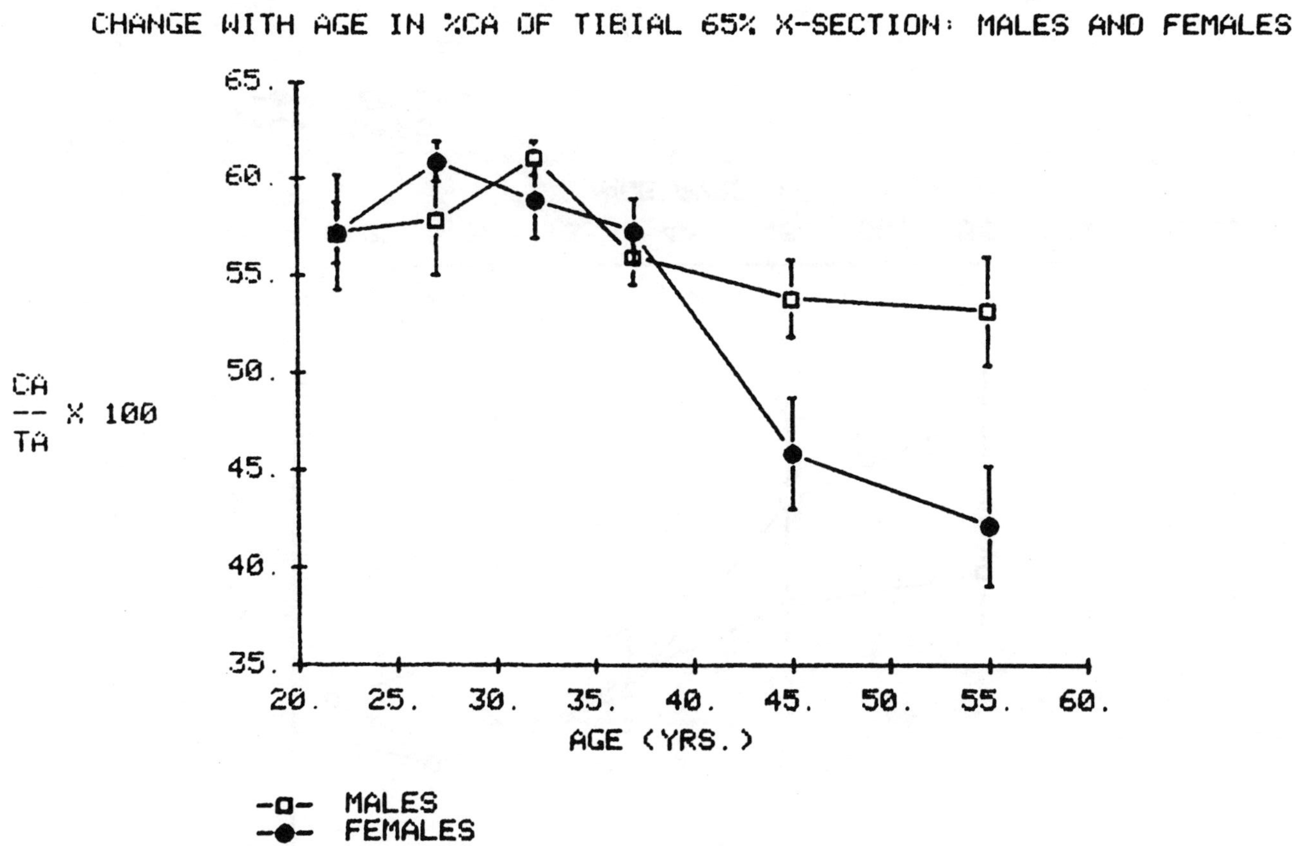
CHANGE WITH AGE IN %CA OF TIBIAL 65% X-SECTION: MALES AND FEMALES
CA
-- X 100
TA
65.
60.
55.
50.
45.
40.
35.
20.
25.
30.
35.
40.
45.
50.
55.
60.
AGE (YRS.)
MALES
FEMALES

Figure 19a

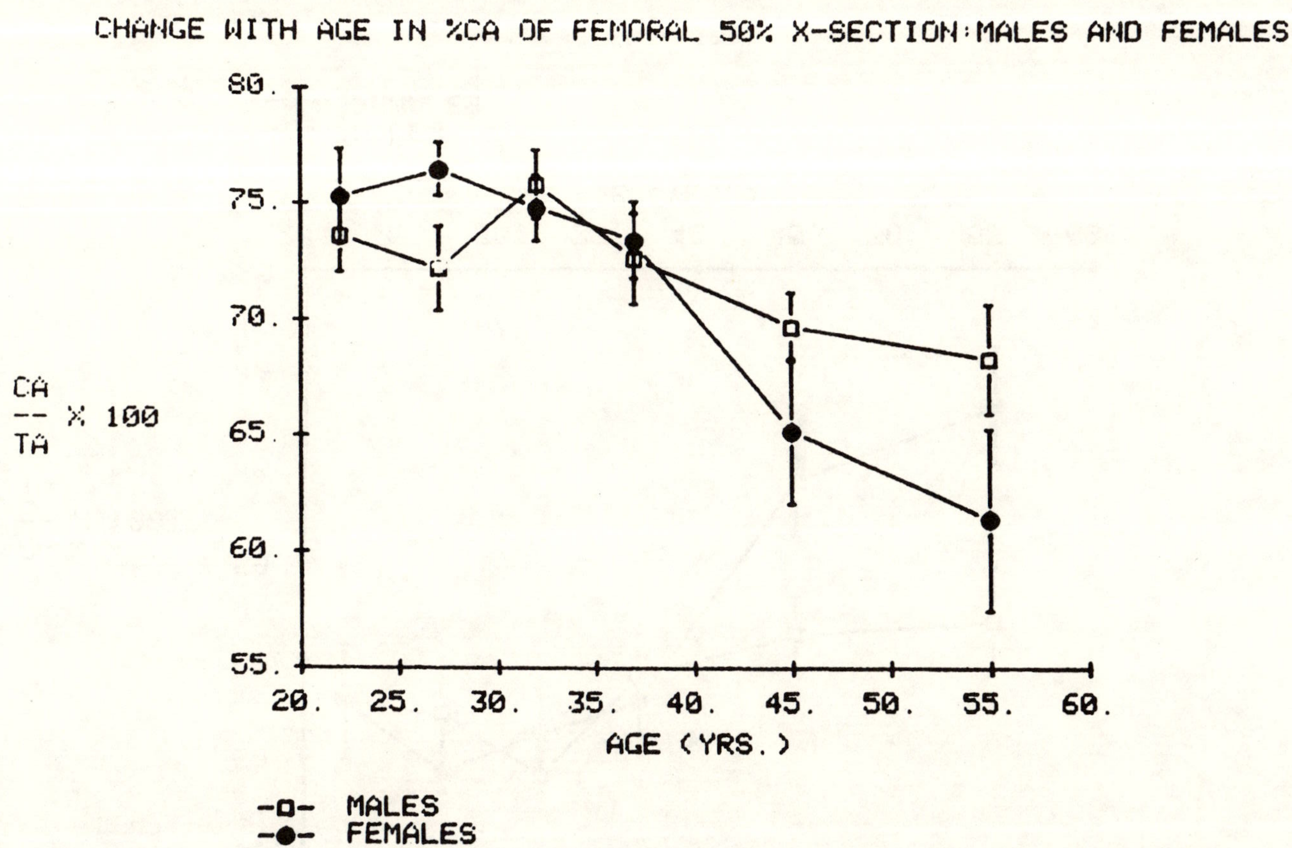
CHANGE WITH AGE IN %CA OF FEMORAL 50% X-SECTION:MALES AND FEMALES
CA
-- X 100
TA
80.
75.
70.
65.
60.
55.
20.
25.
30.
35.
40.
45.
50.
55.
60.
AGE (YRS.)
MALES
FEMALES

Figure 19b

In comparing the figures for bone area changes presented in the present report with those of other studies (almost all extrapolated from radiographic breadths), it should be noted that the upper range of individuals in the Pecos sample is less than that of most studies of the living. Thus, for example, Garn (1970) recorded a total decline with aging of about 28 percentage points in his %CA index among females, while the total decline in %CA among females in the present study ranges in different cross sections from about 15 to 20 percentage points (Figures 19a, b). However, Garn's age range included individuals through the eighth and ninth decades of life. In the present study, while the 50+ year age group probably contains some individuals past 60 years and possibly even past 70 (Todd aged 8% of Pecos adults as over 70 years), the average age of the individuals included in my oldest age group is very likely not more than 60 years (estimated conservatively as 55 years in the 6 age group figures included here, e.g., Figures 19a, b). If a similar age range in Garn's study is used, the age changes observed in bone areas and %CA are quite comparable to those of the present study.

Another factor which could significantly influence the results of the age comparisons reported here is a change in average size of individuals with aging (e.g., due to selective survival), reflected here as differences in bone length. As shown earlier (Table 5 above), while no significant differences in bone length between age groups is present among females, males show a statistically significant, approximately 2% increase in bone length with aging. This increase could conceivably have a significant effect on bone cross-sectional area changes with aging in males, contributing to the observed increase in TA (and MA) and the lack of any decrease in CA. As noted earlier, the elimination of two males--one very large individual in the 50+ year group, and one very small individual in the 30-34 year group--leads to a substantial reduction in the average difference in bone length between the two larger age groupings, dropping from 2% to 1% (tibia and femur combined average). To investigate the effects of differences in bone length on cross-sectional bone area changes, these two individuals were eliminated from the study sample and comparisons between the 20-39 year and 40+ year age groups again carried out on the remaining 57 males. The results of this experiment showed no change in the average difference in CA between age groups (.5%) and an insignificant decrease in the average TA difference between groups (7.5% to 6.9%). In addition, the age difference in the average maximum principal moment, I_{max}, declined only very slightly from 9.6% to 8.8%. Thus, approximately halving the average difference between age groups in bone length produced no significant effect on age differences in cross-sectional properties which might have been expected to reflect "size" changes. This finding is not surprising, since the relationship between bone length and cross-sectional dimensions is a weak one, and use of bone length as a "size" standardizer highly questionable for a study of this kind (see discussion elsewhere). Therefore, differences in bone length between age groups, particularly on the order of only one or two percent, seem unlikely to have significantly affected the pattern of age changes in cross-sectional properties observed. At most, "correcting" for differences in bone length would reduce bone areas (and moments) a few percent in older males relative to younger males, and therefore slightly increase bone "loss" with age in this sex; however, the general sex-related differences in bone aging discussed above will remain unchanged.

One aspect of cross section geometry which was not measured in the present study is cortical thickness, which is probably the most common of all bone volume dimensions to be reported in the literature due to its ease in measurement from radiographs (e.g., Barnett and Nordin, 1960; Exton-Smith et al., 1969; Meema and Meema, 1969; Bloom and Laws, 1970; Garn 1970). However, it is possible to calculate average cortical thickness in a cross section given TA and MA, if the cross section is assumed to be cylindrical (that is, circular with parallel subperiosteal and endosteal borders), using the following formula: cortical thickness = .564 ($\sqrt{TA} - \sqrt{MA}$). Since no cross section in the femur or tibia exactly satisfies this requirement--the middle three femoral diaphyseal sections come closest--the following results must be viewed as only gross approximations of true cortical thicknesses. They do, however, provide some interesting comparisons with data of a similar nature collected by other researchers using different methods.

Using the formula described above, the average differences in cortical thickness between the 20-39 and 40+ age groups in the present study sample are as follows: males, tibia--6.6% loss, femur--5.3% loss; females, tibia--18.6% loss, femur--14.9% loss. (Femoral figures here do not include the femoral neck; it should also be noted that, due to the triangular shape of most of the tibial diaphysis, these results are probably more accurate for the femur than the tibia.) The figures obtained are similar to the age differences in cortical thickness of the second metacarpal reported by Garn (1970: calculated by me as the difference between individuals in his study 20-40 years and 50-70 years), who found an 8% average loss in males and 14% loss in females. In another study which used direct measurement of cut bone cross sections in an archaeological sample of about the same age range as the present study (Carlson et al., 1976), the following figures for average reduction in cortical thickness over 5 femoral diaphyseal sites were obtained: males--11% loss, females--21% loss (my calculations). If only the femoral midshaft is used (the only cross section location which precisely corresponded in this study and the present study), the figures for cortical thickness reductions with age are 21% for females, 8% for males in Carlson et al.'s study, and 18% for females, 6% for males in the present study. These similar findings give some support to the method of calculation of cortical thickness carried out here, and provide some confidence in the resulting values, as least as general indicators of cortical thickness changes with aging. More detailed comparisons of present study results with those of other studies will be presented in chapter 6.

Side differences.

As noted earlier, the differences in bone properties reported here for right and left lower limbs are not truly "bilateral", since right and left sides are represented by different individuals in the sample. As a result, these side differences must be viewed as only approximations which will increase in accuracy as the subsamples of right and left bones approach random samples of right and left sides within the total study sample. Another consequence of this sampling procedure is that the paired t test (Hayes and Winkler, 1970), a much more powerful statistical tool for evaluating the significance of differences between paired data (e.g., Garn et al., 1976; Ruff and Jones, 1981), cannot be used in the present study. Thus, many of the side differences reported here do not reach formal statistical significance (using the standard two sample t test). However, it can be shown that, if the means for right and left subsamples come close to approximating the means for the two sides in the total sample, most of the differences in the paired t statistic and the two sample t statistic will result from the difference in the standard error terms used in the denominator (which will be much larger--hence a smaller t--using the two sample t test). In the present study, 30 individuals of each side and sex (29 male left side) were available, a large enough sample to justify some confidence that the mean values obtained are indeed representative of mean values for the total sample, within sex. Therefore, we may assume that many of the differences not reaching formal statistical significance using the two sample test would in fact be significant if the paired test were available. Thus, somewhat more latitude has been allowed in the following discussions as to what constitutes a "significant" (i.e., meaningful) difference between sides.[12]

---

[12]It is noteworthy in this regard that Latimer and Lowrance (1965) found no statistically significant differences between right and left (paired) bone lengths using the standare two sample t test. Using their data, I recalculated significance levels using the paired t test and found almost all side differences to be highly significant ($p < .001$).

Tables 11a, b summarize side-related differences in bone areas in males and females, respectively.[13] Percentage differences between right and left sides are presented as (right/left)/right, thus, negative percentages indicate larger left side values. On the average, in both sexes, cortical bone areas are larger on the left side, although much variation in this general trend is present among individual cross sections. Females show more asymmetry in CA, averaging 7.8% larger on the left side, while males average only 2.0% larger. This sex difference is particularly marked in the femur, where the left side averages 9.6% larger in CA among females, and only .9% larger among males. The sex difference is not quite as great in the tibia, where females average 5.9% larger and males 3.0% larger on the left.

This left side "dominance" in cortical bone area is brought about by both larger subperiosteal areas and smaller medullary areas on the left side, although the precise combination of these differences varies between bones and sexes. In the tibia, larger left CA values result principally from smaller MA values on the left (3.0% smaller in males, 8.2% smaller in females on average), with TA values approximately equal or slightly larger in right tibiae on average (males .1%, females 1.6% larger on the right side). However, this general pattern varies along the length of the tibial diaphysis. In the distal tibia, both TA and MA tend to be larger on the right (MA more so, thus smaller right side CA), while in the proximal tibia TA and MA are larger on the

---

[13] Significance levels shown are those for the two sample t test. However, because of the arguments presented above, standard errors of right and left side means are not included in these tables, since they are actually of little usefulness in evaluating the true range and distribution of bilateral side differences.

Table 11a

Side Differences in Tibial Bone Areas

CORTICAL AREA

Cross Section	Males			Females		
	Right	Left	%dif [1]	Right	Left	%dif [1]
20%	184.0 [2]	190.6	-3.6	134.8	145.9	-8.2†
35%	253.6	257.4	-1.5	181.2	191.3	-5.6
50%	284.3	290.7	-2.2	202.3	212.5	-5.0
65%	295.0	302.8	-2.6	210.6	220.3	-4.6
80%	265.0	278.5	-5.1	198.2	210.3	-6.1

TOTAL SUBPERIOSTEAL AREA

Cross Section	Males			Females		
	Right	Left	%dif [1]	Right	Left	%dif [1]
20%	461.0 [2]	439.9	4.6	362.3	349.7	3.5
35%	401.9	402.6	-.2	319.3	310.0	2.9
50%	447.0	450.7	-.8	350.4	339.9	3.0
65%	528.1	532.5	-.8	404.9	406.8	-.5
80%	741.6	757.0	-2.1	565.2	582.3	-3.0

MEDULLARY AREA

Cross Section	Males			Females		
	Right	Left	%dif [1]	Right	Left	%dif [1]
20%	277.0 [2]	249.3	10.0*	227.4	203.8	10.4†
35%	148.3	145.2	2.1	138.1	118.7	14.0
50%	162.8	160.0	1.7	148.1	127.4	14.0
65%	233.1	229.7	1.5	194.3	186.5	4.0
80%	476.6	478.5	-.4	366.9	372.0	-1.4

[1] [(Rt-Lt)/Rt] X 100

[2] mm^2

† p<.10 * p<.05 (two sample t test--see text)

Table 11b

Side Differences in Femoral Bone Areas

CORTICAL AREA

Cross Section	Males Right	Males Left	Males %dif[1]	Females Right	Females Left	Females %dif[1]
20%	258.7[2]	264.6	-2.3	202.2	222.0	-9.8*
35%	317.9	318.8	-.3	237.6	266.5	-12.2**
50%	355.3	362.6	-2.1	277.8	307.1	-10.5**
65%	366.0	367.7	-.5	286.9	314.4	-9.6*
80%	350.3	347.7	.7	275.4	291.4	-5.8
Neck	113.8	116.9	-2.7	98.3	95.4	3.0

TOTAL SUBPERIOSTEAL AREA

Cross Section	Males Right	Males Left	Males %dif[1]	Females Right	Females Left	Females %dif[1]
20%	754.9[2]	766.4	-1.5	603.4	623.2	-3.3
35%	541.4	549.1	-1.4	425.7	451.9	-6.2
50%	498.5	500.2	-.3	403.7	424.4	-5.1
65%	525.9	515.6	2.0	425.4	439.7	-3.4
80%	586.2	562.8	4.0†	466.2	474.0	-1.7
Neck	582.5	577.0	.9	461.3	463.5	-.5

MEDULLARY AREA

Cross Section	Males Right	Males Left	Males %dif[1]	Females Right	Females Left	Females %dif[1]
20%	496.2[2]	501.8	-1.1	401.1	401.2	0
35%	223.5	230.3	-3.0	188.0	185.4	1.4
50%	143.2	137.6	3.9	125.9	117.4	6.8
65%	159.9	147.8	7.6	138.5	125.3	9.5
80%	235.9	215.1	8.8*	190.8	182.5	4.4
Neck	468.7	460.1	1.8	363.0	368.0	-1.4

[1] [(Rt-Lt)/Rt] X 100 [2] mm^2

† p<.10 * p<.05 ** p<.01 (two sample t test--see text)

left side (TA more so, thus larger left side CA). In both regions, MA is smaller relative to TA on the left side, thus the CA left "dominance" remains fairly constant along the shaft. However, due to the trend in left tibiae of relative enlargement of both outer and inner bone dimensions in the proximal shaft (displacing the area of cortical bone relatively farther from the section centroid), it will be seen below that left side "dominance" in second moments of area (i.e., resistance to bending stress) increases proximally in the tibia.

In the femur, larger left side values for CA in females are brought about by both larger left TA (3.9%) and smaller left MA (4.4%) values. In male femora, MA is smaller on the left side (3.2%) as in females, but TA is about equal or slightly larger on the right side (.6%), leading to only a small left side "dominance" in CA, as noted above. In general, CA side differences are smaller in the proximal femur (the male 35% section being an exception), due principally to a decrease in left side "dominance" in TA proximally (males are actually larger in TA on the right side in the 65% and 80% sections). As will be shown below, this leads to a decrease in left "dominance" and even a slight right "dominance" in second moments of area in the proximal femur in males. Thus, with respect to external and internal bone dimensions, and consequently bending strength, the left side "dominance" is greatest around the knee joint. The factors which could lead to this type of bilateral asymmetry in the lower lomb are discussed later in chapter 6.

Analyses of side differences in cross-sectional properties were also carried out within different age/sex groups (20-39 and 40+ years).

However, because of the smaller sample sized involved in these comparisons, differences between the two sides are less certain to reflect true bilateral differences within individuals, for the reasons outlined above. Thus, only a brief review of general trends apparent in these age/sex/side comparisons will be presented here.

The most striking result of the side comparisons of bone areas by age group is the change in degree, and in males, direction of asymmetry in CA with aging. These results are apparently brought about by differences in the aging patterns in right and left tibiae and femora, a side/age relationship which is different in the two sexes. In males, except for the distal femur, right side bones gain in CA with aging, while left bones lose CA. Thus, a left "dominance" in younger adults is transformed to a right "dominance" in older adults. In females, exactly the opposite is true--the right side losses *more* CA with aging than the left side, leading to an even more pronounced CA left "dominance" in older adults.[14] This pattern in females is most marked in the proximal femur, where CA actually *increases* slightly with age on the left side, while decreasing on the right (see also Table 11b above). On average, female asymmetry in CA increases from 7.0% to 15.1% (all left side larger). The corresponding figures for male tibial CA are 9.4% larger left side to 7.0% larger right side, femoral CA 2.0% larger left side to 1.1% larger right side. In general, greater losses of bone from one side (right in females, left in males) are due

---

[14] Bone lengths on right and left sides are not significantly dif-derent in all sex/age groups.

to greater increases in MA on that side, caused, presumably, by greater endosteal resorption with aging. However, im most cross sections TA also increases slightly more (but not as much as MA) on the same side. Thus, though CA may decline on one side relative to the other, second moments of area, which reflect both bone area and distribution of area, do not always show the same degree or type of age change in asymmetry as CA values. The age/side comparisons for bone areas also show that females actually do not exhibit more asymmetry than males in tibial CA, as described earlier, if account is taken of age changes in asymmetry. The lower degree of asymmetry for the total male sample is a result of averaging a left side "dominance" in older males. In fact, tibial asymmetry in CA among younger adults is more than twice as great in males as in females, and close to equal in older adults. Asymmetry in CA in the femur is always more marked in females, however, with the sex difference increasing with age.

In an earlier radiographic study by the present author (Ruff and Jones, 1981), females were found to be more asymmetric than males in (extrapolated) CA of the tibia (left side larger), a finding which agrees with those of the present study, and many other studies of lower limb asymmetry (e.g., Ingalls, 1931; Schultz, 1936; Latimer and Lowrance, 1957, 1965). However, in this earlier study, with aging left tibiae in females were found to lose more CA, leading to a small right "dominance" in older females, results which are exactly opposite to what was found in the present study (males showed no significant change in CA asymmetry of the tibia in the earlier study). There are several possible reasons for this apparent contradiction between the two studies, including population-specific differences in asymmetry changes with aging (the previous study sample originated from a Native California coastal population), errors in extrapolating tibial bone areas from radiographs in the previous study, or nonrandom sampling errors in either or both studies. In the earlier study, a physiological "protective" mechanism whereby more bone is lost with aging from the side beginning with more bone in younger adulthood was hypothesized, however, with the exception of male tibiae, this hypothesis is not supported by the present study data.

Orientation of Principal Axes

As discussed earlier in this chapter ("The Biomechanical Significance of Cross Section Geometric Properties"), stresses produced by bending are more likely to cause *in vivo* failure (fracture) of long bones than stresses resulting from pure axial loadings. While actual failure probability under physiological conditions of bending is determined by a number of factors, including the magnitude and direction of loading (see chapter 6) and the bone material properties (see later section this chapter), several geometric properties of long bone cross sections may be identified as having an important bearing on relative bending rigidity and strength: 1) the orientation of the major axis of the section (direction of maximum bending strength), denoted here as phi; 2) the magnitude of the maximum principal second moment of area, I_{max}, which indicates the maximum bending rigidity of the section; 3) the magnitude of I_{min}, the minimum principle moment; the ratio I_{max}/I_{min}, which indicates the relative maximum to minimum rigidity; and 4) the section modulus, Z, an index of bending rigidity at the point of maximum stress in the cross section. Phi, the orientation of the major axis, is considered here first because it aids greatly in the interpretation of the other parameters listed above.

General patterns of variation in phi.

As described earlier, phi in the present study is taken as the angle between the major axis and the transverse, or mediolateral axis of the cross section. This angle is measured, counterclockwise, on the proximal surfaces of the cut bone sections (i.e., as viewed from the proximal end of the bone, or in the case of the femoral neck section, the femoral head), always as a right side section, with left side bone sections reversed to conform to the right hand rule, as described earlier (see chapter 4). Thus, for example, a phi of 90° indicates a cross section whose greatest resistance to bending is in the sagittal or A-P plane, while a phi of 180° (or 0°) indicates greatest bending strength in the frontal, or M-L plane.

In order to better visualize the significance of variation in phi due to sex differences or other factors, a brief description and illustration of the general patterns of variation in phi along the lengths of the tibial and femoral diaphyses, and femoral neck, is presented here. Total ranges of variation and means for the sample have been presented previously in Figures 11 and 12b. Figures 20a, b (following pages) illustrate graphically, using direct output from SLICE and the TEXTRONIX graphics program, the variation in phi in the 11 cross sections from one individual, selected at random from the

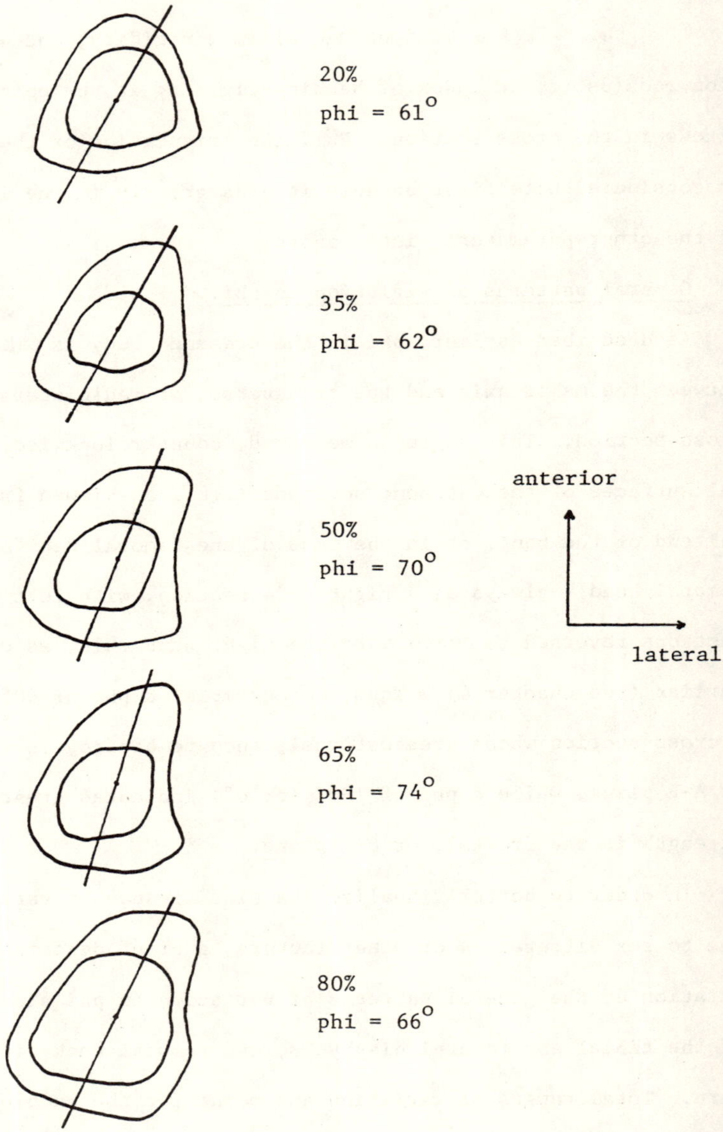

Figure 20a. Illustration of tibial cross section outlines in one individual, showing angle of orientation of major axis, phi (measured counterclockwise from M-L axis). Scale approximately life size.

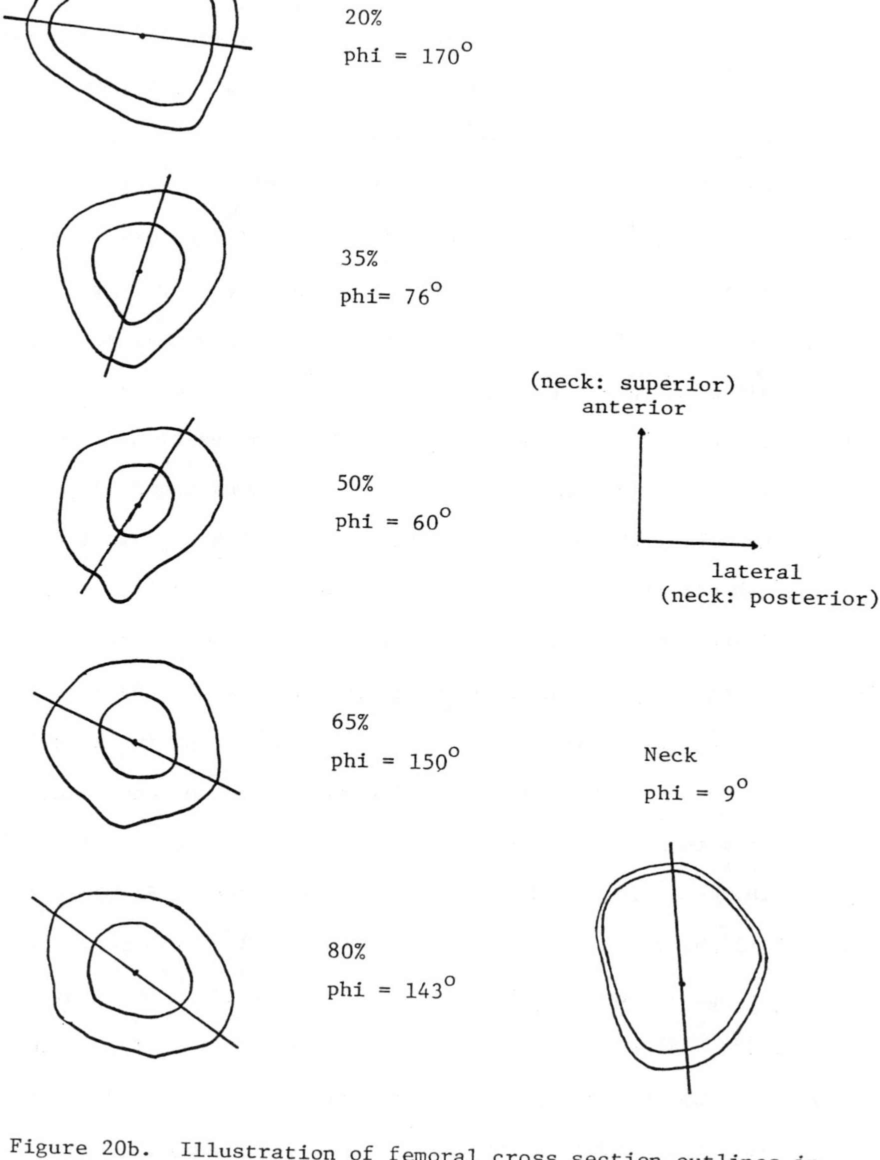

Figure 20b. Illustration of femoral cross section outlines in one individual, showing angle of orientation of major axis, phi (measured counterclockwise from M-L axis, from superior-inferior axis in femoral neck). Scale approximately life size.

study sample (a 30-34 year old male).

Considering first the tibia (Figures 11 and 20a), we find a relatively narrow range of values for phi, both within cross sections (i.e., between individuals) and between cross sections. In all tibial cross sections, phi varies between about 45° and 90°, indicating major axes oriented in the proximal-lateral/distal-medial direction--in the individual illustrated in Figure 20a, phi varies only from 74° to 61°. A general decline in values of phi from the proximal half to the distal half of the tibia is also apparent, with this general trend interrupted by an increase in phi from the 80% to 65% sections. As noted earlier in the presentation of 3-dimensional area centroid positions in the tibia, these changes in phi parallel the sigmoid path of the anterior tibial crest down the length of the diaphysis, except in the most distal section, where this crest is greatly reduced in development. This correspondence is to be expected, since the anterior crest forms the thickest part of the tibial cortex (except in the 20% section) and is also the furthest removed from the section centroid, thus determining to a large extent the direction of maximum bending strength in these sections. The individual illustrated in Figure 20a exhibits tibial phi values representative (within 5°) of the total sample mean except in the 20% (distal-most) cross section, where his phi is about 17° higher than the total sample mean of 44°. Thus, the decline in phi from the proximal to distal ends in most individuals is actually greater than indicated by this figure (see Figure 11, also see below).

As mentioned in chapter 4, some authors (Lovejoy et al., 1976; Miller and Purkey, 1980) have used the transverse and A-P axes of the tibial articulation surface, rather than the proximal tibial articulation used here, to establish reference axes and orient cross sections. To enable comparisons between their results and those of the present study, tibial phi's were also calculated with reference to the transverse axis of the distal tibial articulation, that is, rotated counterclockwise by the magnitude of the tibial "torsion" angle, which averages 18.2° in this sample (Table 4). The results of this transformation are shown in Table 12. As shown here, the "rotated" phi values approach 90° in the proximal tibia, thus indicating a direction of maximum bending strength almost perpendicular to the transverse axis of the distal articulation surface. Similar results were reported in the two studies mentioned above which used the distal reference axis system. Minns et al. (1975), who did not describe which articulation end was used in orientation, also obtained values for phi close to 90° in the proximal tibia (as measured from the M-L axis), while Piziali et al. (1976), who also did not report their specific reference axes orientation procedure, obtained phi's close to my "unrotated" values (thus, presumably Minns used the distal reference, Piziali the proximal reference). Kimura (1974), who used my (proximal) system of orientation, obtained phi values within 10° of mine for comparable cross section locations (my extrapolations from his data). All authors report a relative decline in phi--i.e., a clockwise rotation--in the distal tibia. Most authors also report an increase in the variability of phi values in the distal tibia, which coincides with

Table 13

Tibial Phi's: "Unrotated" and Rotated by the Distal Tibial "Torsion" Angle

MALES

Cross Section	Unrotated [1]	Rotated [2]
20%	45.4 [3]	64.9
35%	56.8	76.4
50%	64.3	83.9
65%	69.2	88.8
80%	66.6	86.2

FEMALES

Cross Section	Unrotated	Rotated
20%	43.2	60.1
35%	58.4	75.3
50%	65.0	81.9
65%	69.4	86.3
80%	65.6	82.5

TOTAL SAMPLE

Cross Section	Unrotated	Rotated
20%	44.3	62.5
35%	57.6	75.8
50%	64.6	82.8
65%	69.3	87.5
80%	66.1	84.3

[1] Values based on proximal articulation centers (standard reference axes)--see text.

[2] Values based on distal articulation surface axes--see text.

[3] degrees

results of the present study (see Figure 11). This increase in variability is attributable to the more circular shape of the distal tibia (Figure 20a),with aless well-defined axis of greatest bending strength (the significance of these observations will be discussed further below and in chapter 6).

The pattern of variation in phi in the femur is more complex than in the tibia (Figures 12b and 20b). Considering first the five diaphyseal sections, we see a range in phi from 140-150° in the proximal femur, to 50-70° in the 50% and 35% sections, to almost 180° in the distal-most section. The individual illustrated in Figure 11 is typical of the sample as a whole (within 10° of the sample mean values). In order to appreciate the significance of these femoral phi's in terms of changes in cross-sectional geometry along the shaft, and loading configurations (see next chapter), the phi's for the distal three sections must be transformed to phi + 180°. If this is done, the pattern of vairation in phi is seen to follow a gradual counterclockwise rotation from proximal to distal ends, beginning at 142° in the 80% section and passing through 146°, 232°, and finally 260° in the 35% section. In the 20% section, because of the increase in mediolateral diaphyseal breadth related to the formation of the femoral condyles at this site, the A-P oriented major axis of the distal femoral shaft has become the minor axis. Therefore, the more appropriate value for comparison with the orientations of the major axes in the 50% and 35% sections is the minor axis of the 20% section. This angle averages 259°, or about equal to the orientation of the major axis of the 35% section. Thus, if allowance is made for the special geometry of the knee joint and its effects on mediolateral "flaring", the counterclockwise rotation of the femoral principal axes, or direction of greatest bending strength, is seen to be continuous from the proximal to distal ends of the diaphysis.

Atkinson and Weatherall (1967) reported a similar "gentle spiral" pattern of bone densities (determined in four cortical quadrants) from the proximal to distal end of the femoral shaft. However, in an extension of the same study to bone geometrical variation (Martin and Atkinson, 1977), the authors do not show the same spiral pattern, claiming only that "most femur cross sections are strongest in the A-P or lateral direction, with relatively few being strongest at some angle in between these 'natural' directions" (ibid.: 228). Setting aside for the moment the question of what constitutes a "natural" direction in this context, the results of the present study clearly show that geometric changes along the length of the femoral diaphysis also follow a counterclockwise spiral or rotation from the junction of the shaft with the femoral neck to the distal articulation of the femur. This is apparent not only from values of phi at different cross section locations, but even subjectively from the shapes of the cross section outlines. Observing these outlines in Figure 12b, definite "bulges" in cortical contours are apparent, beginning as anterior-medial/posterior-lateral protrusions in the 80% section and gradually rotating counterclockwise to reach a posterior-medial/proximal-lateral orientation in the 50% and 35% sections. The relatively small differences in average phi values for the 80% and 65% sections (see Figure 12b) may be due partly to the presence in the 80% outline of a distal-medial ridge formed by the initial slope of the lesser trochanter, which was inadvertantly included in this cross section in many individuals (see Figure 20b). This may tend to rotate the major axis of this section closer to the transverse axis, thereby increasing the value of phi and decreasing the difference between the two proximal sections orientations. If this ridge were eliminated (e.g., by moving the location of the 80% section slightly distally), the phi for this section might average 10-20° less (i.e., 120-130°), resulting in a more evenly-spaced distribution of phi values within the general counterclockwise rotation pattern. As discussed further in chapter 6, the "uneveness" of the rotation pattern in femoral diaphyses may also be brought about by the competing (or combining) of different bending loads (e.g., A-P and M-L) at different levels of the shaft.

The femoral neck cross section outline, as noted above, is illustrated in Figure 20b as it appears viewed from the medial-superior end, i.e., the femoral head. Since this cross section lies in a plane perpendicular to the cervical axis, not the longitudinal axis of the diaphysis (see chapter 4), the reference axes for this section do not correspond to the A-P and M-L axes of the femoral diaphyseal sections. The y axis of the femoral neck cross section is defined by the intersection of the frontal plane (of the femur and body as a whole) and the plane of the cross section surface. Thus, it forms the superior-inferior axis of the femoral neck. Deviations of the major axis from this axis indicate the degree of "twisting" of the femoral neck from a superior-inferior direction of maximum bending strength (for a fuller discussion of these points, see Backman, 1957).

As Figures 12b and 20b show, the major axis of the femoral neck cross section does in fact lie close to the superior-inferior axis, but is rotated somewhat counterclockwise (i.e., to a superior-anterior/inferior-posterior position). When calculated from the cortical shell area alone, this angle averages about 6°; when calculated from the total subperiosteal area (considered to be of a homogeneous material) the angle averages about 8°.[15] The total range of variation in the sample is from 38° to -28° (Figure 12b). Backman (ibid.: 29) reports a somewhat greater average value of 15° for his corresponding angle, measured on 126 adult femora. However, not having access to the measurement techniques available in the present study, he was forced to rely on an external indicator to determine this parameter, using the inclination of the flattened ventral subperiosteal border of the neck to derive his angle. As is apparent from Figure 20b, this method will overestimate the magnitude of this angle. In fact, applying Backman's technique to the femoral neck outline illustrated here, which had a calculated phi of 8.6° (cortical shell property), yields an estimate of approximately 15°, or precisely the average value obtained by Backman for his total sample. The use of the outer ventral border alone to derive phi does not take into account the rounded contour of the medial border, which tends to rotate the direction of maximum bending strength closer to the superior-inferior axis. This

[15] Figure 12b plots phi values calculated from the total subperiosteal area of the femroal neck cross section, Figure 20b phi's calculated from the area of the cortical shell alone. The difference in results obtained using the two methods of measurement, as indicated in the text, is neglible.

medial "bulge", easily descernible on an unsectioned femur, may have an important bearing on the bending strength of the femoral neck in directions other than the superior-inferior direction, and on torsional strength. In any event, if account is taken of the differences in measurement technique, Backman's results and those of the present study concerning the direction of maximum bending strength of the femoral neck are probably in fairly close agreement.

It is interesting to compare ranges of variation and distributions curves for phi values in different femoral cross sections (Figure 12b). The femoral neck and 80% and 20% diaphyseal sections exhibit a relatively narrow distribution of values for this property (disregarding the few outliers for the 20% section), indicating a well-defined axis of greatest bending strength in these sections. In contrast, the three middle femoral diaphyseal sections show wide ranges of variation in phi, with only a vaguely defined direction of maximum bending strength, particularly in the 65% and 35% sections. As in the tibia, these differences are partly a function of the relative circularity of cross-sectional shape (for example, if "circularity" is measured by the ratio of the maximum to the minimum prinicipal moments, over the total sample the three middle sections of the femoral diaphysis average closer to 1.0 than any other section--see below).

But "circularity" is itself a function of either no strongly-defined direction of in vivo bending, or perhaps more likely of a loading pattern consisting of the simultaneous application of two or more forces or bending moments from different directions.[16] Thus, both mean values and relative dispersion of values for phi can provide clues to the types of biomechanical stresses bones are subjected to under physiological loading conditions. These issues are explored further in the context of other cross section geometric properties presented below.

Sex differences.

Differences in male and female values for tibial and femoral phi's are presented in Table 13. As shown in this table, phi's for tibial cross sections show no significant or near-significant differences between the sexes. In fact, tibial phi's are remarkably similar in males and females, with mean differences of at most only 1 or 2 degrees at any cross section site. This indicates very similar orientations of greatest bending strength, and thus, presumably, similar directions of applied bending moments in the tibia in the two sexes. If tibial phi's are compared across sex with respect to the distal articulation axes (i.e., rotated counterclockwise through the "torsion" angle as described above), sex differences increase slightly, since males exhibit a tibial "torsion" angle almost 3° greater on average than females (see Table 4 above). Thus, with respect to the transverse axis

---

[16]This assumes no torsional loadings, which also would select for a relatively circular cross section, and is an important factor in the distal tibia and possibly the femoral neck.

Table 13

Sex Differences in Tibial and Femoral Phi's

TIBIA

Cross Section	Male Mean	Male SE	Female Mean	Female SE	Male - Female
20%	45.4[1]	2.0	43.2	2.0	2.2
35%	56.8	1.1	58.4	1.1	-1.6
50%	64.3	1.0	65.0	1.0	-.7
65%	69.2	.8	69.4	.8	-.2
80%	66.6	.8	65.6	.7	1.0

FEMUR

Cross Section	Male Mean	Male SE	Female Mean	Female SE	Male - Female
20%	168.0[1]	3.4	170.8	1.2	-2.8
35%	75.1	3.2	84.0	5.7	-8.9
50%	55.6	2.3	47.8	3.5	7.8[†]
65%	147.4	1.3	144.0	3.3	3.4
80%	141.1	1.3	143.4	1.2	-2.3
Neck-CA	6.5	1.5	6.0	1.4	.5
Neck-TA	8.8	1.4	7.8	1.2	1.0

[1] degrees

[†] $p<.10$

of the distal articulation surface, male tibial phi's fall somewhat closer to 90° than female phi's. However, even with this transformation, the differences between the sexes are not statistically significant, and in no case are greater than 5°.

No statistically significant sex-related differences in femoral phi's are present either, however, male and female values for phi appear to show a greater degree of divergence and less close correspondence in the femur than in the tibia. The greatest differences between the sexes occur in the 50% and 35% sections, averaging 8° and 9°, respectively (statistically near-significantly different in the 50% section). In the midshaft cross section, the major axis is somewhat more A-P oriented in males, while in the 35% section, females are more A-P oriented. However, much of this apparent difference between the sexes may be due to sampling error and the method of transforming phi's in the femur, described earlier in this chapter. This is particularly true with respect to the 35% section (see Figure 12b above). As noted earlier, the distribution of phi values for this cross section is skewed to the left, and it was considered during transformations whether to change individuals with 35% cross section phi's of 150°-180° to -30°-0°. In fact, most of the left-positioned outliers (i.e., 150°-180° range) in this distribution are females, including the four largest values of phi. If the above transformation had been carried out, the female mean for the 35% section would actually have fallen <u>below</u> (to the right of in Figure 12b) the male mean, rather than above the male average as reported here (in this connection, note the very large standard error associated with the female 35% cross section phi mean). Thus, at least with respect to the 35% section, the apparent differences in phi between the sexes may be more an artifact of the measurement and transformation procedures employed than a real difference. The statistically near-significant difference in phi at the 50% section may be more important, since the distribution of phi values for this cross section is much less skewed (see Figure 12b--also note smaller standard error terms in Table 13), and thus less subject to artifacts of the kind described above. If this represents a true structural difference, it might indicate somewhat more A-P oriented bending moments in the midshaft region of the femur in males. In the proximal and distal diaphysis and the femoral neck, however, the directions of greatest bending strength do not appear to vary appreciably in the two sexes.

Age differences.

Differences in phi values between the 20-39 year and 40+ year age groups are shown in Tables 14a, b.

A number of interesting age-related differences in phi are apparent. In the tibia, phi shows a consistent increase with age, more so in males. Thus, older adults, especially males, have an orientation of greatest bending strength in the tibia closer to the A-P axis. The age-related rotation of the major axes increases from the proximal to the distal end of the tibial diaphysis, reaching a maximum of almost 9° in the male 20% section. As discussed earlier, this redistribution of cross-sectional area, reflected in changes in phi, also produces a displacement of the area centroids of the tibial diaphysis, moving them closer to the longitudinal axis of

Table 14a

Age Differences in Tibial and Femoral Phi's: Males

TIBIA

Cross Section	20-39 yrs Mean	SE	40+ yrs Mean	SE	40+ yrs - 20-39 yrs	F [1]
20%	42.4[2]	2.1	51.1	3.9	8.7*	2.36*
35%	54.1	1.3	62.1	1.4	8.0***	3.16**
50%	61.7	1.1	69.5	1.3	7.8***	4.03**
65%	67.2	.8	73.1	1.2	5.9***	3.53**
80%	64.4	.9	70.9	1.2	6.5***	3.76**

FEMUR

Cross Section	20-39 yrs Mean	SE	40+ yrs Mean	SE	40+ yrs - 20-39 yrs	F [1]
20%	169.6[2]	3.5	165.0	7.5	-4.6	1.23
35%	78.9	4.5	67.6	3.0	-11.3†	.75
50%	56.3	3.1	54.1	3.0	-2.2	.22
65%	143.1	5.9	155.8	8.9	12.7	.49
80%	140.4	1.8	142.4	1.4	2.0	.29
Neck-CA	7.1	1.6	5.3	3.2	-1.8	.74
Neck-TA	9.7	1.5	7.2	3.1	-2.5	1.08

[1] Results of 6 age group ANOVA--see text.

[2] degrees

† $p<.10$ * $p<.05$ ** $p<.01$ *** $p<.001$

Table 14b

Age Differences in Tibial and Femoral Phi's: Females

TIBIA

Cross Section	20-39 yrs Mean	SE	40+ yrs Mean	SE	40+ yrs - 20-39 yrs	F [1]
20%	41.9[2]	2.3	46.0	3.8	4.1	.69
35%	56.7	1.3	61.8	1.7	5.1*	1.57
50%	63.6	1.2	67.9	1.6	4.3*	2.06†
65%	68.6	1.0	71.0	1.4	2.4	1.01
80%	65.1	.9	66.6	1.2	1.5	.77

FEMUR

Cross Section	20-39 yrs Mean	SE	40+ yrs Mean	SE	40+ yrs - 20-39 yrs	F [1]
20%	170.4[2]	1.5	171.7	2.1	1.3	.55
35%	92.3	6.7	67.2	10.0	-25.1*	1.39
50%	49.4	4.2	44.6	6.5	-4.8	.71
65%	142.9	4.5	146.2	4.4	3.3	1.15
80%	140.9	1.3	148.5	2.0	7.6**	2.57*
Neck-CA	3.8	1.6	10.5	2.7	6.7*	1.22
Neck-TA	5.3	1.4	12.8	2.0	7.5*	1.39

[1] Results of 6 age group ANOVA--see text.

[2] degrees

† $p<.10$ * $p<.05$ ** $p<.01$

the bone, again more markedly so in males (see Figures 14a, 15a above). Since the longitudinal axis of the tibia is defined as the straight line between the proximal and distal centers of articulation of the tibia (see chapter 4), redistribution of bone with aging in the tibia, primarily involving relocation of the anterior crest, not only produces a more A-P oriented direction of greatest bending strength, but also reduces the mediolateral eccentricity of the diaphysis, relative to the articulation centers.

Interestingly, this change in the orientation of greatest bending strength is only very weakly reflected in changes in the tibial distal "torsion" angle with aging (Table 5 above). The degree of distal "torsion" does decrease slightly with age (i.e., the A-P and M-L reference axes of the distal articulation surface rotate closer to the A-P and M-L reference axes of the shaft), but by an average of only .9°, far less than the 4°-9° age change observed in distal tibial cross section phi's. (This poor correspondence between variation in the tibial "torsion" angle and variation in cross section phi's is also underscored by low correlations between these properties: the highest correlations between the "torsion" angle and tibial cross section phi's, occurring at the 20% site, are only .239 and .380, for males and females, respectively.) Thus, with aging the major axes of the tibial diaphyseal sections are brought closer not only to the A-P axis of the diaphysis, but also to the A-P axis of the distal articulation surface. In older males, the direction of greatest bending strength in the three proximal tibial sections rotates to within 2° of the A-P axis of the distal articulation (in females, to within 7°). The significance of this observation will explored further in succeeding segments of this report.

The most marked changes with age in femoral phi's occur among females. Most age-related changes in this bone involve the proximal femur, and appear to be related to changes in the femoral antetorsion angle with age. As noted earlier (see Table 5), the antetorsion angle of the femoral head and neck decreases by almost 7° in older females (by more than 8° if one individual is eliminated), and by 1.6 in older males. Changes in the orientation of the major axes of the proximal femoral diaphysis, particularly the 80% section, parallel the decline with age in the antetorsion angle (note that the "decline" in the antetorsion angle is actually equivalent to an increase in phi, since phi is measured counterclockwise, and the antetorsion angle, following tradition, is measured clockwise). This is clearly shown in Figure 21, which plots the six age group means for the antetorsion angle and 80% and 65% section phi's in females. The closeness of the relationship between the antetorsion angle and the direction of greatest bending strength in the proximal femoral diaphysis is further illustrated by comparing changes with age in values of the antetorsion angle and the 80% cross section phi in the two sexes: in females, the antetorsion angle changes 6.8°, the 80% phi 7.6°; in males, the corresponding changes are 1.6° and 2.0°, respectively. Correlation coefficients between the antetorsion angle and the direction of the major axis of the 80% section (phi) are also relatively high: .708 in females, .593 in males. Thus, there is strong evidence that the geometry of the proximal femoral diaphysis is largely determined by the relative position of the femoral head and neck (the antetorsion angle), i.e., by the *in vivo* forces and moments transmitted through them. Based on the correlation

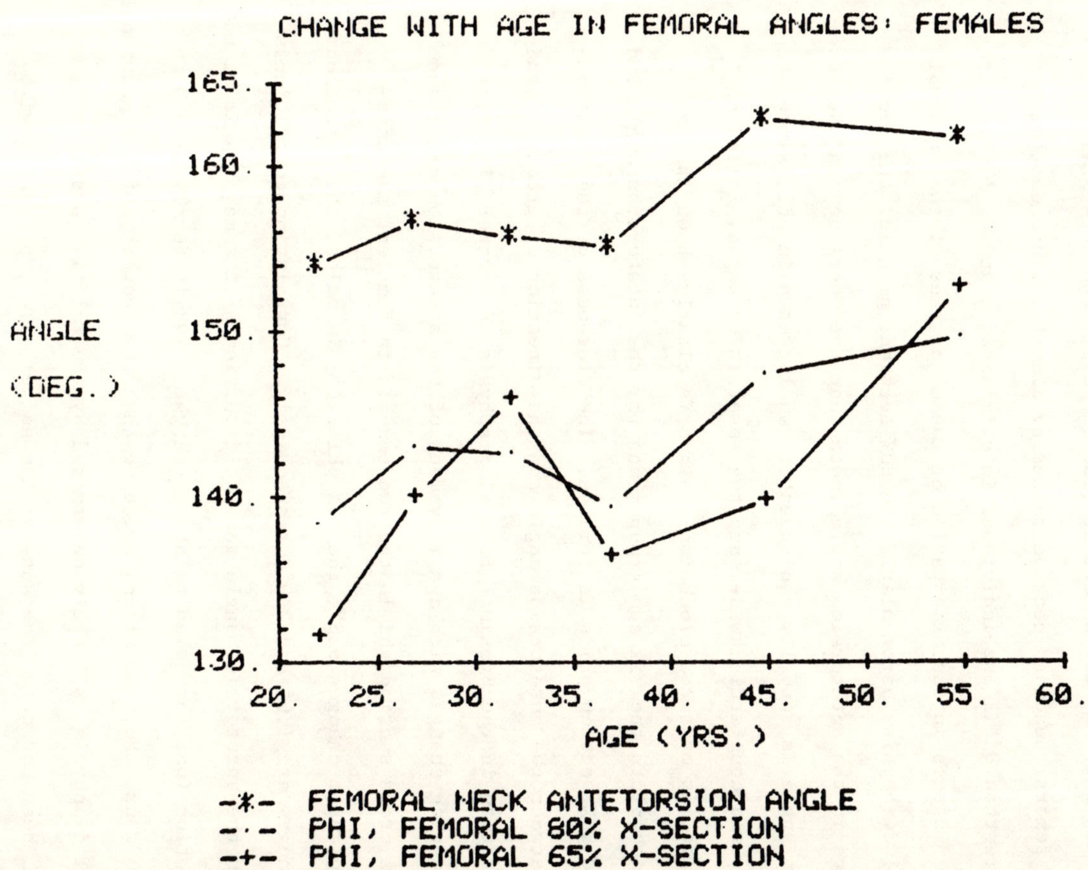
CHANGE WITH AGE IN FEMORAL ANGLES: FEMALES
ANGLE
(DEG.)
165.
160.
150.
140.
130.
20.
25.
30.
35.
40.
45.
50.
55.
60.
AGE (YRS.)
-*- FEMORAL NECK ANTETORSION ANGLE
-·- PHI, FEMORAL 80% X-SECTION
-+- PHI, FEMORAL 65% X-SECTION

Figure 21

coefficients quoted above, there is also evidence that this relationship is stronger among females. This may be due to relatively higher M-L bending moments transmitted through the femoral head in females, leading to a closer correspondence between structural elements in the proximal femur, i.e., the relative position of the head and the direction of greatest bending strength in the proximal shaft (see chapter 6).[17]

The angle of the major axis of the femoral neck itself also undergoes a significant change with age in females, specifically, an increase in the degree of departure from the superior-inferior axis of the neck. In older females, the superior surface of the femoral neck is tilted further anteriorly, and the inferior surface further posteriorly, than in younger females. Males show no significant age change in this property. Interestingly, the change in females is very close to the change with age in the antetorsion angle in this sex: about 7°. While at this point it is difficult to formulate specific hypotheses regarding the precise biomechanical relationships between the various angles of the proximal femur, the three variables--femoral antetorsion angle, proximal diaphyseal and femoral neck orientations of greatest bending strength-do appear to form a kind of functional complex whose elements vary in concordance with one another. This functional interrelationship is underscored by the presence of significant correlations between all of the various elements. Besides the correlations between the antetorsion angle and proximal diaphyseal phi's,

[17]Higher correlations in females may also be a result of the greater variation in these properties among females, e.g., age-related variation.

correlations between the antetorsion angle and femoral neck phi's in females are r = .5 (neck cortical shell phi) and r = .6 (total subperiosteal area phi). Interestingly, males show only low correlations between these two variables of r = .2-.3, which again may reflect a weaker functional relationship between these structural features, as noted above (see also footnote 17 above). An interpretation of this functional complex, and possible explanations for its observed sex/age variation, will be presented in chapter 6.

Some changes with age in phi's for other femoral cross sections are also shown in Table 14. Again,the changes are more marked in females, but appear to some extent in both sexes. These involve primarliy the 35% section and to a lesser degree the 50% section, and entail a counterclockwise rotation of the major axis, moving it closer to the transverse axis. In the female 35% section, this rotation occurs through more than 15°, in the same section in males, about 11°. As detailed later in chapter 6, these changes in the mid-distal femur may also be related to age differences in the proximal femoral complex described above, and may also involve a more general trend towards preserving (or relatively enhancing) mediolateral bending strength in the femoral shaft among older adults, especially females.

Side differences.

Side differences in phi are shown in Table 15. It should be recalled that, as explained earlier, statistical tests of significance between sides here are not particularly powerful, due to the unavailability of paired data. Thus, side differences which do not reach formal significance may still represent true bilateral differences in structure.

Table 15

Side Differences in Tibial and Femoral Phi's

TIBIA

Cross Section	Males			Females		
	Right	Left	Rt-Lt	Right	Left	Rt-Lt
20%	42.1[1]	48.7	-6.6†	40.6	45.9	-5.9
35%	54.3	59.4	-5.1*	57.7	60.1	-3.4
50%	63.0	65.7	-2.7	64.4	65.7	-1.3
65%	68.3	70.1	-1.8	70.6	68.2	-2.4
80%	65.8	67.4	-1.6	66.7	64.5	-2.2

FEMUR

Cross Section	Males			Females		
	Right	Left	Rt-Lt	Right	Left	Rt-Lt
20%	174.8[1]	160.8	-14.0*	169.4	172.3	-2.8
35%	68.2	82.2	-14.0*	87.6	80.3	7.3
50%	52.7	58.5	-5.8	50.2	45.4	4.8
65%	144.7	150.2	-5.5	139.2	148.9	-9.7
80%	139.3	143.0	-3.7	141.6	145.2	-3.6
Neck-CA	5.1	8.0	-2.9	2.8	9.1	-6.3*
Neck-TA	6.8	10.8	-4.0	5.2	10.4	-5.2*

[1] degrees

† $p<.10$ * $p<.05$ (two sample t test--see text)

In both the tibia and femur, most side-related differences in phi can be related to parallel differences in the antetorsion angle of the femur and the tibial distal "torsion" angle. As shown earlier (Table 6 above), at both proximal and distal ends of the lower limb, bones from the left side are less "twisted", or closer to the A-P/M-L reference axes, than right bones. This same side difference is apparent in the direction of major axis of tibial and proximal femoral diaphyseal cross sections. In the tibia, a gradual "untwisting" of the tibial diaphysis occurs from the proximal to the distal end on the left side relative to the right. This relative rotation of the major axis reaches its maximum in the 20% section, averaging 6°-7°, which is similar to the 5°-10° average side difference in the tibial "torsion" angle (Table 6). Thus, the direction of maximum bending strength in the left tibia is more A-P oriented, and corresponds to a more A-P oriented distal articulation.

In the proximal femoral diaphysis, side differences in phi correspond to differences in the femoral antetorsion angle, and indicate a direction of maximum bending strength closer to the M-L axis in left femora. Phi values for the 65% and 80% femoral section average 4°-10° greater (closer to the M-L axis) on the left, comparable to the 4°-8° more M-L oriented cervical axis (i.e., antetorsion angle) on the same side (Table 6). In the mid and distal femur, a more complex pattern of side differences in phi exists. In the 50% through 20% cross sections, the major axis is more A-P oriented in males on the left side, while in females the opposite pattern holds true, except in the 20% section, where right and left sides average very close together. It is difficult here to explain each of these sex/side-related differences, since they probably depend on a complex interaction between relative position of the femoral head and other factors, discussed further in chapter 6. It should be noted, however, that A-P bending stresses are probably relatively greater than M-L bending stresses in the distal femur, so that the more A-P oriented major axes among males in this region may reflect greater bending stresses on the left side. The reverse direction of side differences in females in the 50% and 35% sections may be due partly to nonrandom sampling and the method of transforming the phi values for the 35% sections, and/or to differences in loadings of the femur in the two sexes (e.g., greater use of the left lower limb produces relatively greater A-P stress in the distal femur in males and greater M-L stress in the distal femur in females).

The side-related differences in phi in the femoral neck cross section, as part of the "proximal femoral functional complex" described above, vary in concordance with other proximal femoral angular differences. Thus, as with age changes, as the antetorsion angle (and major axis angle of the proximal femoral diaphysis) moves closer to the transverse axis in left femora, the major axis of the femoral neck cross section rotates further from the superior-inferior axis of the neck. The magnitude of this relative rotation is greater in females, who also show a greater side difference in the antetorsion angle (8°, versus 4° in males).

Side comparisons of phi values by age group were also carried out. In no case does the direction of right-left differences in phi change with age, however, the magnitude of side differences does vary by age group. In the femur, side differences in phi generally decrease with age, particularly in males. In the tibia, side differences in phi increase with age in the distal sections, showing very little change in the proximal sections, with this age pattern again more marked in males. It should be noted in this connection that the side-related differences in thedistal tibial torsion angle decrease somewhat with age in males, remaining about the same in females. Left tibiae in the older male age group exhibit directions of greatest bending strength quite close to the A-P axis of the distal articulation--within 1° for the 50%, 65% and 80% sections. The side-related difference in the antetorsion angle of the femur remains about the same with aging in females, while decreasing somewhat in males.

Principal Moments

Sex differences.

As described earlier in this chapter, the principal second moments of area, I_{max} and I_{min}, are critical geometric properties in determining the bending strength of a beam (or long bone) cross section. Based on the formula $\sigma = \frac{M\,y}{I}$ (see page 166), the greater the value of I, the lower the stress, given equivalent bending moments (the effects of differences in y--the distance from the centroid to the outermost fiber of the cross section--will be considered in the next section of this chapter). Thus, the maximum value of I, I_{max}, will indicate the minimum bending stress, and thus the greatest bending strength of the section. Similarly, I_{min} indicates the minimum bending strength of the section. Since I_{max} and I_{min} always lie in directions perpendicular to each other, the ratio of I_{max}/I_{min} can be used not only to provide an idea of the relative difference between maximum and minimum bending strengths, but also an appreciation of the general distribution of cross-sectional volume within the cross section (e.g., oval or rectangular vs. circular or square). Thus, this index may serve as a biomechanical interpretation of cross-sectional "shape", a component of long bone morphology traditionally evaluated using A-P/M-L breadth ratios (e.g., Bass, 1971).

Tables 16a, b summarize sex differences in I_{max}, I_{min}, and I_{max}/I_{min} for the 11 tibial and femoral cross sections. As in the presentation of bone area results, to facilitate male-female comparisons, in additon to male and female means and standard errors,the ratio of male/female values for each cross section and property are given.

Although important sex-specific differences in these properties exist, the general pattern of variation in principal moments along the lengths of the tibial and femoral diaphyses are similar in the two sexes. Thus, in the tibia, I_{max} increases in magnitude from the distal end to a peak in the most proximal section in both sexes. Since the major axis of the tibia is directed anteroposteriorly, particularly in the proximal tibia, this indicates an increase in A-P bending strength in the tibia from the distal to proximal ends. It should also be noted that the large principal moments at the 80% section are related in large part to the flaring of the tibial condyles and development of the tibial tuberosity just proximal to this site. I_{min} in the tibia also reaches a maximum at the 80% level, but unlike I_{max}, shows a distinct minimum at the 35% level. Thus, the minimum bending strength of

Table 16a

Sex Differences in Principal Moments: Tibia

MAXIMUM PRINCIPAL MOMENT (I_{max})

Cross Section	Male Mean	Male SE	Female Mean	Female SE	Male - Female[1]	Male/Female X 100
20%	12887[2]	380	7558	236	5329	170.5
35%	17831	510	9996	372	7835	178.4
50%	25118	708	13266	469	11852	189.3
65%	33512	969	17646	650	15866	189.9
80%	44142	1267	24121	828	20021	183.0

MINIMUM PRINCIPAL MOMENT (I_{min})

Cross Section	Male Mean	Male SE	Female Mean	Female SE	Male - Female[1]	Male/Female X 100
20%	9153[2]	269	5511	204	3642	166.1
35%	7337	258	4525	193	2812	162.1
50%	8251	308	5059	204	3192	163.1
65%	10216	374	6184	237	11032	165.2
80%	16349	559	10086	356	6263	162.1

I_{max} / I_{min}

Cross Section	Male Mean	Male SE	Female Mean	Female SE	Male - Female[1]	Male/Female X 100
20%	1.42	.03	1.40	.03	.02	101.4
35%	2.51	.06	2.28	.05	.23	110.1
50%	3.15	.08	2.70	.07	.45	116.7
65%	3.37	.08	2.93	.08	.44	115.0
80%	2.77	.06	2.45	.07	.32	113.1

[1] All male-female differences significant at .001, except I_{max}/I_{min} of 35% section, significant at .01 level, and 20% section, nonsignificant.

[2] mm^4

Table 16b

Sex Differences in Principal Moments: Femur

MAXIMUM PRINCIPAL MOMENT (I_{max})

Cross Section	Male Mean	Male SE	Female Mean	Female SE	Male - Female[1]	Male/Female X 100
20%	30664[2]	915	21343	630	9321	143.7
35%	22892	543	13770	460	9122	166.2
50%	22434	554	14359	480	8075	156.2
65%	22510	616	15641	464	6869	143.9
80%	30691	710	20513	627	10178	149.6
Neck-CA	11538	300	7593	259	3945	152.0
Neck-TA	34630	1114	21999	767	12631	157.4

MINIMUM PRINCIPAL MOMENT (I_{min})

Cross Section	Male Mean	Male SE	Female Mean	Female SE	Male - Female[1]	Male/Female X 100
20%	22986[2]	538	13888	411	9068	165.5
35%	17123	435	11678	331	5445	146.6
50%	15334	386	11172	322	4162	137.2
65%	17592	441	11744	395	5848	149.8
80%	16414	435	10902	400	5512	150.6
Neck-CA	6974	224	4748	168	2226	146.9
Neck-TA	21813	747	13895	507	7918	157.0

I_{max} / I_{min}

Cross Section	Male Mean	Male SE	Female Mean	Female SE	Male - Female[1]	Male/Female X 100
20%	1.33	.02	1.55	.03	-.22	85.8
35%	1.36	.03	1.18	.02	.18	115.2
50%	1.48	.03	1.29	.02	.19	114.7
65%	1.28	.02	1.36	.02	-.08	94.1
80%	1.90	.04	1.93	.04	-.03	98.4
Neck-CA	1.69	.04	1.63	.03	.06	103.7
Neck-TA	1.61	.03	1.60	.03	.01	100.6

[1] All male-female differences significant at .001 level, except I_{max}/I_{min}, 65% section, significant at .05 level, 80% and two femoral neck sections, nonsignificant.

[2] mm^4

the tibia decreases from the 20% to 35% cross section before increasing to its greatest value in the proximal section. This minimum at the 35% level corresponds to an M-L narrowing of the shaft at this point, easily detectable on an intact tibia. Again, the relatively large value of I_{min} in the 80% section should be interpreted as partially a result of the broadening of the diaphysis as it approaches the metaphyseal region. In this connection, it should be noted that while total subperiosteal area for the tibia reached a maximum at the 80% level, cortical area peaked in the 50% and 65% sections of the tibia (see Table 8a above), indicating an expanded, but relatively thin-walled cortical shell at this most proximal level. The ratio of I_{max}/I_{min} in the tibia is greatest in the 65% section, where the tibial shaft is most oblong or triangular in shape, and smallest in the 20% section, where the shaft is most nearly circular (see also Figure 20a above).

In the femoral diaphysis, the greatest values of I_{max} in both sexes are found at the 20% and 80% levels, both sites where the shaft is relatively mediolaterally broadened (see Figure 20b above). I_{min} is greatest in the 20% cross section and smallest in the 50% section. Again it should be remarked that, as in the tibia, the relatively large principal moments in the section closest to the knee joint (femoral 20%) are partly due to the expansion of the diaphysis here leading to the development of the femoral condyles. TA is also greatest in the femur at the 20% level, while CA reaches its maximum in the 50% and 65% sections. The I_{max}/I_{min} ratio is greatest at the 80% level, where the shaft is most anteroposteriorly flattened (or mediolaterally broadened) and smaller in the mid and distal femur. The femoral neck also exhibits a relatively high ratio, reflecting a somewhat A-P flattened (superior-inferior broadened) shape. However, the femur as a whole is characterized by much smaller I_{max}/I_{min} ratios than the tibia, with the exception of the tibia 20% section. This reflects the generally more cylindrical shape of the femoral diaphysis (and neck), easily observed on intact specimens. While a more detailed consideration of loadings of these bones will be presented in the next chapter, this observation in and of itself indicates a general difference in the loading configurations of the two bones, with tibiae designed principally to withstand A-P bending moments (and some torsion in the distal diaphysis), and femora designed to withstand both A-P and M-L bending moments.

As was observed for bone areas, all male values for principal moments at the 11 cross sections are significantly greater than female values for corresponding cross sections, due to sexual dimorphism in size. Of more interest is the relative magnitude of these sex-related differences at different cross section sites and for different properties (I_{max}, I_{min}, or their ratio). These differences are indicated by the male/female rations shown in the last columns of Tables 16a, b, and plotted in Figures 22a, b (following pages).

In the tibia, a difference in the overall degree of sexual dimorphism in I_{max} and I_{min} is immediately apparent (Figure 22a). The male/female ratio for I_{max} in tibial sections averages 1.82, for I_{min} only 1.64. This difference is reflected in the I_{max}/I_{min} ratio, with a more A-P oriented distribution of bone volume in males (using traditional osteometric terminology, male tibiae of this sample would be said to

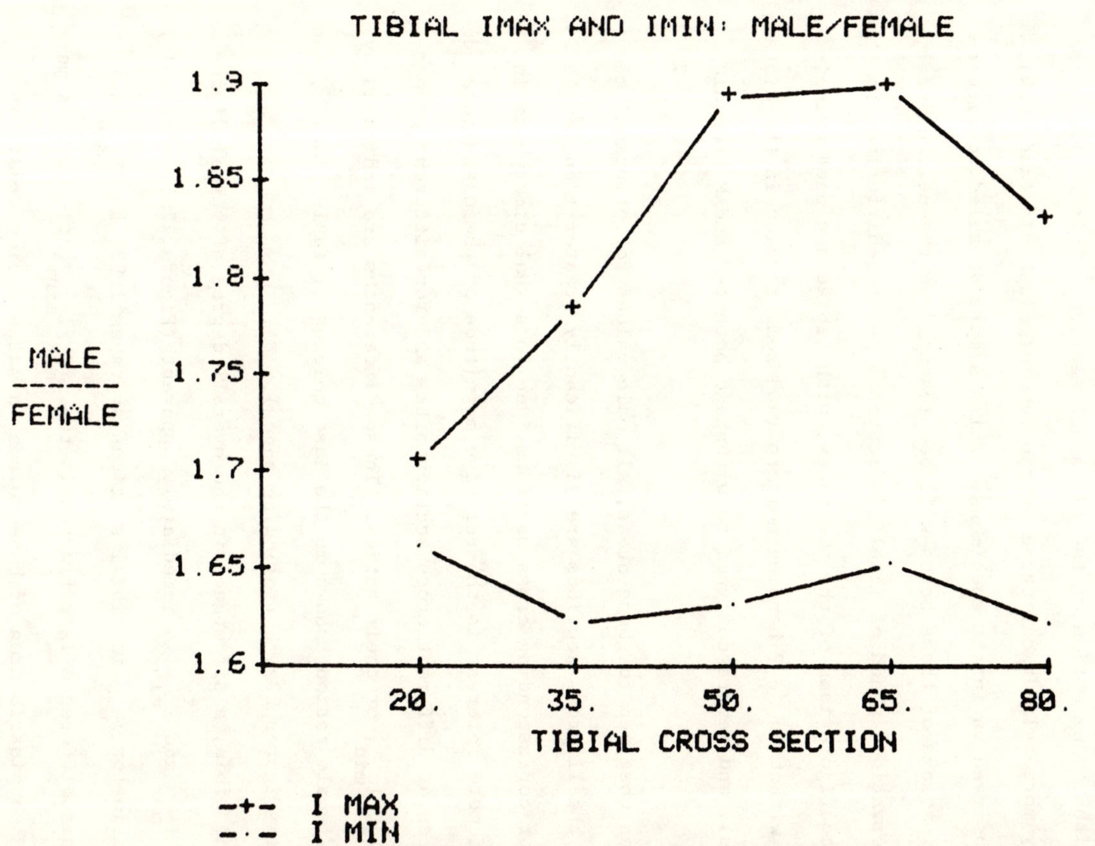
TIBIAL IMAX AND IMIN: MALE/FEMALE
MALE / FEMALE
1.9
1.85
1.8
1.75
1.7
1.65
1.6
20.
35.
50.
65.
80.
TIBIAL CROSS SECTION
-+- I MAX
-·- I MIN

Figure 22a

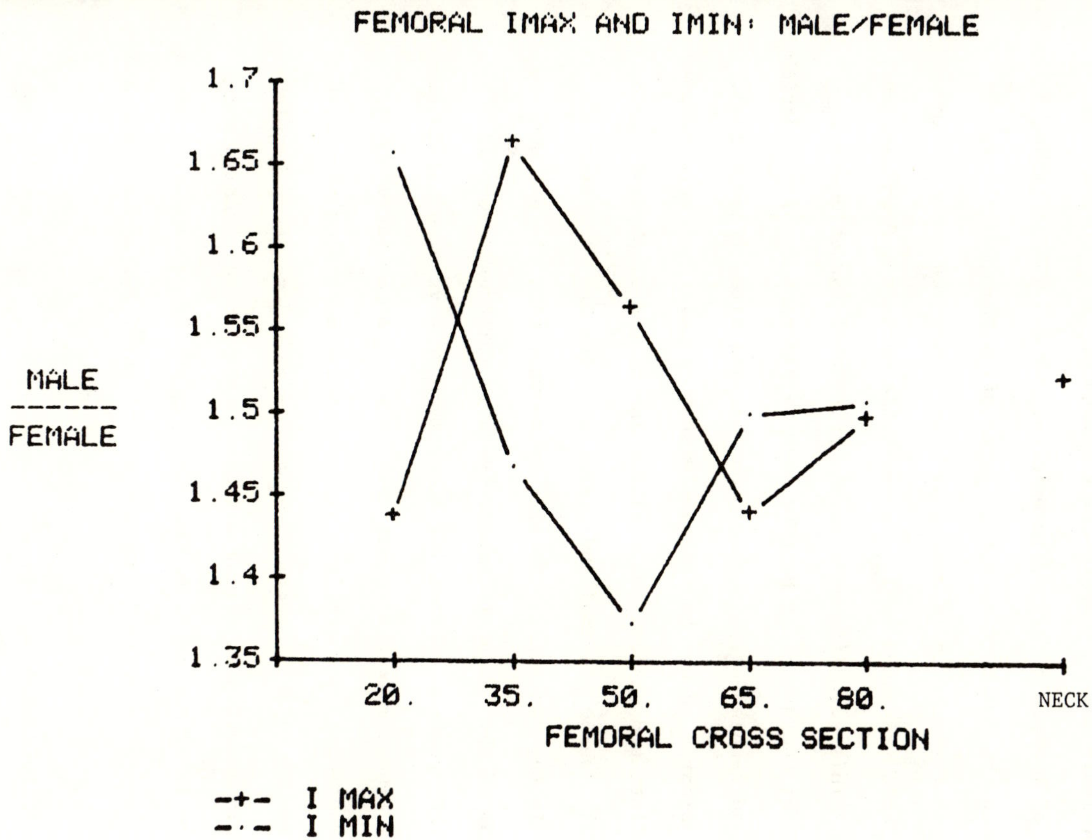
FEMORAL IMAX AND IMIN: MALE/FEMALE
MALE

FEMALE
1.7
1.65
1.6
1.55
1.5
1.45
1.4
1.35
20.
35.
50.
65.
80.
NECK
FEMORAL CROSS SECTION
-+- I MAX
-·- I MIN

Figure 22b

be more "platycnemic" than female tibiae).

Sexual dimorphism in the shape of the tibial shaft, reflected in the male/female principal moment ratios, also varies by level along the diaphysis (Figure 22a). Male-female differences in I_{max} rise from the distal end of the tibia to a maximum in the 50% and 65% sections, then fall again in the 80% section (it is interesting that CA follows the same pattern of variation in the tibia--see Figure 16a above). In contrast, sex differences in I_{min} show little variation from section to section. This sex difference in the pattern of variation of I_{max} and I_{min} along the shaft can best be seen by plotting I_{max}/I_{min} ratios for males and females at different cross section sites (Figure 23a). This graph clearly shows the relatively less circular and more medio-laterally flattened shape of male tibiae in the mid-diaphyseal region (50% and 65%), while proximal and distal sections are more similar in "shape" in the two sexes. As will be shown later, the 65% and 50% sections of the tibia are precisely those which are probably subjected to the greatest A-P bending moments under physiological loading conditions. Thus, male tibiae not only show a general pattern of greater A-P bending strength, but also demonstrate particularly high bending strength in those sections under the most pronounced bending stress. This constitutes strong evidence for a marked sexual dimorphism in the *in vivo* loading of the tibia in the Pecos sample, a phenomenon which may be explained on the basis of several factors, discussed further in chapter 6.

Principal moments of the femoral cross sections show a different pattern of sexual dimorphism (Table 16b and Figure 22b). First, it

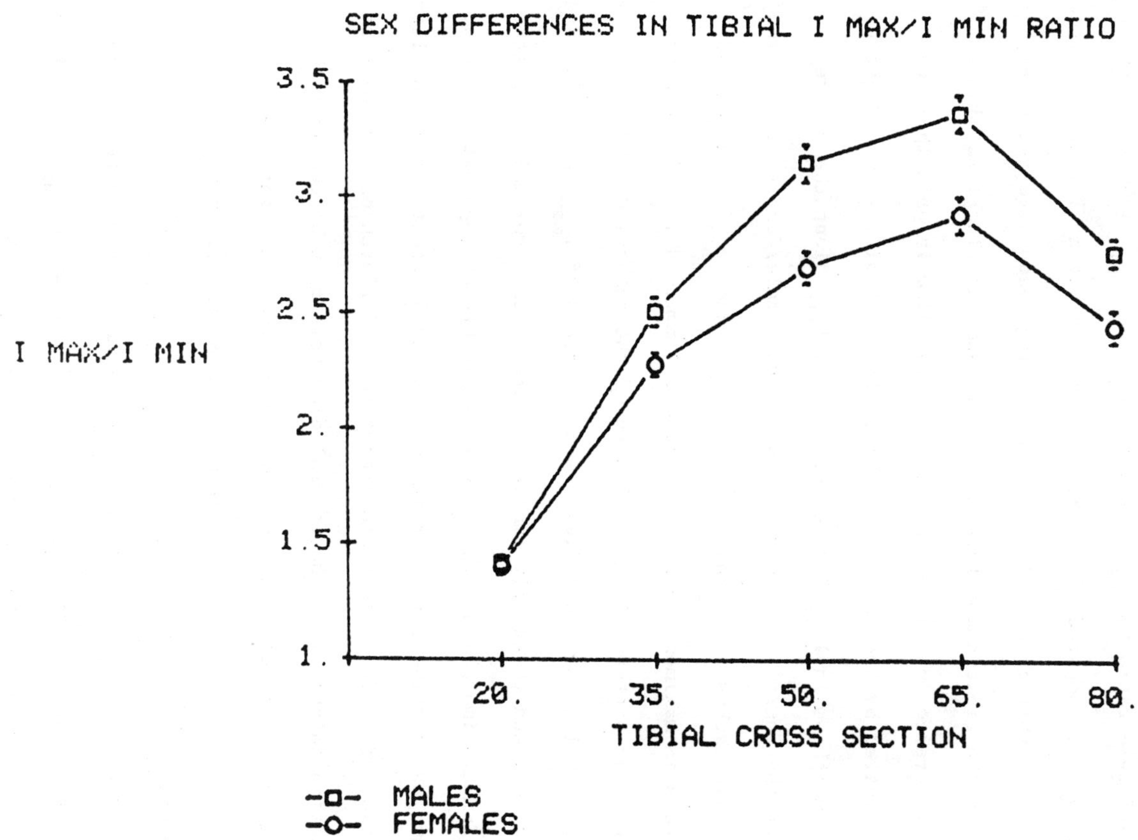
SEX DIFFERENCES IN TIBIAL I MAX/I MIN RATIO
I MAX/I MIN
3.5
3.
2.5
2.
1.5
1.
20.
35.
50.
65.
80.
TIBIAL CROSS SECTION
MALES
FEMALES

Figure 23a

should be noted that, as with bone areas, sexual dimorphism in second moments of area is smaller in the femur as a whole than in the tibia, the male/female ratio in I_{max} averaging 1.52 and I_{min} 1.50 (diaphyseal sections only), compared with 1.82 and 1.64, respectively, in the tibia. This suggests that as a whole, _in vivo_ bending loads were relatively smaller in the tibia among females of this sample. There is also a marked difference in the type of sexual dimorphism in principal moments exhibited by the femur and tibia. In the tibia, as noted above, sexual dimorphism in I_{max} is greater than in I_{min} along the entire length of the shaft. In contrast, the average male/female ration along the femoral diaphysis is about equal for I_{max} and I_{min}. Sex-related differences in maximum and minimum principal moments vary in an apparently complex pattern along the length of the femoral diaphysis (Figure 22b). Males have relatively greater values of I_{max} in the 35% and 50% sections, and relatively smaller values at the proximal and distal ends. Sexual dimorphism in I_{min} follows exactly the opposite pattern, with females relatively greater in the 35% and 50% sections and smaller at the two ends. In order to interpret these differences, the pattern of variation in phi along the femoral diaphysis must be considered (see Figure 20b above). Considering the two factors together--the magnitudes and orientations of the principal moments--it is apparent that males exhibit relatively greater values of I_{max} (and smaller values of I_{min}) in regions of the femur where I_{max} is more A-P oriented, while females are relatively greater in I_{max} (and smaller in I_{min}) in regions where the major axis of the cross section is more M-L oriented. In other words, along the entire length of the femoral diaphysis, males are more adapted for A-P bending stress, females for M-L bending stress.

This point is also made clear by considering the variation in I_{max}/I_{min} ratios along the femoral shaft in the two sexes (Figure 23b). This graph shows that male cross sections are less circular (i.e., more specialized for bending in one particular plane) in the 35% and 50% sections, while females are less circular in the 20%, 65% and 80% cross sections. As will be discussed in more detail in the next chapter, the 35% and 50% sections of the femur may be subjected to the greatest A-P bending moments under physiological loading conditions, while the 65% and 80% sections almost certainly undergo the greatest M-L bending forces. Thus, female femora are not only relatively more robust over-all than female tibiae, but they are especially strong for withstanding the particular stresses which would result from M-L bending of the fe-mur. In contrast, A-P bending stress appears to be a relatively more important in male femora, as it is in male tibiae.

Sexual dimorphism in the femoral neck cross section for I_{max} and I_{min} falls close to the average male/female values for the femoral diaphysis as a whole (Figure 22b and Table 16b). Considering proper-ties of the cortical shell alone, males tend to have relatively greater values of I_{max} than I_{min} (compared to females), and are somewhat less circular in shape (Figure 23b). If the entire subperiosteal area is used to determine these properties, male/female values for the two principal moments are very close, as is their "shape". If we consider the true bending strength of the femoral neck, in terms of cross-sec-tional geometric parameters, to fall somewhere between these two

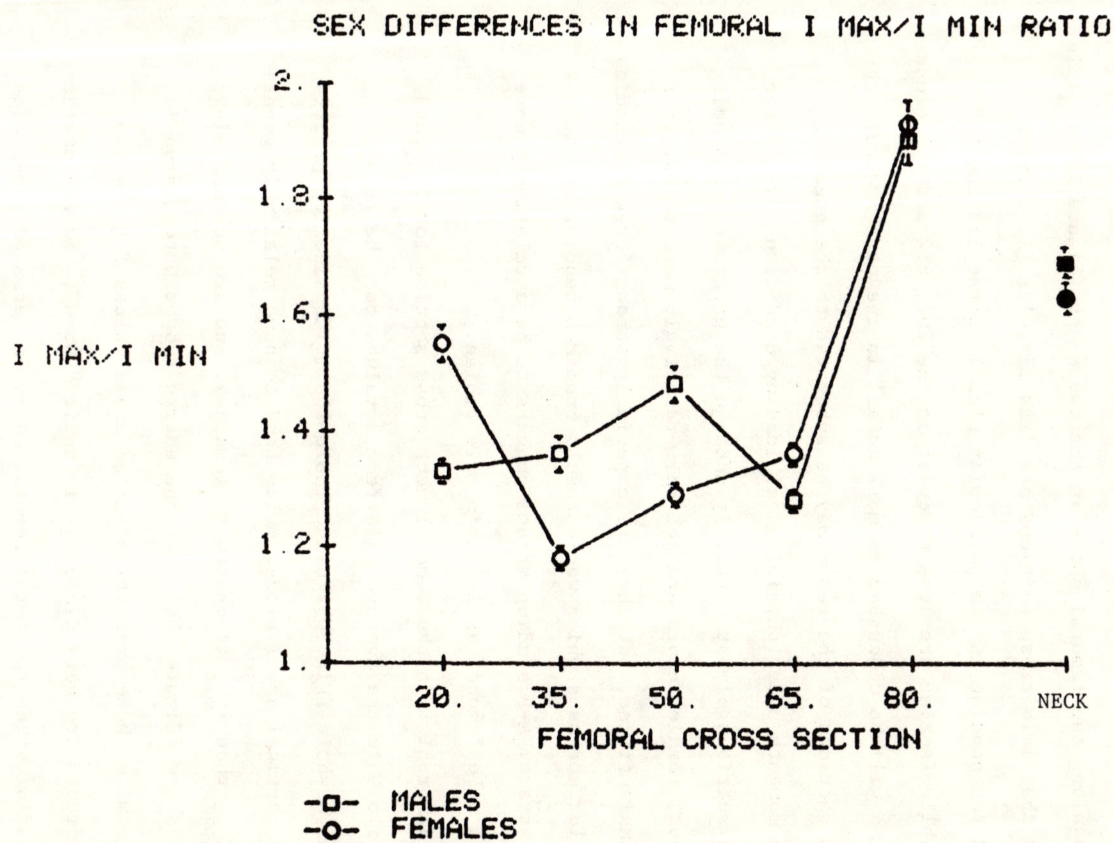
SEX DIFFERENCES IN FEMORAL I MAX/I MIN RATIO
I MAX/I MIN
2.
1.8
1.6
1.4
1.2
1.
20.
35.
50.
65.
80.
NECK
FEMORAL CROSS SECTION
MALES
FEMALES

Figure 23b

results, it appears that relatively little sexual dimorphism in the relative bending strengths around the major and minor axes of the femoral neck exists. If any trend is apparent, it is that males are relatively slightly stronger with respect to superior-inferior bending of the femoral neck (i.e., bending in the superio-inferior plane produced by an M-L bending moment around the femoral head). This is somewhat surprising, since with the hypothesized greater M-L bending moments in female femora (see earlier discussions and chapter 6), one would expect relatively greater bending stresses in the female femoral neck. Two factors might explain this apparent discrepancy: a) stresses other than those produced by superior-inferior bending of the neck (e.g., A-P bending, torsion) are also critical in determining cross-sectional geometry in this section; and b) other structural properties of the femoral neck, such as trabecular density or orientation, are more important than gross cross-sectional geometry or distribution of compact cortical bone in determining bending strength in this section. (It should also be pointed out, as discussed earlier, that the beam model used in the present study is probably quite imprecise for estimating the true bending stresses in an area as complex in geometry and as close as articulation endpoints as the femoral neck.) More detailed analyses of both trabecular configurations and stress distributions under simulated loading conditions, perhaps using finite element analysis (Hayes et al., 1978), will be necessary before hypotheses such as these can be fully evaluated.

Age differences.

Age differences in I_{max}, I_{min} and I_{max}/I_{min} are summarized in Tables 17a, b (tibia) and 18a, b (femur). As in the earlier presentation of bone areas, specific mean and standard error values are shown for only the two larger age groupings: 20-39 and 40+ years. F values for a 1-way ANOVA of each property on age, using the 6 smaller age groups (sexes separate), are included in the last column of each table for reference, as was done earlier with bone areas.

Perhaps the single most important observation to be drawn from these data is that maximum and minimum second moments of area increase with age at most sites. This increase is characteristic of both males and females, although the degree of age-related change varies by bone and by cross section site. The percentage increase is greater in males in the tibia, averaging 13% for both I_{max} and I_{min}, compared with changes of 5% and -2% (decrease) for I_{max} and I_{min}, respectively, in females. However, females show greater average percentage gains in the femur: 10% and 11% for I_{max} and I_{min}, respectively, versus 6% for both principal moments in males (diaphyseal sections only). This bone/sex-related difference in age change of principal moments is not unexpected, given the earlier observations with regard to bone area age changes in the two sexes--namely that females gain more in TA and lose relatively less in CA with age in the femur than in the tibia, while males show smaller age increases in TA in the femur than the tibia (see Tables 9a, b 10a, b above).

Table 17a

Age Differences in Tibial Principal Moments: Males

MAXIMUM PRINCIPAL MOMENT (I_{max})

Cross Section	20-39 yrs Mean	20-39 yrs SE	40+ yrs Mean	40+ yrs SE	40+ yrs - 20-39 yrs	%dif	F[1]
20%	12489[2]	421	13663	748	1174	9.4	1.59
35%	16921	536	19605	988	2684**	15.9	2.43*
50%	23935	817	27425	1220	3490*	14.6	2.66*
65%	31848	1166	36758	1522	4909*	15.4	2.52*
80%	42694	1525	46967	2179	4274	10.0	.86

MINIMUM PRINCIPAL MOMENT (I_{min})

Cross Section	20-39 yrs Mean	20-39 yrs SE	40+ yrs Mean	40+ yrs SE	40+ yrs - 20-39 yrs	%dif	F[1]
20%	8917[2]	301	9612	531	695	7.8	1.96†
35%	6838	236	8310	504	1472**	21.5	2.89*
50%	7730	330	9266	590	1536*	19.9	1.89
65%	9694	452	11234	617	1540*	15.9	1.63
80%	16435	677	16181	1014	-254	-1.5	.36

I_{max} / I_{min}

Cross Section	20-39 yrs Mean	20-39 yrs SE	40+ yrs Mean	40+ yrs SE	40+ yrs - 20-39 yrs	%dif	F[1]
20%	1.42	.03	1.44	.05	.02	--	.76
35%	2.55	.08	2.42	.10	-.13	--	1.03
50%	3.20	.10	3.06	.12	-.14	--	.95
65%	3.37	.09	3.37	.14	0	--	.78
80%	2.66	.07	2.98	.12	.32*	--	1.69

[1] Results of 6 age group ANOVA--see text.

[2] mm^4

† $p<.10$ * $p<.05$ ** $p<.01$

Table 17b

Age Differences in Tibial Principal Moments: Females

MAXIMUM PRINCIPAL MOMENT (I_{max})

Cross Section	20-39 yrs Mean	SE	40+ yrs Mean	SE	40+ yrs - 20-39 yrs	%dif	F[1]
20%	7596[2]	290	7482	415	-114	-1.5	.32
35%	9672	477	10644	566	972	10.0	.39
50%	12827	571	14143	807	1316	10.3	.55
65%	17597	828	17744	1056	147	.8	.68
80%	23676	965	25011	1579	1335	5.6	1.71

MINIMUM PRINCIPAL MOMENT (I_{min})

Cross Section	20-39 yrs Mean	SE	40+ yrs Mean	SE	40+ yrs - 20-39 yrs	%dif	F[1]
20%	5567[2]	255	5399	346	-168	-3.0	.56
35%	4405	246	4765	309	360	8.2	.16
50%	4991	249	5194	364	203	4.1	.07
65%	6394	311	5764	335	-630	-9.8	.63
80%	10386	430	9488	630	-898	-8.6	.89

I_{max} / I_{min}

Cross Section	20-39 yrs Mean	SE	40+ yrs Mean	SE	40+ yrs - 20-39 yrs	%dif	F[1]
20%	1.39	.03	1.43	.06	.04	--	1.19
35%	2.26	.06	2.33	.12	.07	--	.34
50%	2.63	.06	2.85	.15	.22	--	.74
65%	2.81	.08	3.16	.16	.35*	--	1.24
80%	2.31	.06	2.74	.15	.43**	--	3.08*

[1] Results of 6 age group ANOVA--see text.

[2] mm^4

* $p<.05$ ** $p<.01$

Table 18a

Age Differences in Femoral Principal Moments: Males

Cross Section	20-39 yrs Mean	20-39 yrs SE	40+ yrs Mean	40+ yrs SE	40+ yrs - 20-39 yrs	%dif	F [1]
			MAXIMUM PRINCIPAL MOMENT (I_{max})				
20%	31222[2]	1273	29575	1064	-1647	-5.3	1.11
35%	22219	686	24205	826	1986†	8.9	.67
50%	21538	658	24183	914	2645*	12.3	1.49
65%	21931	841	23637	744	1706	7.8	1.00
80%	29930	840	32175	1272	2245	7.5	.93
Neck-CA	11399	396	11810	438	411	3.6	.37
Neck-TA	33425	1411	36979	1724	3554	10.6	1.39
			MINIMUM PRINCIPAL MOMENT (I_{min})				
20%	22688[2]	737	23568	676	880	3.9	.62
35%	16978	594	17406	566	428	2.5	.28
50%	14986	535	16013	437	1027	6.8	1.50
65%	16884	559	18972	618	2088*	12.4	2.17†
80%	16152	568	16926	649	774	4.8	.33
Neck-CA	7016	269	6893	413	-123	-1.8	.28
Neck-TA	21045	838	23311	1452	2266	10.8	.46
			I_{max} / I_{min}				
20%	1.37	.03	1.26	.03	-.11*	--	2.48*
35%	1.33	.03	1.41	.06	.08	--	.52
50%	1.46	.03	1.52	.06	.06	--	.64
65%	1.30	.03	1.25	.03	-.05	--	1.97
80%	1.89	.05	1.91	.05	.02	--	.54
Neck-CA	1.65	.04	1.77	.06	.12	--	.74
Neck-TA	1.60	.04	1.62	.04	.02	--	.62

[1] Results of 6 age group ANOVA--see text.

[2] mm^4

† p<.10 * p<.05

Table 18b

Age differences in Femoral Principal Moments: Females

MAXIMUM PRINCIPAL MOMENT (I_{max})

Cross Section	20-39 yrs Mean	20-39 yrs SE	40+ yrs Mean	40+ yrs SE	40+ yrs - 20-39 yrs	%dif	F [1]
20%	21790 [2]	798	20450	1008	-1340	-6.1	.62
35%	13352	588	14607	709	1255	9.4	1.28
50%	13912	609	15253	750	1341	9.6	1.60
65%	14665	511	17594	792	2929**	20.0	3.37**
80%	19443	688	22654	1164	3210**	16.5	2.34*
Neck-CA	7465	312	7849	470	384	5.1	.39
Neck-TA	21067	962	23864	1189	2797†	13.3	1.39

MINIMUM PRINCIPAL MOMENT (I_{min})

Cross Section	20-39 yrs Mean	20-39 yrs SE	40+ yrs Mean	40+ yrs SE	40+ yrs - 20-39 yrs	%dif	F [1]
20%	14039 [2]	534	13585	629	-454	-3.2	.70
35%	11303	417	12428	516	1125	9.9	1.80
50%	10620	369	12275	556	1655**	15.6	3.59**
65%	11075	474	13082	623	2007*	18.1	3.04*
80%	10350	447	12006	756	1656*	16.0	1.55
Neck-CA	4914	208	4417	278	-497	-10.1	.53
Neck-TA	13636	682	14413	678	777	5.7	.46

I_{max} / I_{min}

Cross Section	20-39 yrs Mean	20-39 yrs SE	40+ yrs Mean	40+ yrs SE	40+ yrs - 20-39 yrs	%dif	F [1]
20%	1.57	.03	1.51	.05	-.06	--	.65
35%	1.18	.02	1.18	.04	0	--	.21
50%	1.30	.03	1.25	.04	-.05	--	.65
65%	1.35	.03	1.36	.04	.01	--	2.45*
80%	1.92	.04	1.93	.06	.01	--	.35
Neck-CA	1.53	.03	1.81	.07	.28***	--	3.89**
Neck-TA	1.57	.03	1.66	.05	.11†	--	1.27

[1] Results of 6 age group ANOVA--see text.

[2] mm^4

† $p<.10$ * $p<.05$ ** $p<.01$ *** $p<.001$

Age trends in the 6 smaller age groups clearly support the 2 age group results presented above. As an illustration of these trends, values for I_{max} in the femoral midshaft section are plotted for the 20-24 year through 50+ year age groups in Figure 24. Although some slight irregularities in these age curves are apparent, particularly in the younger age groups, the general trend is of increasing values with age over the entire adult life span. Note that there is no clear "peak" in values of I_{max} in the younger age groups (25-39 years) corresponding to that observed in CA (see Figures 17a, b and 18a, b above). In general, the pattern shown in Figure 24 is more similar to the age pattern in TA (see same earlier figures). There is, however, a tendency for females to "level off" somewhat in age increases in I_{max} in the middle part of the age range sampled here (25-49 years), while males show a more even and consistent pattern of increase over all age groups, a difference which may be related to the greater loss of cortical bone among females beginning in the fourth or fifth decades.

The fact that both sexes exhibit increases with age in principal moments, particularly I_{max}, is of critical importance in evaluating the effects of age-related remodeling of the femur and tibia in this sample. This indicates that even with endosteal resorption of bone, and in females, some decline in compact cortical bone volume, the diaphyseal cortex may be remodeled in such a way (principally through subperiosteal expansion) so that bending strength is actually increased with respect to geometrical parameters (age change in bone material properties are considered in a later subdivision of this chapter). Previous to this study, the only evidence for this kind of age-related

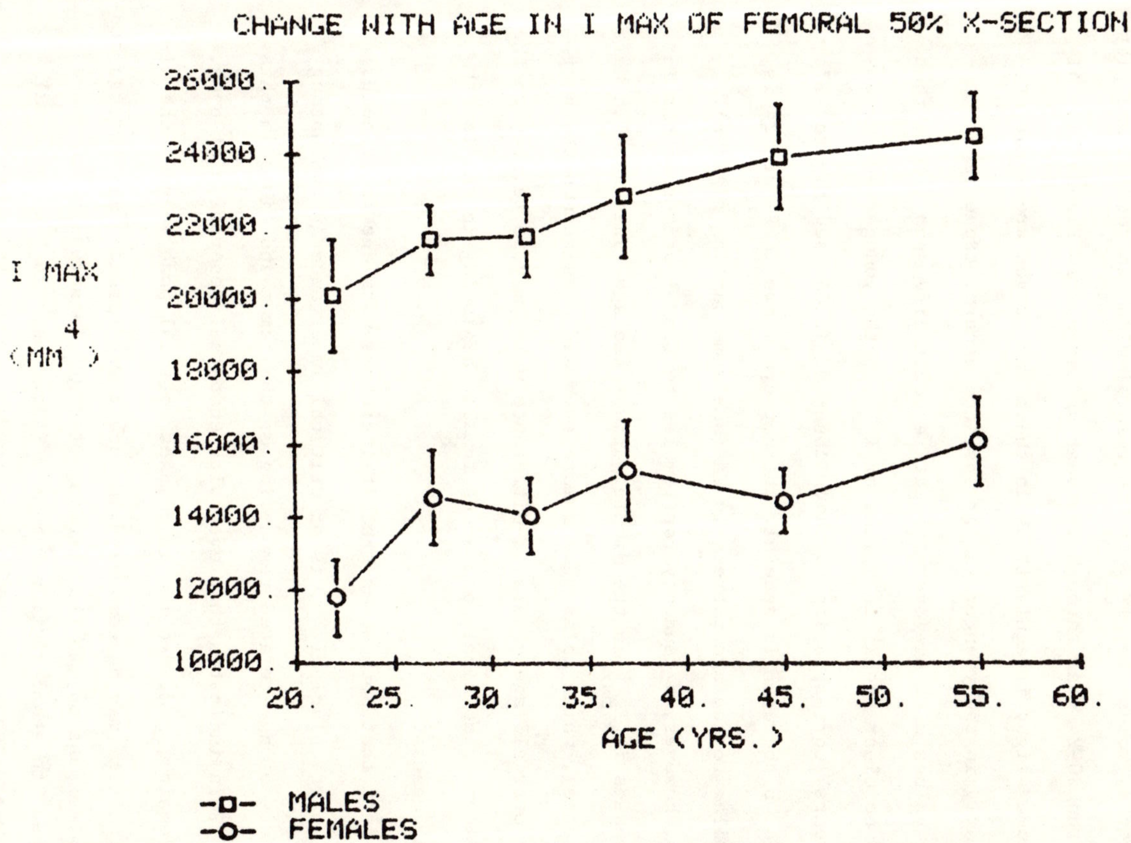
CHANGE WITH AGE IN I MAX OF FEMORAL 50% X-SECTION
I MAX
4
(MM)
26000.
24000.
22000.
20000.
18000.
16000.
14000.
12000.
10000.
20.
25.
30.
35.
40.
45.
50.
55.
60.
AGE (YRS.)
MALES
FEMALES

Figure 24

change came from radiographic investigations (e.g., Smith and Walker, 1964), with their concomitant problems of extrapolation from 2 to 3 dimensions (see chapter 2), and the single direct measurement study of the femur by Martin and Atkinson (1977). However, Martin and Atkinson found that second moments of area, while increasing with age in males, decreased with age in females. They interpreted this evidence to imply a fundamental difference between the sexes in type of biomechanically "compensatory" mechanism available to each as response to loss of bone (and thus increased stress) with aging. The present study results indicate that both males and females share in this "compensatory" mechanism, although the degree to which they do so varies by bone and by site. The implications of these conflicting data from the two investigations in terms of characteristics of the study samples and study designs used will be explored in greater depth in the next chapter.

Site-specific differences in age changes in I_{max} and I_{min} are of great interest in that they may indicate the relative importance of maintaining bending strength in different areas of the lower limb. For example, the overall greater increases in principal moments of the femur relative to the tibia in females is indicative of the relative importance of maintaining bending strength in the female femur. Similarly, greater increases with age among males in tibial principal moments may indicate a relatively greater need for maintaining bending strength in the tibia among older males. Several site-specific trends <u>within</u> bones are also apparent from the data presented in Tables 17a, b and 18a, b. In the tibia, percentage increases with age in I_{max} are greater in the middle cross sections than at the two ends. This is illustrated graphically in Figure 25a, which plots the ratio of I_{max} in the 40+ year age group to I_{max} in the 20-39 year age group for the 5 tibial cross sections. As can be seen from this figure, the relative difference in I_{max} between older and younger age groups is greater among males in the 35%, 50%, and 65% tibial sections, and in females in the 35% and 50% sections. Age differences in I_{min} in the tibia follow basically the same pattern (Tables 17a, b), although in females I_{min} shows smaller gains (or greater reductions) with age than I_{max} over the entire diaphysis. This is particularly true in the proximal tibial diaphysis, where the I_{max}/I_{min} ratio increases significantly with age in females. A similar "shape" change occurs in males in the 80% section, although the 35% and 50% sections follow an opposite pattern, tending to become more circular with age in males. In fact, although males still maintain a relatively less circular and more A-P oriented distribution of bone volume, the cross-sectional shape of the tibial diaphysis does become somewhat similar overall in the two sexes in the older age group (also see Figure 23a). This may indicate more similar types of bending stress in the tibia in older males and females than in younger adults (see next chapter). In any event, the greater increases in principal moments, particularly I_{max}, in the middle region of the tibial shaft in both sexes is consistent with the model of greater *in vivo* A-P bending stress in this region presented earlier. The peak increase in I_{min} in the 35% section in both sexes may also be interpreted as an adaptation to torsional stress in this area.

In the femur, site-specific patterns of age change in principal moments vary between the sexes. In males, the greatest increase in

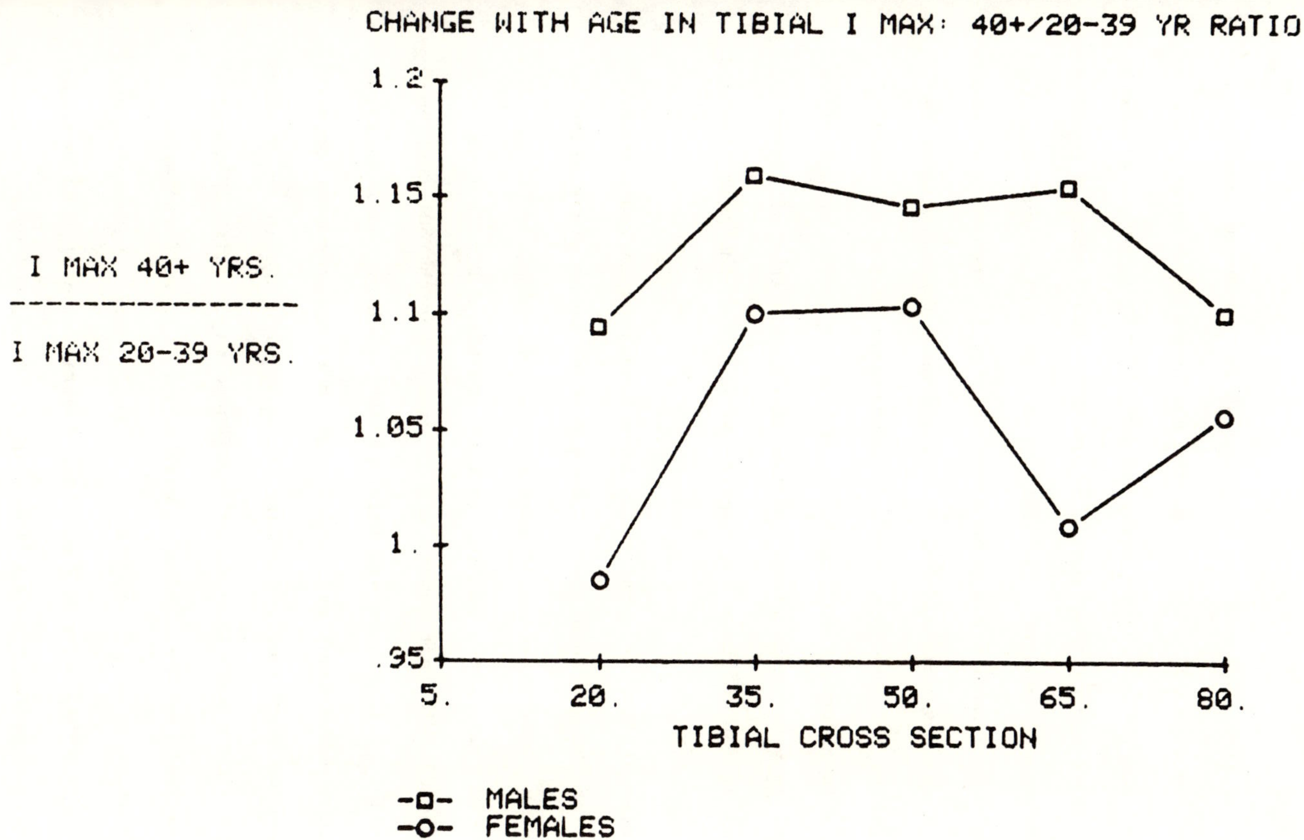
CHANGE WITH AGE IN TIBIAL I MAX: 40+/20-39 YR RATIO
I MAX 40+ YRS.
I MAX 20-39 YRS.
1.2
1.15
1.1
1.05
1.
.95
5.
20.
35.
50.
65.
80.
TIBIAL CROSS SECTION
MALES
FEMALES

Figure 25a

I_{max} occurs at the 50% location. In females the greatest increase in I_{max}--amounting to a 20% increase in the 65% section--occurs in the two proximal locations. These patterns are clearly shown in Figure 25b, which plots age ratios (as in Figure 25a) for femoral I_{max} values. The sex differences support the view, presented earlier and discussed below, that male femora are subjected to relatively greater A-P bending stress (at a maximum near midshaft), while female femora undergo relatively greater M-L bending stress (at a maximum in the proximal diaphysis). In effect, bone remodeling differences with aging reflect differential _in vivo_ stresses in the two sexes, and thus differential preservation of (or increase in) bending strength at the most critical sites in each sex.

It is also of interest that males show a maximum increase with age in I_{min} at the 65% location, a cross section whose _minor_ axis orientation in older males is at an angle of approximately 66° to the transverse axis, i.e., relatively close to the A-P axis (see Table 18a above). In fact, among males, except for the 80% section, where I_{max} and I_{min} increases are fairly close, each of the femoral cross sections whose major axis is oriented more anteroposteriorly (35%, 50%) shows a greater percentage increase with age in I_{max}, while each section with a more transversely oriented major axis (20%, 65%) shows a relatively greater increase with age in I_{min}. These age changes are also reflected in changes in the I_{max}/I_{min} ratio of these sections, which show a more A-P oriented distribution of bone cross-sectional volume in older males (Table 18a). Thus, the male femur as a whole becomes relatively better adapted with aging for withstanding

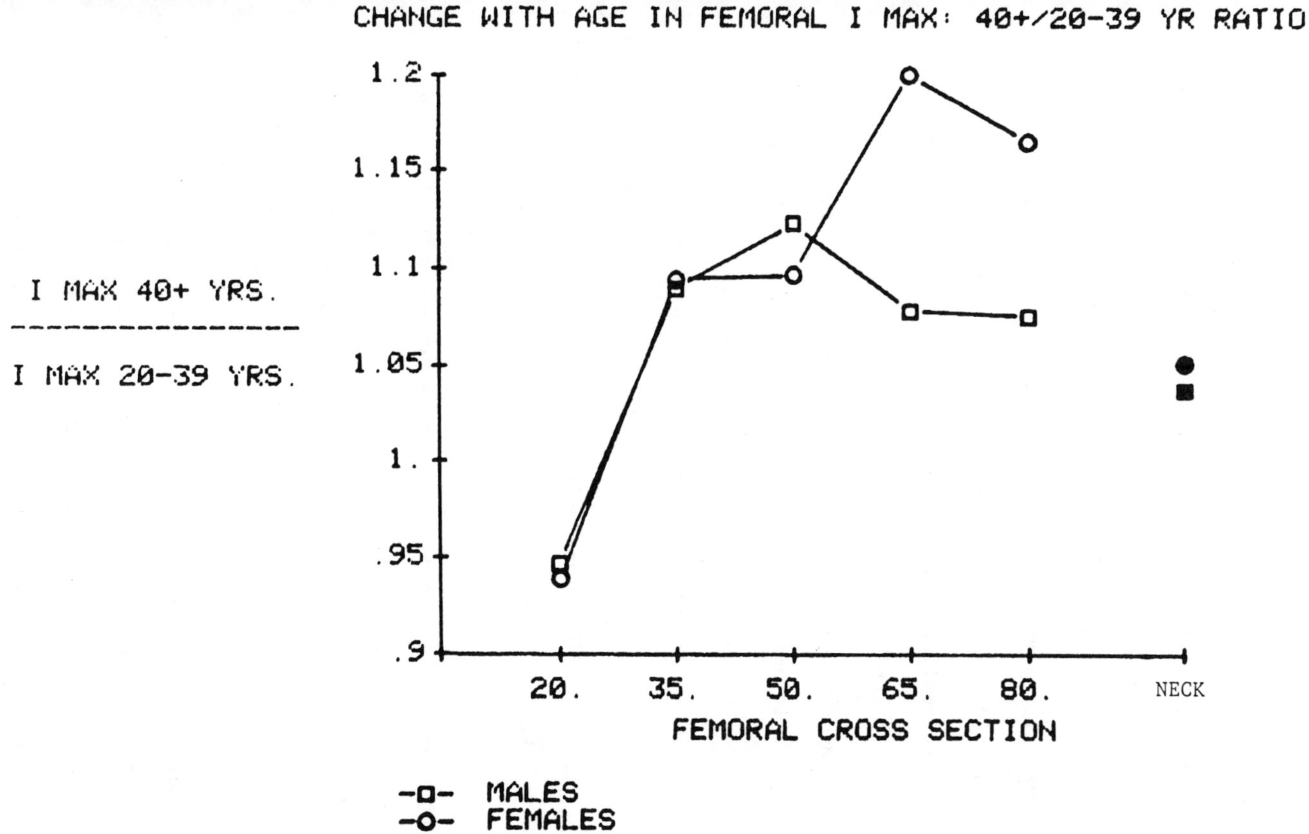
CHANGE WITH AGE IN FEMORAL I MAX: 40+/20-39 YR RATIO
I MAX 40+ YRS.
I MAX 20-39 YRS.
1.2
1.15
1.1
1.05
1.
.95
.9
20.
35.
50.
65.
80.
NECK
FEMORAL CROSS SECTION
MALES
FEMALES

Figure 25b

A-P bending stress, particularly near midshaft. In contrast, the midshaft section of the female femur shows a greater percentage increase in I_{min} with aging than I_{max}, leading to a more circular rather than A-P oriented cross-sectional shape in this region. Thus, throughout the entire proximal and midshaft regions of the femoral diaphysis, female femora maintain or increase M-L bending strength with aging, while any age-related remodeling of male femora leads to a relative increase in A-P bending strength.

The femoral neck shows a very interesting pattern of age change in I_{max} and I_{min}. In both sexes, I_{max} of this section increases with age, while I_{min} decreases (or in the case of properties determined on the basis of total subperiosteal area, I_{min} increases less). Thus, particularly with respect to cortical shell determined properties, the ratio of I_{max}/I_{min} increases with age, indicating a less circular cross-sectional shape, with more bone volume distributed along the superior-infereior axis. This pattern is much more marked among females, where even the subperiosteal area properties show a near-significant change in "shape" with age (indicating a remodeling of the entire cross-sectional geometry, including the trabecular bone area). Thus, bending strength in the superior-inferior plane of the femoral neck, at least in terms of cross-sectional geometry, increases with age, more so in females, who would be expected to have particularly high levels of bending stress in this area. The fact that females show more marked changes with age in the magnitudes of principal moments of the femoral neck is also in concert with their more pronounced changes in several other proximal femoral characteristics, i.e., the femoral antetorsion angle and angles of major axes of the proximal diaphysis and femoral neck. The age-related variation in I_{max} and I_{min} of the femoral neck cross section is therefore probably best viewed as another aspect of the "proximal femoral functional complex" introduced above. The implications of these age changes in terms of loadings of the femur are discussed further in chapter 6.

<u>Side differences.</u>

Side differences in principal moments of the tibia and femur are presented in Tables 19a, b. As in the presentation of bone area results earlier, only means for each side and percentage differences ((rt-lt)/rt) are given, for the reasons discussed in that section (see footnote 14).

The tibia shows a consistent pattern, present in both sexes, of greater values for I_{max} on the left side, averaging 6% larger than right side values over all five cross sections (in both sexes). This left side "dominance" in I_{max} increases from the distal to proximal ends, reaching a maximum in the 50% to 80% sections. I_{min} shows relatively little side difference in the tibia, averaging 2% larger on the right side in males, and equal on right and left sides in females. Because of the increase in I_{max}, left tibiae are markedly less circular and more A-P oriented in distribution of cross-sectional bone volume than right tibiae. This is particularly evident in the asymmetry in I_{max}/I_{min} ratios on right and left sides (Table 19a). A moderate right side "dominance" in I_{min} in the middle region of the tibial diaphysis leads to an even greater asymmetry in this ratio for these sections, reaching a peak difference between sides in the 50% cross section in both sexes. It should also be noted from earlier

Table 19a

Side Differences in Tibial Principal Moments

MAXIMUM PRINCIPAL MOMENT (I_{max})

Cross Section	Males Right	Males Left	Males %dif[1]	Females Right	Females Left	Females %dif[1]
20%	13002[2]	12768	1.8	7541	7576	-.5
35%	17315	18364	-6.0	9865	10128	-2.7
50%	24095	26177	-8.6	12973	13558	-4.5
65%	32578	34480	-5.8	16962	18329	-8.1
80%	42297	46051	-8.9	22690	25551	-12.6†

MINIMUM PRINCIPAL MOMENT (I_{min})

Cross Section	Males Right	Males Left	Males %dif[1]	Females Right	Females Left	Females %dif[1]
20%	9324[2]	8976	3.7	5420	5603	-3.4
35%	7468	7202	3.6	4594	4457	3.0
50%	8423	8073	4.2	5203	4914	5.6
65%	10272	10157	1.1	6176	6192	-.3
80%	15981	16730	-4.7	9849	10332	-5.0

I_{max} / I_{min}

Cross Section	Males Right	Males Left	Males Rt-Lt	Females Right	Females Left	Females Rt-Lt
20%	1.41	1.44	-.03	1.42	1.39	.03
35%	2.39	2.63	-.24†	2.20	2.36	-.16
50%	2.98	3.33	-.35*	2.58	2.82	-.24†
65%	3.25	3.50	-.25†	2.84	3.01	-.17
80%	2.73	2.81	-.08	2.36	2.54	-.18

[1] [(Rt-Lt)/Rt] X 100

[2] mm^4

† p<.10 * p<.05 (two sample t test--see text)

Table 19b

Side Differences in Femoral Principal Moments

MAXIMUM PRINCIPAL MOMENT (I_{max})

Cross Section	Males			Females		
	Right	Left	%dif [1]	Right	Left	%dif [1]
20%	29615[2]	31749	-7.2	20201	22485	-11.3
35%	23240	22533	3.0	12601	14939	-18.6**
50%	22590	22274	1.4	13279	15439	-16.3*
65%	22970	22033	4.1	14834	16449	-10.9†
80%	31670	29679	6.3	19956	21070	-5.6
Neck-CA	11599	11476	1.1	7682	7503	2.3
Neck-TA	35125	34118	2.9	22123	21876	1.1

MINIMUM PRINCIPAL MOMENT (I_{min})

Cross Section	Males			Females		
	Right	Left	%dif [1]	Right	Left	%dif [1]
20%	22745[2]	23235	-2.2	13277	14498	-9.2
35%	16491	17777	-7.8	10832	12524	-15.6**
50%	14932	15751	-5.5	10522	11821	-12.3*
65%	17754	17424	1.8	11062	12427	-12.3†
80%	16870	15942	5.5	10474	11330	-8.2
Neck-CA	6762	7194	-6.4	4786	4711	1.6
Neck-TA	21853	21772	.4	13681	14109	-3.1

I_{max} / I_{min}

Cross Section	Males			Females		
	Right	Left	Rt-Lt	Right	Left	Rt-Lt
20%	1.30	1.37	-.07	1.53	1.58	-.05
35%	1.42	1.28	.14*	1.17	1.19	-.02
50%	1.51	1.44	.07	1.26	1.31	-.05
65%	1.30	1.27	.03	1.37	1.34	.03
80%	1.90	1.89	.01	1.95	1.90	.05
Neck-CA	1.76	1.62	.14*	1.64	1.62	.02
Neck-TA	1.64	1.57	.07	1.63	1.57	.06

[1] [(Rt-Lt)/Rt] X 100 [2] mm^4

† p<.10 * p<.05 ** p<.01 (two sample t test--see text)

observations that the direction of the major axis of greatest bending strength in the tibia is more A-P oriented in left tibiae (see Table 15 above). Thus, with respect to the direction, magnitude, and location along the shaft of maximum bending strength, left tibiae are adapted to greater A-P bending loadings than right tibiae. This pattern in present in both sexes, but is more marked among males (Tables 11a, 15, 19a).

The femur also displays side-related asymmetry in principal moments, but the nature of the asymmetry appears to be different in males and females. Female side differences in the femur are similar to those of the tibia, and indicate an even more pronounced left side "dominance" in bending strength. Female left femora average 12.5% greater values of I_{max}, and 11.5% greater values of I_{min} than right femora (diaphyseal sections only), with peak levels of asymmetry in the 35% and 50% sections (Table 19b). The ratio of I_{max}/I_{min} does not vary greatly between the two sides, although there is a tendency for I_{max} to be relatively greater on the left side in regions where the major axis is M-L oriented (65%, 80%)--the 20% section is an exception. Thus, there is some evidence for a relatively more A-P oriented distribution of cross-sectional volume in left female femora, as in left tibiae.

Male femora display much less side-related asymmetry in principal moments than female femora, averaging over the five diaphyseal sections a slight (1.5%) right "dominance" in I_{max}, and a slight (1.6%) left "dominance" in I_{min} (this same sex difference in asymmetry was also found in CA--see above). The right side "dominance" in I_{max} increases from the distal end (where the left side I_{max} is actually larger) to the proximal end, while I_{min} exhibits its peak "dominance" in the 35% and 50% sections, changing to a right "dominance" in the proximal two sections. Side differences among males in the I_{max}/I_{min} ratio indicate a relatively more A-P oriented distribution of cross-sectional bone volume in right femora in the mid-distal shaft (particularly the 35% section), a pattern of "shape" asymmetry opposite to that found in female femora (with the exception of the 20% section in females).

Thus, based on magnitudes of the two principal moments and their ratio, female femora appear to be more adapted for A-P bending on the left side, and male femora on the right side. However, side-related differences in the orientation of principal moments, indicated by differences in phi values, must also be considered here (see Table 15). Concentrating first on the 35% and 50% sections--those that show the difference between the sexes in terms of side-related "shape" asymmetry--we see that males have more A-P oriented major axes on the left side, while female major axes are more A-P oriented on the right side. Therefore, the side-related differences in direction of the principal axes tend to cancel out the side-related differences in relative magnitude of I_{max} or I_{min}, in terms of establishing a more A-P oriented or M-L oriented distribution of bone area in these sections. In fact, if second moments of area are calculated about x and y reference axes rather than principal axes, males are found to have relatively more cross-sectional bone volume distributed in an A-P direction (about the M-L axis) in left femora than females, due primarily to the differences in phi asymmetry in the two sexes. Thus, the orientation and relative magnitude of principal moments in the mid-distal femoral diaphysis appear to vary in a fairly complex manner, which may be due to the combination of different bending loads in this area (see chapter 6).

Variation in other femoral characteristics, such as the femoral antetorsion angle, which shows more side-related asymmetry in females than in males (see Table 4 above), probably also plays a role in producing these sex/side differences in the femoral shaft.

The femoral neck cross section principal moments do not appear to follow the same side-related trends as diaphyseal properties in the two sexes. Thus, based on cortical shell determined properties, females exhibit a slight (1-2%) <u>right</u> side "dominance" in both I_{max} and I_{min}, while males exhibit a slight (1%) right "dominance" in I_{max} but relatively large (6.4%) left "dominance" in I_{min}. Using total subperiosteal area determined properties for this cross section tends to increase right side "dominance" in males, but decrease right "dominance" in females. Using either set of properties, femoral neck cross sections in both sexes tend to be less circular, with a more superior-inferiorly distribution of bone volume in right femora. Although this would seem to indicate greater resistance to bending in right femoral necks, it should be noted that female femoral necks (on either side), which might be expected to experience greater M-L bending moments, are more circular with a less superior-inferiorly distributed bone volume than males. As discussed above, this may indicate that bending moments applied to the femoral head are translated into stresses other than superior-inferior bending stresses in the femoral neck. Therefore, more circular left femoral neck cross sections may still be consistent with a female left femoral diaphysis adapted for greater bending stresses.

In summary, it may be said that both sexes exhibit left side "dominance" in terms of bending strength in the tibia: females also exhibit left "dominance" in the femur as a whole, but males show little overall asymmetry, or a slight right "dominance" in the femur. Interpretation of femoral asymmetry in principal moments is made more difficult, however, by side/sex-related variation in the orientation of principal axes and the lack of strongly-defined directions of principal axes in some sections of the femur.

As with other geometric properties, age differences in side-related asymmetry of principal moments were also investigated. In general, females tend to increase in left side "dominance" with age, both in terms of magnitude of I_{max} and I_{min}, and in terms of bone "shape" in the tibia (i.e., a more A-P oriented distribution of bone area on the left side in older females). Males show less distinct and consistent changes with age in asymmetry, although there is a tendency for male tibiae to become somewhat less asymmetric with age in terms of principal moment magnitudes (but not "shape"), and in the femoral 80% section, right "dominance" in both I_{max} and I_{min} increases with age. In the femoral neck cross section, both sexes show an increase in both principal moments with age on the right side, while the left side increases less or decreases in the magnitudes of these properties. Thus, the right "dominant" femoral neck observed above only develops among older adults, with younger adults showing an approximately symmetric or slightly left "dominant" femoral neck section. The "shape" asymmetry of the femoral neck cross section, whereby right sections are less circular and more A-P oriented in distribution of bone volume, also only develops among females in the older age group, although males show this side difference in both age groups.

Section Moduli

As described earlier in this chapter, the maximum stress in a beam cross section in bending will occur at the most distant point in the section from the neutral axis. If this distance is denoted c, the section modulus I/c, or Z, will be inversely proportional to the maximum stress at this point given a particular applied bending moment. Thus, Z may be used as an indicator of maximum stress in a cross section under bending loads, with larger values of Z indicating smaller maximum bending stresses.

There are several problems in interpreting the section modulus, however, one being that it is only strictly applicable to bending moments applied in the same plane (i.e., applied in a plane running through the cross section which contains the distant-most point from the neutral axis). Also, values of Z will depend upon both maximum distance from the neutral axis (c) and the second moment of area corresponding to this direction (I). Thus, calculation of the true maximum and minimum values of Z for a particular cross section is complex, theoretically involving separate determinations of c and I and comparisons of I/c for each point along the outer perimeter.

In the present study, section moduli were calculated for points along each of the principal axes, with maximum distances along the major and minor axes (i.e., to their intersections with the subperiosteal boundary) designated c_{maj} and c_{min}, respectively. Thus, Z_{maj} =

I_{max}/c_{maj}, and $Z_{min} = I_{min}/c_{min}$. While, for the reasons stated above, these values may not include the absolute maximum and minimum values of Z for the section, they do give some indication of the maximum stress endountered along each of the principal axes. Thus, Z_{maj} and Z_{min} can be interpreted as indices of the minimum stress in a cross section along its directions of maximum and minimum bending strength, respectively. Also, since c_{maj} is in most cases a good estimate of the true maximum distance from the section centroid, given that the major axis of a bone cross section usually passes through the subperiosteal surface close to its maximum diameter (see Figures 20a, b above), Z_{maj} may be taken as a good estimate of Z calculated for the most distant point in the cross section from the section centroid.

A summary of Z_{maj} and Z_{min} by sex and age group are shown in Tables 20a (males) and 20b (females). The general patterns of variation in these properties along the tibial and femoral shafts, in both sexes, are very similar to those observed in the corresponding principal moments (see Tables 16a, b). That is, relative increases or decreases in Z_{max} and Z_{min} values at different cross section sites closely parallel changes in I_{max} and I_{min}, respectively. This is not surprising, since variation in the magnitude or distribution of cross-sectional area will have a relatively greater (fourth power) effect on the second moment of area of the section, and a relatively smaller (first power) effect on linear distances within the cross section. Thus, the magnitudes of the I_{max}/c_{maj} and I_{min}/c_{min} section modulus ratios will be largely determined by variation in I_{max} and I_{min}.

Table 20a

Section Moduli, Tibia and Femur: Males

TIBIA

Cross Section	Property	Total Mean	Total SE	20-39 yrs Mean	20-39 yrs SE	40+ yrs Mean	40+ yrs SE	40+ yrs - 20-39 yrs
20%	Zmaj[1]	.94	.02	.92	.02	.98	.05	.06
	Zmin[2]	.77	.02	.76	.02	.78	.04	.02
35%	Zmaj	1.19	.03	1.15	.03	1.28	.05	.13**
	Zmin	.70	.02	.66	.02	.78	.04	.12**
50%	Zmaj	1.53	.03	1.46	.04	1.66	.06	.20**
	Zmin	.84	.02	.81	.03	.89	.04	.07†
65%	Zmaj	1.86	.04	1.78	.05	2.02	.08	.24**
	Zmin	.96	.03	.93	.03	1.03	.04	.10†
80%	Zmaj	2.18	.05	2.13	.06	2.27	.10	.14
	Zmin	1.24	.03	1.26	.04	1.21	.06	.05

FEMUR

Cross Section	Property	Total Mean	Total SE	20-39 yrs Mean	20-39 yrs SE	40+ yrs Mean	40+ yrs SE	40+ yrs - 20-39 yrs
20%	Zmaj	1.72	.04	1.74	.05	1.69	.04	.05
	Zmin	1.50	.03	1.50	.04	1.50	.04	0
35%	Zmaj	1.55	.03	1.52	.03	1.60	.04	.08
	Zmin	1.32	.02	1.30	.03	1.34	.03	.04
50%	Zmaj	1.56	.03	1.52	.04	1.64	.04	.12*
	Zmin	1.28	.02	1.26	.03	1.33	.02	.07
65%	Zmaj	1.59	.03	1.56	.05	1.66	.04	.10
	Zmin	1.38	.03	1.34	.04	1.46	.04	.12*
80%	Zmaj	1.88	.03	1.85	.04	1.95	.06	.10
	Zmin	1.34	.03	1.32	.04	1.37	.04	.05
Neck-CA	Zmaj	.73	.02	.74	.03	.71	.03	-.03
	Zmin	.54	.02	.55	.02	.52	.03	-.03
Neck-TA	Zmaj	2.16	.07	2.13	.09	2.20	.11	.07
	Zmin	1.69	.05	1.65	.06	1.77	.10	.12

[1] $Z_{maj} = (I_{max}/c_{maj}) \times 1000$

[2] $Z_{min} = (I_{min}/c_{min}) \times 1000$

† $p<.10$ * $p<.05$ ** $p<.01$

Table 20b

Section Moduli, Tibia and Femur: Females

TIBIA

Cross Section	Property	Total Mean	Total SE	20-39 yrs Mean	20-39 yrs SE	40+ yrs Mean	40+ yrs SE	40+ yrs - 20-39 yrs
20%	Zmaj[1]	.63	.02	.64	.02	.61	.03	-.03
	Zmin[2]	.53	.02	.54	.02	.49	.03	-.05†
35%	Zmaj	.77	.02	.76	.03	.79	.04	.03
	Zmin	.49	.02	.48	.02	.50	.03	.02
50%	Zmaj	.93	.02	.90	.03	.98	.04	.08
	Zmin	.57	.02	.58	.02	.56	.03	-.02
65%	Zmaj	1.12	.03	1.13	.04	1.11	.06	-.02
	Zmin	.66	.02	.68	.02	.61	.03	-.07†
80%	Zmaj	1.36	.04	1.36	.04	1.37	.08	.01
	Zmin	.87	.02	.90	.03	.81	.04	-.09†

FEMUR

Cross Section	Property	Total Mean	Total SE	20-39 yrs Mean	20-39 yrs SE	40+ yrs Mean	40+ yrs SE	40+ yrs - 20-39 yrs
20%	Zmaj	1.29	.03	1.31	.04	1.25	.05	-.06
	Zmin	1.05	.03	1.09	.03	.98	.04	-.11*
35%	Zmaj	1.05	.03	1.04	.03	1.08	.04	.04
	Zmin	.97	.02	.96	.03	1.00	.04	.04
50%	Zmaj	1.11	.03	1.09	.03	1.16	.05	.07
	Zmin	.98	.02	.96	.03	1.04	.04	.08†
65%	Zmaj	1.20	.03	1.16	.03	1.29	.05	.13*
	Zmin	1.02	.03	.99	.03	1.10	.04	.11*
80%	Zmaj	1.38	.03	1.33	.03	1.46	.06	.13*
	Zmin	.98	.03	.94	.03	1.04	.05	.10†
Neck-CA	Zmaj	.52	.02	.50	.02	.57	.04	.07†
	Zmin	.42	.01	.44	.01	.38	.03	-.06*
Neck-TA	Zmaj	1.50	.06	1.38	.05	1.74	.12	.36**
	Zmin	1.21	.03	1.20	.04	1.23	.05	.03

[1] $Z_{maj} = (I_{max}/c_{maj}) \times 1000$

[2] $Z_{min} = (I_{max}/c_{maj}) \times 1000$

† $p<.10$ * $p<.05$ ** $p<.01$

However, while not changing the general patterns of variation in bending strength described earlier, division of principal moments by c_{maj} and c_{min} does tend to reduce difference in relative benuing strength between different cross section sites and between the directions of the major and minor axes. For example, values of I_{max} increase by more than three times from the distal to the proximal tibia (see Table 16a above). Values of Z_{maj} for these same cross sections increase by just over two times (Tables 20a, b). Similarly, in the femoral diaphysis, I_{max} varies between different cross sections by 31% (males) and 43% (females); Z_{maj} in the femoral diaphysis varies by only 16% (males) and 27% (females). The ratio of Z_{maj}/Z_{min} also shows less variation at different sites than I_{max}/I_{min}, and is invariably closer to 1.0, that is Z_{maj} and Z_{min} are always more equal in magnitude than I_{max} and I_{min}. As would be expected, the effect of dividing by a distance from the section centroid to the outermost fiber (c_{maj} and c_{min}) is greater in less circular cross sections. For example, in the <u>least</u> circular cross section--themale tibial 65% section--I_{max}/I_{min} averages 3.37, and Z_{maj}/Z_{min} 1.94, a 42% reduction, while in the <u>most</u> circular cross section--the female femoral 35% section--I_{max}/I_{min} averages 1.18, and Z_{maj}/Z_{min} 1.08, only an 8% reduction. Some other comparisons of these ratios include the femoral 80% cross section--I_{max}/I_{min} about 1.9 in both sexes, Z_{maj}/Z_{min} about 1.4--and the femoral neck cross section (both cortical shell and total subperiosteal area calculations)--I_{max}/I_{min} about 1.6-1.7, Z_{maj}/Z_{min} about 1.2-1.3.

These results are not unexpected, since with increasing values of I_{max} or I_{min}, c_{maj} and c_{min} also increase, leading to a "dampening" effect on differences between cross section sites or between different directions within the same cross section. What this means in terms of bending stress is that the maximum stress in the distant-most fibers of different cross sections or different planes of bending will be closer in magnitude than the total sum of stresses throughout the cross section in that plane of bending (i.e., as reflected in second moments of area). With regard to bending stress at the most distant fiber of cross section, then, and keeping in mind the cautions expressed above, the "maximum" bending strength of the tibia (Z_{maj}) may be said to be about twice as great in the proximal shaft than in the distal shaft and the "minimum" bending strength (Z_{min}) somewhat less than twice as great in the proximal as the distal shaft, reaching a minimum at the 35% section site (Tables 20a, b). Similar patterns of variation for "maximum" and "minimum" section moduli for the tibia have been reported by Lovejoy et al. (1976). The femoral shaft sections vary much less in these properties (Tables 20a, b). However, definite patterns of variation, such as the gradual decline in Z_{maj} in female femoral from the 80% to 35% sections, and the lack of any such decline in values of Z_{min} over the same cross sections, are clearly apparent, perhaps even more so than in comparisons of principal second moments of area at different sites (Tables 16a, b).

The tendency for equalization of bending stresses at the outermost fibers of different cross sections or planes of bending noted above, is also apparent in comparisons of section modulus values between the sexes. The mean male/female ration in I_{max} is 1.82 and 1.52, respectively, for the tibia and femur (diaphysis only), in Z_{maj}, 1.59 and

1.38. Similar reductions in male-female differences are also apparent in comparisons of average I_{min} and Z_{min} values: tibia--1.64 to 1.44; femur--1.50-1.36. The femoral neck shows a mean reduction in the sex ratio (average of cortical shell and total subperiosteal area calculations) of 1.55 to 1.42 (Z_{maj}) and 1.52 to 1.34 (Z_{min}). Following the same line of reasoning as that presented above, this means that maximum bending stressses in the outermost fibers of bone cross sections in the lower limb are more similar in males and females than the total sum of stresses throughout the cross section. In other words, if the magnitude of stress at the point of maximum stress in a cross section is taken as an index of bending strength, males and females are closer in bending strength than would be indicated by a comparison of second moments of area.

Age comparisons (between 20-39 year and 40+ year age groups) are also shown in Tables 20a, b. Again, as with site and sex variation in these properties, age changes Z_{maj} and Z_{min} generally parallel those observed earlier for I_{max} and I_{min}: a general increase with age (except in I_{min} and Z_{min} of female tibiae), most marked in the midshaft region of the tibia in malès and proximal femur in females. However, some modifications of these general patterns are also apparent. First, the age-related increase in Z_{maj} of mid-tibial sections in males is relatively greater than the corresponding increase in I_{max}, and the increase in Z_{min} of these sections relatively less than that of I_{min}. Thus, the Z_{maj}/Z_{min} ratio increases with age, particularly in the 50% and 65% sections. This was not found to be true of principal moments in male tibiae (see Table 17a above), where age increases were about equal or relatively greater in $\underline{I_{min}}$ in the middle of the shaft. In contrast to males, female tibiae show a relatively <u>smaller</u> increase with age in the Z_{maj}/Z_{min} ratio than in the I_{max}/I_{min} ratio (Table 17b). Over all five tibial sites, with aging males gain an average of .15 units in Z_{maj} and .07 units in Z_{min} (compared to 13% increases in both I_{max} and I_{min}); the corresponding figures for female Z_{maj} and Z_{min} are .01 and -.04 (loss) (compared to 5% and -2% in I_{max} and I_{min}). Thus, based on the section modulus criterion of bending strength (stress in the outermost fiber), male tibiae become relatively more adapted for A-P bending with age than female tibiae. The difference between male and female age increases in both section moduli are also greater than in the corresponding principal moments (see figures quoted above). This is probably due to the fact that females undergo more cortical thinning with aging than males; any increases with age in principal moments among females are thus due entirely to relatively greater subperiosteal expansion, i.e., remodeling of the (thinner) cortex to a position more distant overall from the section centroid. Males undergo some cortical thinning (see earlier section), but not as marked as in females, due to less endosteal resorption of bone with aging. Increase in subperiosteal area in males is also relatively smaller than in females, especially in the femur. Thus, division of principal moments by linear distance to the outermost fiber along that axis (i.e., derivation of section moduli) will tend to produce less of an effect on age increases in males than in females.

Femoral section moduli show the same effect of dividing c_{maj} and c_{min} on age differences in males and females. It was observed earlier that female femoral exhibit greater increases with age in principal moments on average than male femora (see Tables 18a, b). However, comparison of age changes in section moduli averaged over the five femoral diaphyseal sections reveal very little sex difference, and in fact a slight male advantage: increases with age in Z_{maj} average .09 units in males versus .06 units in females; increases in Z_{min} average .06 units in males and .04 units in females. Females still gain slightly more in Z_{maj} in the 65% and 80% sections, however, the sex difference here is not nearly so marked as in comparisons of I_{max} (see above). Thus, in terms of maximum stress in the outermost cross section fiber, age changes in bending strength of male and female femoral sections are fairly equal.

The only femoral site where consideration of distance to the outermost fiber (the section modulus) does not lead to a relative increase in male bending strength and a decrease in female bending strength is the femoral neck. Here, using both cortical shell and total subperiosteal area methods of calculation, but especially with regard to subperiosteal properties, females gain significantly more in Z_{maj} with aging than males, much more so than was observed in I_{max} sex/age comparisons. Females still show smaller gains (or greater losses) than males in Z_{min}, as was observed in I_{min}. However, this only adds to the increase in the Z_{maj}/Z_{min} ratio with age in females, especially using the subperiosteal area properties, where the ratio increases with age from 1.15 to 1.41. In males, the same ratio <u>decreases</u> with age from 1.29 to 1.24. Thus, in terms of maximum stress, only females show a relative increase with age in superior-inferior bending strength of the femoral neck (based on cross section geometry, and assuming no loss in bone density). This is particularly interesting in light of sexual dimorphism in loading of tne hip joint, as described briefly above and discussed further in the next chapter. If section moduli are used as the criterion of bending strength, a clear pattern of decline in female gains relative to male gains from the proximal to distal end of the lower limb is apparent: females gain relatively most in bending strength in the femoral neck, followed by the proximal femoral diaphysis, after which male gains in section moduli are always greater, especially in the mid-tibia (Tables 20a, b).

Right and left side values for section moduli were also compared. The results of these comparisons are very similar to those of the principal moments (see Table 15 above), so are not shown here. As with second moments of area, section moduli reflect a general left "dominance" in lower limb, except in male femora. Interestingly, however, the relative "dominance" of the left side Z_{maj} values in the tibia is greater in males (.08 units on average) than in females (.04 units), unlike I_{max}, which averaged 6% larger on the left side in both male and female tibiae. Thus, according to the section modulus criterion, male tibiae are relatively more asymmetric in bending strength than female tibiae. Sex/side differences in femoral section moduli are quite similar to corresponding differences in principal moments, however: females average .13 units greater in Z_{maj} on the left side (compared to 13% greater I_{max} values), peaking in the 35% and 50% sections, while males show almost no average side differences in Z_{maj} (.01 units larger on the right side, compared to 1.5% larger right side values of

I_{max}). Z_{min} side differences in the femoral diaphysis are also very similar to those in I_{min}: females .09 units larger left side (11.5% larger in I_{min}), and males .01 units larger left side (1.6% larger in I_{min}). The femoral neck shows no side asymmetry in cortical shell section moduli in either sex, with the only significant side difference occurring in male Z_{maj} values calculated from the total subperiosteal area, where the left side is .11 units larger (versus 2.9% larger in I_{max} values). The Z_{maj}/Z_{min} ratio for the femoral neck (either method of calculation) is about the same in males as the I_{max}/I_{min} ratio--right side somewhat greater--but differs in females from the principal moment ratio, averaging slightly greater on the left side (compared with I_{max}/I_{min}, which is slightly greater in right female femora). Thus, left "dominance" in terms of minimizing the maximum stress in the principal (superior-inferior) plane of bending, is present among females in the femoral neck as well as in the femoral diaphysis.

BONE MINERAL MEASUREMENTS

General Issues

In the preceding sections of this chapter, bone strength has been assessed as a function of cross-sectional geometric variation. However, bone strength under loading depends upon a number of factors other than geometrical configuration of compact cortical bone, including bone material properties (density, modulus of elasticity), chemical composition, microstructure and ultrastructure, and trabecular morphology in regions of cancellous bone (see chapter 2). Of these other factors, bone material properties are probably the most critical in assessing changes in bone strength with aging, since chemical composition of bone remains essentially unchanged during aging (Mueller et al., 1966), and microstructural and ultrastructural changes are largely reflected in variation in bone density (Carter et al., 1976).

The modulus of elasticity (E), which describes the relative stiffness of a material under loading (stress/strain) prior to plastic deformation, is critical in evaluating material strength (Reilly and Burstein, 1974). However, it has been demonstrated that a direct relationship between bone density and E over a wide range of density values exists (Carter and Hayes, 1977). Also, a direct relationship between bone density and compressive strength, whereby strength varies by the squared power of density, has also been empirically demonstrated (Carter and Hayes, 1976). Thus, differences in bone density should provide a good approximation of relative differences in bone strength due to variation in material properties.

As documented in chapter 2, variation in bone "density" has been assessed in several different ways, including dividing whole bone weights by total external bone volumes (Trotter et al., 1959), measurement of radiographic density (Meema and Meema, 1965), use of photon absorptiometry (Johnston et al., 1968), and direct measurement of weight and volume of cut bone sections (Atkinson and Weatherall, 1967). In fact, only the last of these methods--direct measurment of cut bone samples--actually measures true tissue density. Division of intact bone weights by total subperiosteal volumes does not take into account changes at the endosteal surface of long bones, i.e., endosteal resorption, and thus may actually reflect volumetric changes (cortical thinning) rather than true tissue density differences (see below). Radiographically determined bone densities are subject to a number of inaccuracies (Cameron and Sorenson, 1963), and are not sensitive to small changes in bone density. Photon absorptiometry (ibid.) has been shown to be more accurate and reproducible as a means of measuring bone mineral content (BMC), but is limited in that it provides only a summative measure of BMC through a total scan cross section, with no account taken of variability in bone volume. In an effort to standardize BMC readings by an index of bone size (i.e., volume), a bone mineral index, BMI, may be derived by dividing BMC by bone subperiosteal width across the scan path (BW), which can also be accurately determined from bone mineral scanner data (Cameron and Sorenson, 1968). However, like the total bone weight/bone volume method, this still cannot control for changes occurring at the endosteal surface, or for deviations in cross-sectional shape from a cylindrical model. Thus, differences in BMI may again be largely determined by volumetric rather than true tissue density variation.

In the present study, as described in chapter 4, an ^{125}I Bone Mineral Analyzer was used to determine BMC and BMI indices at all 11 cross section sites in 40 individuals (10 of each sex 20-39 years and 40+ years of age), and at the femoral midshaft and neck sections in all 119 individuals. Because precise bone cross-sectional volumes (CA) were also later determined at these same sections using SLICE, theoretically it should be possible to control for differences in bone cortical volume when comparing bone mineral analyzer results, i.e., by dividing BMC by CA. In the following presentations, BMC/CA and BMC/TA (a kind of "total subperiosteal area density" similar to Trotter et al.'s density index) are compared with traditional photon absorptiometric indices to test the relative effectiveness of BMC and BMI in assessing sex, age, and side variation in bone density.

In addition to these data, directly measured wet bone densities for bone cores removed from the femoral midshaft are also available for 39 individuals (see below). Thus, three different methods of assessing bone density: bone mineral analyzer data, bone mineral analyzer data controlled for volumetric variation using SLICE results, and direct estimates of true tissue density at one site, may be compared in these individuals.

Bone Mineral Analyzer Data

Sex differences.

Sex differences in BMC, BW, and BMI (BMC/BW) for the 11 cross sections measured in 40 individuals are shown in Tables 21a,b. In both bones, BMC averages higher values among males, with greater sexual dimorphism in the tibia (mean male/females ratio=1.42) than in the femur (mean male/females ratio=1.30, not including the femoral neck). BW is also, expectedly, greater among males, with the sex ratio averaging 1.14 in the tibia and 1.10 in the femoral diaphysis. (It should be noted that, with the exception of the femoral neck, these refer to mediolateral total subperiosteal breadths, since bone mineral scans were taken across the M-L diameter; in the femoral neck, a superior-inferior scan was taken--see chapter 4). Because sexual dimorphism in BMC is greater than in BW, the bone mineral index ratio, BMI, is also higher in males, again more marked in the tibia (mean difference=.23 units) than in the femur (mean difference=.19 units).

This last finding might imply greater bone "density" in male tibiae and femora in this sample, except that, as noted above, differences in bone volume as well as tissue density may account for variation in BMI. To test this, sex differences in BMC were compared with sex differences in CA at the same sites (see Tables 8a,b above). Before carrying out this comparison, however, it was first determined whether the subsample of 40 individuals with all 11 bone mineral scans is representative of the sample as a whole, since the total sample (n=119) was used to derive the bone areas shown in

Table 21a

Sex Differences in Bone Mineral Analyzer Data: Tibia

BMC (gm/cm)

Cross Section	Male[1] Mean	Male[1] SE	Female[2] Mean	Female[2] SE	Male - Female[3]	Male/Female X 100
20%	2.25	.06	1.64	.09	.64	137.2
35%	2.77	.07	1.95	.08	.82	142.0
50%	3.02	.08	2.15	.09	.87	140.5
65%	3.06	.07	2.11	.08	.95	145.0
80%	3.09	.09	2.14	.10	.95	144.4

BW (cm)

Cross Section	Male Mean	Male SE	Female Mean	Female SE	Male - Female[3]	Male/Female X 100
20%	2.52	.04	2.24	.04	.28	112.5
35%	2.26	.03	2.02	.03	.24	111.9
50%	2.32	.04	2.03	.03	.27	114.3
65%	2.37	.03	2.08	.03	.29	113.9
80%	2.97	.04	2.58	.04	.39	115.1

BMI (gm/cm^2)

Cross Section	Male Mean	Male SE	Female Mean	Female SE	Male - Female[3]	Male/Female X 100
20%	.89	.02	.73	.03	.16	--
35%	1.23	.03	.97	.04	.26	--
50%	1.31	.03	1.06	.04	.25	--
65%	1.29	.03	1.02	.04	.27	--
80%	1.04	.03	.84	.04	.20	--

[1] n = 20

[2] n = 20

[3] All male-female differences significant at .001 level.

Table 21b

Sex Differences in Bone Mineral Analyzer Data: Femur

BMC (gm/cm)

Cross Section	Male[1] Mean	Male[1] SE	Female[2] Mean	Female[2] SE	Male − Female[3]	Male/Female X 100
20%	3.14	.10	2.41	.10	.73	130.3
35%	3.41	.08	2.53	.08	.88	134.8
50%	3.66	.08	2.84	.09	.82	128.9
65%	3.79	.07	3.00	.09	.79	126.3
80%	3.70	.08	2.91	.09	.79	127.1
Neck	2.36	.07	1.90	.09	.46	124.2

BW (cm)

Cross Section	Male Mean	Male SE	Female Mean	Female SE	Male − Female[3]	Male/Female X 100
20%	3.48	.07	3.19	.06	.29	109.1
35%	2.67	.04	2.45	.03	.22	109.0
50%	2.57	.03	2.38	.03	.19	108.0
65%	2.72	.04	2.46	.04	.26	110.6
80%	2.93	.03	2.64	.04	.29	111.0
Neck	3.07	.04	2.72	.04	.35	112.9

BMI (gm/cm^2)

Cross Section	Male Mean	Male SE	Female Mean	Female SE	Male − Female[3]	Male/Female X 100
20%	.91	.03	.76	.04	.15	--
35%	1.28	.03	1.03	.03	.25	--
50%	1.43	.02	1.20	.04	.23	--
65%	1.39	.02	1.22	.04	.17	--
80%	1.26	.02	1.11	.04	.15	--
Neck	.78	.02	.70	.04	.08	--

[1] n = 20 [2] n = 20

[3] All male-female differences significant at .001 level, except BMI 20% and 80% sections, significant at .01 level.

Tables 8a,b. Although it is impossible to determine the exact degree to which this subgroup represents a random sample of bone mineral measurements in the total study sample, a general idea of its representativeness may be derived by comparing means calculated from the subsample (n=40) with means for the total sample (n=119) at sites where all individuals in the total sample were scanned, i.e, the femoral midshaft and femoral neck. Table 22 presents bone mineral analyzer data for these sections in the total sample. Comparing Tables 21b and 22, it is apparent that the 40 individual subsample is slightly atypical of the sample as a whole in terms of overall "size", specifically, males are slightly larger on average than in the total sample. This is particularly apparent in the femoral midshaft section, where BMC averages 5.2% greater in the male subsample than in the total male sample ($p<.10$). CA and bone length are also somewhat greater in the male subgroup (data not shown). In contrast, female BMC in the smaller sample averages about the same as in the total sample (this is also true for CA and bone length). Therefore, as an approximate "correction" factor for sampling error in the smaller subsample, 5% was subtracted from the mean male values for BMC at each cross section site in the subsample, and sex ratios again calculated. The resulting average sex ratios in BMC--1.35 for the tibia, 1.23 for the femoral diaphysis--are almost precisely the same as the average sex ratios in CA for these bones--1.36 for the tibia, 1.23 for the femur (see earlier results). In addition, variation in sexual dimorphism in BMC at different sites within the same bone parallels site variation in CA (see Tables 8a,b).

Table 22

Sex and Age Differences in Bone Mineral Analyzer Data for Total Sample (N=119) at Femoral Midshaft and Iemoral Neck Cross Sections

SEX DIFFERENCES

Cross Section	Property	Male Mean	Male SE	Female Mean	Female SE	Male - Female[1]	$\frac{M-F}{M}$X100
50%	BMC	3.48	.05	2.86	.04	.62	121.7
	BW	2.50	.02	2.31	.02	.19	108.2
	BMI	1.39	.02	1.24	.02	.15	--
Neck	BMC	2.33	.04	1.90	.04	.43	122.6
	BW	3.00	.02	2.71	.02	.29	110.7
	BMI	.77	.01	.70	.02	.07	--

AGE DIFFERENCES--MALES

Cross Section	Property	20-39 yrs Mean	20-39 yrs SE	40+ yrs Mean	40+ yrs SE	40+ yrs - 20-39 yrs	%dif
50%	BMC	3.45	.07	3.52	.06	.07	2.0
	BW	2.47	.03	2.56	.02	.09*	3.6
	BMI	1.40	.02	1.38	.02	-.02	--
Neck	BMC	2.39	.05	2.21	.07	-.18*	-7.5
	BW	2.99	.05	3.07	.05	.09†	3.0
	BMI	.80	.02	.72	.02	-.08**	--

AGE DIFFERENCES--FEMALES

Cross Section	Property	20-39 yrs Mean	20-39 yrs SE	40+ yrs Mean	40+ yrs SE	40+ yrs - 20-39 yrs	%dif
50%	BMC	2.91	.05	2.75	.08	-.16†	5.5
	BW	2.26	.02	2.41	.04	.15***	6.6
	BMI	1.29	.02	1.14	.03	-.15***	--
Neck	BMC	1.96	.05	1.77	.06	-.19*	-9.7
	BW	2.69	.03	2.75	.04	.06	2.2
	BMI	.73	.02	.65	.03	-.08**	--

[1] All male-female differences significant at .001 level.

† p<.10   * p<.05   ** p<.01   *** p<.001

Thus, the observed sexual dimorphism in BMC (a mineral mass measurement) is almost completely accounted for by sexual dimophism in CA (a volume measurement). Consequently, since bone mineral density = bone mineral mass/bone volume, and the differences between the sexes in mass and volume are about the same, no sex-related difference in bone mineral density can be demonstrated for this sample.

This is shown even more clearly by comparing BMC/CA ratios in males and females (all data for the 40 individual subsample), presented in Tables 23a,b. With the exception of the femoral neck, no BMC/CA ratio is significantly ($p<.05$) different in the two sexes. BMC/TA ratios are also presented here for comparison with whole bone weight/bone volume determinations made by other authors (see above). These also show no significant sex-related differences. It should be noted that BMC/CA ratios overestimate actual tissue density in the femoral neck due to the presence of trabecular bone, which is included in BMC measurement but not in CA (the same is true to a lesser extent in the femoral 20% and tibial 80% sections). The significantly greater BMC/CA ratio for the female femoral neck in this (n=40) sample may be due to sampling error, since no significant sex difference in this index exists in the total sample (n=119). Table 24 presents BMC/CA and BMC/TA ratios for the femoral midshaft and neck sections in the total sample, analyzed by sex and age. The sex comparisons shown here bear out the general conclusion of no sex-related differences in these density indices derived from the smaller subsample.

The problems inherent in using BMI to estimate bone "density"

Table 23a

Sex Differences in BMC/CA and BMC/TA: Tibia

BMC/CA

Cross Section	Male[1] Mean	SE	Female[2] Mean	SE	Male - Female
20%	1.18	.03	1.24	.06	-.06
35%	1.06	.02	1.07	.02	-.01
50%	1.01	.02	1.06	.02	-.05†
65%	.99	.02	1.05	.03	-.06
80%	1.11	.03	1.12	.03	-.01

BMC/TA

Cross Section	Male Mean	SE	Female Mean	SE	Male - Female
20%	.48	.02	.47	.02	.01
35%	.67	.02	.62	.03	.05
50%	.65	.02	.63	.03	.02
65%	.55	.01	.53	.02	.02
80%	.40	.01	.39	.03	.01

[1] $n = 20$

[2] $n = 20$

† $p<.10$

Table 23b

Sex Differences in BMC/CA and BMC/TA: Femur

BMC/CA

Cross Section	Male [1] Mean	SE	Female [2] Mean	SE	Male - Female
20%	1.19	.03	1.17	.04	.02
35%	1.05	.02	1.03	.01	.02
50%	1.00	.01	.99	.02	.01
65%	.99	.01	1.03	.03	-.04
80%	1.04	.02	1.08	.03	-.04
Neck	2.02	.05	2.27	.06	-.25**

BMC/TA

Cross Section	Male Mean	SE	Female Mean	SE	Male - Female
20%	.41	.02	.40	.02	.01
35%	.60	.01	.57	.02	.03
50%	.70	.01	.68	.02	.02
65%	.70	.01	.68	.03	.02
80%	.63	.01	.63	.03	0
Neck	.40	.01	.41	.02	-.01

[1] n = 20

[2] n = 20

** p<.01

Table 24

Sex and Age Differences in BMC/CA and BMC/TA for Total Sample (N=119) at Femoral Midshaft and Femoral Neck Cross Sections

SEX DIFFERENCES

Cross Section	Property	Male Mean	Male SE	Female Mean	Female SE	Male - Female
50%	BMC/CA	.97	.01	.98	.01	-.01
	BMC/TA	.70	.01	.70	.01	0
Neck	BMC/CA	2.04	.04	2.00	.04	.04
	BMC/TA	.40	.01	.42	.01	-.02

AGE DIFFERENCES--MALES

Cross Section	Property	20-39 yrs Mean	20-39 yrs SE	40+ yrs Mean	40+ yrs SE	40+ yrs - 20-39 yrs
50%	BMC/CA	.96	.01	.98	.01	.02
	BMC/TA	.71	.01	.68	.02	-.03
Neck	BMC/CA	2.07	.05	1.98	.07	-.09
	BMC/TA	.42	.01	.37	.01	-.05***

AGE DIFFERENCES--FEMALES

Cross Section	Property	20-39 yrs Mean	20-39 yrs SE	40+ yrs Mean	40+ yrs SE	40+ yrs - 20-39 yrs
50%	BMC/CA	.98	.01	.99	.02	.01
	BMC/TA	.73	.01	.62	.02	-.11***
Neck	BMC/CA	2.01	.05	1.98	.08	-.03
	BMC/TA	.43	.01	.38	.02	-.05**

** $p<.01$ *** $p<.001$

in irregularly shaped cross sections are made even more apparent by examining relative degrees of sexual dimorphism in BMI at different cross section sites (Tables 21a,b). Since BMI is derived by dividing BMC by bone subperiosteal breadth across the scan path, departures from a cylindrical cross section which distribute relatively more bone volume along the scan path (i.e., in this study, A-P flattening or M-L broadening) will tend to decrease the index by increasing BW, while for the same reasons a disproportionate distribution of bone volume in a direction perpendicular to the scan path will increase the index. For example, BMI values in the tibia rise from the distal end to the midshaft section, a result of both an increase in BMC and a reduction in BW due to the progressively more flattened cross-sectional shape in the midshaft region (see earlier results). However, BMC/CA values in these same sections show no such rise, and even a decline from the 20% to 35% sections (this decline may be partially caused by the presence of trabecular bone at the 20% site--see above). In the femoral diaphysis, the BMI index is lowest in the 20% and 80% sections, both relatively broad across the M-L scan path, and higher in the middle regions of the diaphysis, where cortical volume is more A-P distributed; again BMC/CA shows no such pattern of variation. Where sex differences in cross-sectional shape occur (see Tables 16a,b above), these are also reflected in BMI sexual dimorphism. For example, relatively greater sexual dimorphism in BMI in the middle three tibial sections is brought about by more A-P distributed ("platycnemic") cortical bone volume in male tibiae in this region (see earlier results), not by denser bone tissue in males (Table 23a). Thus, serious misinterpretations may result from using BMI to estimate sex and/or site differences in bone "density", if **concomitant** changes in bone shape are not taken into consideration.

Age differences.

Age differences in BMC, BW, and BMI are shown in Tables 25a,b and 26a,b, and age differences in BMC/CA and BMC/TA in Tables 27a,b and 28a,b. Both BMC and BMI show significant decreases with age in females in both bones, and nonsignificant changes in males. As with sex-related variation, however, these age changes appear to be related to bone volume changes rather than mineral density changes. This is indicated by the analyses of age variation in BMC/CA in Tables 26a,b and 27a,b, where only one differences between 20-39 year and 40+ year age groups reaches statistical significance (a gain in BMC/CA in the tibia 65% section in females). Thus, while females exhibit an average loss with age in BMI, this change merely parallels the age loss in CA among females (see Tables 9b, 10b above), with no age changes in bone mineral mass standardized for volumetric changes (i.e., no mineral density changes).

It is interesting that the BMC/TA index, analogous to the Trotter et al. "whole bone density" measurements described above, shows about a 15-20% reduction with age among females, similar to the "density" declines documented by Trotter. (It is also interesting that in her White female group, femora showed less change in "density" than tibiae or other long bones--see Chart I, Trotter et al., 1960: 53. Smaller losses in BMC/TA in the femur than in the tibia were also observed

Table 25a

Age Differences in Tibial Bone Mineral Analyzer Data: Males

BMC (gm/cm)

Cross Section	20-39 yrs[1] Mean	SE	40+ yrs[2] Mean	SE	40+ yrs - 20-39 yrs	%dif
20%	2.18	.05	2.31	.11	.13	6.0
35%	2.75	.10	2.79	.10	.04	1.4
50%	3.04	.10	3.01	.14	-.03	-1.0
65%	3.02	.09	3.10	.12	.08	2.6
80%	3.00	.08	3.18	.16	.18	6.0

BW (cm)

Cross Section	20-39 yrs Mean	SE	40+ yrs Mean	SE	40+ yrs - 20-39 yrs	%dif
20%	2.45	.06	2.59	.03	.14†	5.7
35%	2.23	.05	2.28	.04	.05	2.2
50%	2.31	.06	2.33	.06	.02	.9
65%	2.37	.05	2.38	.04	.01	.4
80%	2.98	.06	2.97	.06	-.01	-.3

BMI (gm/cm^2)

Cross Section	20-39 yrs Mean	SE	40+ yrs Mean	SE	40+ yrs - 20-39 yrs	%dif
20%	.89	.02	.89	.04	0	--
35%	1.24	.04	1.22	.04	-.02	--
50%	1.32	.05	1.29	.05	-.03	--
65%	1.28	.04	1.30	.05	.02	--
80%	1.01	.03	1.07	.05	.06	--

[1] n = 10

[2] n = 10

† $p<.10$

Table 25b

Age Differences in Tibial Bone Mineral Analyzer Data: Females

BMC (gm/cm)

Cross Section	20-39 yrs[1] Mean	SE	40+ yrs[2] Mean	SE	40+ yrs - 20-39 yrs	%dif
20%	1.92	.12	1.37	.08	-.55***	-28.6
35%	2.21	.06	1.70	.10	-.51***	-23.1
50%	2.41	.07	1.89	.12	-.52**	-21.6
65%	2.35	.07	1.87	.11	-.48**	-20.4
80%	2.42	.11	1.87	.12	-.55**	-22.7

BW (cm)

Cross Section	20-39 yrs Mean	SE	40+ yrs Mean	SE	40+ yrs - 20-39 yrs	%dif
20%	2.25	.07	2.22	.06	-.03	-1.3
35%	2.02	.04	2.01	.05	-.01	-.5
50%	2.02	.05	2.03	.04	.01	.5
65%	2.08	.04	2.09	.05	.01	.5
80%	2.57	.04	2.59	.07	.02	.8

BMI (gm/cm^2)

Cross Section	20-39 yrs Mean	SE	40+ yrs Mean	SE	40+ yrs - 20-39 yrs	%dif
20%	.85	.04	.61	.03	-.24**	--
35%	1.09	.03	.84	.04	-.25***	--
50%	1.19	.03	.93	.04	-.26***	--
65%	1.14	.03	.90	.05	-.24***	--
80%	.94	.04	.73	.05	-.21**	--

[1] n = 10

[2] n = 10

** $p<.01$ *** $p<.001$

Table 26a

Age Differences in Femoral Bone Mineral Analyzer Data: Males

BMC (gm/cm)

Cross Section	20-39 yrs[1] Mean	SE	40+ yrs[2] Mean	SE	40+ yrs - 20-39 yrs	%dif
20%	3.20	.14	3.09	.15	-.11	-3.4
35%	3.46	.12	3.36	.12	-.10	-2.9
50%	3.75	.12	3.57	.09	-.18	-4.8
65%	3.87	.11	3.70	.09	-.17	-4.4
80%	3.79	.12	3.60	.12	-.19	-5.0
Neck	2.46	.09	2.27	.10	-.19	-7.7

BW (cm)

Cross Section	20-39 yrs Mean	SE	40+ yrs Mean	SE	40+ yrs - 20-39 yrs	%dif
20%	3.52	.11	3.43	.10	-.09	-2.6
35%	2.71	.06	2.63	.05	-.08	-3.0
50%	2.59	.05	2.55	.03	-.04	-1.5
65%	2.76	.06	2.69	.04	-.07	-2.5
80%	2.96	.05	2.91	.04	-.05	-1.7
Neck	2.99	.05	3.07	.05	.08	2.7

BMI (gm/cm^2)

Cross Section	20-39 yrs Mean	SE	40+ yrs Mean	SE	40+ yrs - 20-39 yrs	%dif
20%	.92	.05	.91	.05	.01	--
35%	1.28	.04	1.28	.05	0	--
50%	1.45	.03	1.40	.04	-.05	--
65%	1.40	.02	1.38	.03	-.02	--
80%	1.28	.03	1.23	.03	-.05	--
Neck	.83	.03	.74	.03	-.11[†]	--

[1] n = 10 [2] n = 10

[†] $p<.10$

Table 26b

Age Differences in Femoral Bone Mineral Analyzer Data: Females

BMC (gm/cm)

Cross Section	20-39 yrs[1] Mean	SE	40+ yrs[2] Mean	SE	40+ yrs - 20-39 yrs	%dif
20%	2.65	.11	2.18	.13	-.47**	-17.7
35%	2.72	.08	2.34	.11	-.38**	-14.0
50%	3.06	.09	2.62	.12	-.44**	-14.4
65%	3.19	.10	2.81	.13	-.38*	-11.9
80%	3.13	.10	2.69	.12	-.44**	-14.0
Neck	2.14	.10	1.67	10	-.47**	-22.0

BW (cm)

Cross Section	20-39 yrs Mean	SE	40+ yrs Mean	SE	40+ yrs - 20-39 yrs	%dif
20%	3.25	.08	3.13	.09	-.12	-3.7
35%	2.44	.04	2.46	.05	.02	.8
50%	2.34	.03	2.42	.05	.08	3.4
65%	2.40	.04	2.52	.05	.12†	5.0
80%	2.53	.04	2.75	.06	.22**	8.7
Neck	2.69	.05	2.75	.07	.06	2.2

BMI (gm/cm^2)

Cross Section	20-39 yrs Mean	SE	40+ yrs Mean	SE	40+ yrs - 20-39 yrs	%dif
20%	.82	.04	.71	.05	-.11†	--
35%	1.12	.03	.95	.05	-.17**	--
50%	1.30	.03	1.09	.05	-.21***	--
65%	1.33	.03	1.12	.05	-.21**	--
80%	1.24	.03	.98	.05	-.26***	--
Neck	.80	.04	.61	.05	-.19**	--

[1] n = 10 [2] n = 10

† p<.10 * p<.05 ** p<.01 *** p<.001

Table 27a

Age Differences in Tibial BMC/CA and BMC/TA: Males

BMC/CA

Cross Section	20-39 yrs [1] Mean	SE	40+ yrs [2] Mean	SE	40+ yrs - 20-39 yrs
20%	1.17	.04	1.19	.04	.02
35%	1.06	.02	1.06	.02	0
50%	1.04	.02	.99	.06	-.05
65%	.97	.03	1.00	.03	.03
80%	1.08	.04	1.13	.04	.05

BMC/TA

Cross Section	20-39 yrs Mean	SE	40+ yrs Mean	SE	40+ yrs - 20-39 yrs
20%	.50	.02	.47	.02	-.03
35%	.70	.02	.63	.02	-.07*
50%	.69	.01	.61	.03	-.08*
65%	.57	.01	.53	.02	-.04†
80%	.40	.02	.40	.02	0

[1] n = 10

[2] n = 10

† $p<.10$ * $p<.05$

Table 27b

Age Differences in Tibial BMC/CA and BMC/TA: Females

BMC/CA

Cross Section	20-39 yrs[1] Mean	SE	40+ yrs[2] Mean	SE	40+ yrs - 20-39 yrs
20%	1.31	.11	1.16	.06	-.15
35%	1.08	.03	1.06	.03	-.02
50%	1.04	.02	1.09	.04	.05
65%	.99	.04	1.11	.05	.12*
80%	1.11	.05	1.12	.04	.01

BMC/TA

Cross Section	20-39 yrs Mean	SE	40+ yrs Mean	SE	40+ yrs - 20-39 yrs
20%	.55	.03	.38	.02	-.17***
35%	.73	.03	.52	.03	-.21***
50%	.72	.03	.54	.03	-.18***
65%	.60	.03	.47	.03	-.13**
80%	.45	.03	.34	.03	-.11*

[1] n = 10

[2] n = 10

* p<.05 ** p<.01 *** p<.001

Table 28a

Age Differences in Femoral BMC/CA and BMC/TA: Males

BMC/CA

Cross Section	20-39 yrs[1] Mean	SE	40+ yrs[2] Mean	SE	40+ yrs - 20-39 yrs
20%	1.18	.04	1.22	.06	.04
35%	1.05	.02	1.05	.02	0
50%	1.00	.02	.99	.02	-.01
65%	.99	.02	1.00	.01	.01
80%	1.02	.02	1.06	.02	.04
Neck	2.02	.07	2.03	.09	.01

BMC/TA

Cross Section	20-39 yrs Mean	SE	40+ yrs Mean	SE	40+ yrs - 20-39 yrs
20%	.42	.02	.39	.02	-.03
35%	.61	.02	.59	.02	-.02
50%	.73	.01	.67	.02	-.06**
65%	.72	.01	.68	.01	-.04*
80%	.65	.02	.37	.01	-.04**
Neck	.43	.02	.37	.01	-.06**

[1] n = 10

[2] n = 10

* p<.05 ** p<.01

Table 28b

Age Differences in Femoral BMC/CA and BMC/TA: Females

BMC/CA

Cross Section	20-39 yrs[1] Mean	SE	40+ yrs[2] Mean	SE	40+ yrs - 20-39 yrs
20%	1.14	.07	1.19	.02	.05
35%	1.02	.02	1.05	.02	.03
50%	.96	.02	1.02	.03	.06†
65%	1.00	.02	1.05	.05	.05
80%	1.07	.03	1.09	.05	.02
Neck	2.32	.09	2.22	.09	-.10

BMC/TA

Cross Section	20-39 yrs Mean	SE	40+ yrs Mean	SE	40+ yrs - 20-39 yrs
20%	.44	.08	.36	.03	-.08*
35%	.62	.03	.52	.03	-.10*
50%	.73	.02	.62	.04	-.11*
65%	.75	.02	.61	.04	-.14**
80%	.70	.02	.56	.04	-.14**
Neck	.46	.02	.35	.03	-.09**

[1] n = 10

[2] n = 10

† $p<.10$ * $p<.05$ ** $p<.01$

among females in the present study, due to smaller losses of cortical bone volume in the female femur--see Tables 9a,b,10a,b). Males also show a small (about 5-10%) decline in BMC/TA. These declines in both sexes in this index are due to both cortical thinning and subperiosteal expansion with aging, i.e., decline in %CA (see Figures 19a,b above) (subperiosteal expansion will also create a greater decline in BMC/BW, or BMI).

The data from the femoral midshaft and neck for the total study sample, shown in Table 24 bear out these general observations on the smaller sample--essentially no change with age in BMC/CA in either sex (possibly a slight but nonsignificant decline in the femoral neck) and a decline in BMC/TA due to the geometric changes noted above. Again, it should be noted that use of BMI to estimate bone "density" changes with age in these sections is not only misleading in a general sense (i.e., by indicating an overall loss in "density" with age), but also misrepresents relative age changes in "density" at different sites undergoing different kinds of geometric age change. For example, the female femur shows relatively greater declines in BMI with age than might have been expected (Table 26b), given the smaller age losses in BMC (and CA) in this bone relative to the tibia. However, the female femur and tibia undergo different kinds of geometric change with age, with cross-sectional volume in the femur becoming relatively more M-L distributed, and in the tibia more A-P distributed (see Tables 17b, 18b above). Given mediolateral bone mineral scans, for the reasons stated earlier, this will lead to an inflation of tibial BMI values in the older female age group, and an underestimate of "density" losses in the tibia, while having exactly the opposite effect on female femora, leading to an overestimate of "density" losses with age in this bone.

No marked site-specific age changes in BMC/CA in either bone or sex are apparent in Tables 27a,b or 28a,b. The only possible site-related differences are in the female tibia, which shows a (nonsignificant) decline in this index in the distal two sections, and an increase in the midshaft and proximal sections (significant in the 65% section). However, since the sample sizes are quite small here (n=10 for each sex/age group), these sites differences may be attributable to sampling error, and in any case are relatively minor compared to site-specific differences in geometric changes with age (see earlier sections).

Analyses of age variation in smaller age groupings were also carried out for bone mineral analyzer data. Because of the small sample sizes available for all 11 sections (n=20 for each sex), individuals in this subsample were grouped into 3 age groups--20-29, 30-39, and 40+ years (with n's of 5, 5, and 10, respectively)--rather than 6 age groups for ANOVA tests. The femoral midshaft and neck sections, where data for all 119 individuals were available, were also analyzed using the 6 age group divisions. The results of these analyses confirm the observations presented above: no significant age-related change in BMC/CA occurs in either sex, i.e., bone mineral density does not vary significantly between five and ten-year age categories from the third through six decades. Any suggestion of an age trend in this index indicates, as in the two age group analyses (see Tables 27a,b; 28a,b), a slight increase in BMC/CA with aging, although inconsistent and never more than a few percent.

Age variation in BMC closely parallels that in CA, rising from the third to fourth decades, then falling slightly in males and more markedly in females. Again, this demonstrates the dependence of BMC on volumetric changes, and not necessarily on density changes with age.

Side differences.

Side differences in the various bone mineral properties discussed in this section were also investigated in both the smaller sample for all 11 cross sections sites, and in the full sample for the femoral midshaft and neck sections. The results of these comparisons are briefly summarized here.

Interestingly, side-related variation in BMC does not always parallel or equal side differences in CA (see Tables 11a,b above). Cortical bone area was found to be larger on the left side in general, more marked among females, especially in the distal femur. BMC shows a variable pattern of side "dominance" in the tibia, but averages a slight (1-2%) right "dominance" in both sexes, while the femur shows a left "dominance" in BMC, especially among females, where it averages almost 5% (which is still only about half the left "dominance" in female CA in this bone). BMC/CA shows a clear right "dominance" in male tibiae and female femora (averaging 6% in both), and a varaible pattern in female tibiae and male femora.

Analysis of side differences in principal moments (Tables 19a,b above) showed males to be somewhat more left "dominant" in terms of geometrical bending strength in the tibia, and females much more left "dominant" in bending strength in the femur. In other words, in comparisons between sides, the geometrically weakest bones (right tibia in males, right femur in females) are also those which show higher bone mineral densities (see above). Thus, there is at least some evidence for a negative relationship between bone material and geometric properties, with density and cross-sectional geometry possibly partially offsetting or "compensating" each other in terms of strength. However, it should be noted that these observations are based on only small samples (n=10 for each sex/side group), using unpaired data, so that these conclusions must remain very tentative until further data can be collected (for example, the full (n=119) sample side comparisons for the femoral midshaft show only a 3% right "dominance" in BMC/CA, compared to a 6% difference in the samller sample--see also results of bone core analyses below). The femoral neck shows a 5% left "dominance" in BMC/CA in females (using the full sample data), and a 5% right "dominance" in BMC/CA among males. Neither sex shows any side difference in BMC/TA for the femoral neck.

Bone Core Densities

As noted earlier (chapter 4), bone cores 4 mm in diameter were removed from the anterior quadrant of the femoral diaphysis just proximal to the 50% cross section location in 39 individuals. These were used to calculate wet bone tissue density by grinding the two ends of the core down to yield approximately "perfect" cylinders, weighing wet (stored in 10% formalin), and dividing bone weight by calculated core volume ($\pi r^2 h$) (Thompson, 1980).[22] Of the 39 core densities available, 21 are from males, 18 from females. Both sexes are about equally divided between younger (20-39 year) and older (40+ year) adults: n=11 and 10, respectively, for males, and n=9 and 9 for females in the two age groups. Because of the limited size of the sample, 6 age group analyses were not carried out, however, 3 age group differences were tested with the sample divided into 20-29, 30-39, and 40+ year categories (with n's of 4, 7, and 10, respectively, for males, and 5, 4, and 9 for females). Side differences were also tested , with n's for right and left sides of 12 and 9 for males, and 10 and 8 for females, respectively. Thus, the sample sizes by sex/age group and sex/side for these measurements are small, but should still provide useful independently derived data on bone density to compare with the bone mineral data presented above.[23]

Results of sex, age, and side comparisons for bone core densities are shown in Table 29. Before discussing these differences, it should be pointed out that the general range of values obtained for bone tissue density in this sample from Pecos is remarkably similar to those obtained for modern cadaver derived samples. The mean density over all 39 individuals in the present sample is 1.826 gm/cm^3, with a

---

[22] Courtesy David Thompson, University of Connecticut. All individuals in the sample will eventually be cored and analyzed for density and histological changes with age.

[23] Of the 39 individuals with bone core density data, 34 are included in the sample of 40 individuals with bone mineral data at all 11 cross section sites; bone mineral data for the femoral 50% section is, of course, available for all cored individuals since these determinations were carried out on the entire sample of 119 individuals.

Table 29

Summary of Bone Core Density Data [1]

I. Sex Differences

Male Mean	Male SE	Female Mean	Female SE	Male - Female	Male/Female X 100
1.852	.023	1.796	.024	.056†	103.1

II. Age Differences--2 Age Groups

MALES

20-39 yrs Mean	20-39 yrs SE	40+ yrs Mean	40+ yrs SE	40+ yrs - 20-39 yrs	%dif
1.848	.031	1.855	.035	.077	.4

FEMALES

20-39 yrs Mean	20-39 yrs SE	40+ yrs Mean	40+ yrs SE	40+ yrs - 20-39 yrs	%dif
1.844	.029	1.749	.033	-.091	-4.9*

III. Age Differences--3 Age Groups

MALES

20-29 yrs Mean	20-29 yrs SE	30-39 yrs Mean	30-39 yrs SE	40+ yrs Mean	40+ yrs SE	F
1.845	.052	1.850	.042	1.855	.035	.01

FEMALES

20-29 yrs Mean	20-29 yrs SE	30-39 yrs Mean	30-39 yrs SE	40+ yrs Mean	40+ yrs SE	F
1.814	.037	1.881	.044	1.747	.033	2.94†

IV. Side Differences

Males Right	Males Left	Males (Rt-Lt)/RtX100	Females Right	Females Left	Females (Rt-Lt)/RtX100
1.890	1.801	4.9*	1.795	1.797	-.1

[1] All data gm/cm^3. Total n = 39; see text for n's in individual comparisons.

† $p<.10$ * $p<.05$

range from 1.601 to 2.008 gm/cm^3. This compares to a mean cortical bone density from the same site determined using the same method in 98 US White cadavers of 1.847 gm/cm^3 (Thompson, 1980). Atkinson and Weatherall (1967), also using modern cadaver material, obtained mean (dry) cortical bone densities for the femoral cortex in the younger-middle adult range (20-60 years) of approximately 1.80 gm/cm^3 (my calculation), with a range from 1.58 gm/cm^3 to 2.16 gm/cm^3. Thus, the rewetted bone density values obtained here for an archaeological sample some 500 years old (see chapter 2) are certainly well within the range of modern, perfectly preserved material, and indicate that very little bone tissue was lost during the hundreds of years of burial at Pecos Pueblo.

Sex differences in bone core densities in the sample are relatively small, averaging about 3% greater among males ($p<.10$) (Table 29, I). Comparison of males and females by two age groups (Table 29, II) shows that the slightly lower values for females as a whole are brought about by a loss in density among older females, with no parallel loss among males. Among younger (20-39 year) adults, male and female densities are almost identical--about 1.84 to 1.85--while among older adults (40+ years) females are about 6% lower in density than males ($p<.05$, between sex difference). Males show a very slight, nonsignificant gain in density with aging, while females exhibit a significant, approximately 5% loss.

Analysis of age differences in core densities using the three age group divisions (Table 29, III) reveals further sex-related differences in the pattern of density changes with age. Males show no significant differences between any age groups in core density, with only a very small (.5%) increase from the youngest through oldest age groups. Females, on the other hand, show an interesting pattern which seems to parallel bone volume changes with aging: first an increase in core density from the third to fourth decades (3.6% gain), then a relatively large decrease from the fourth to the fifth and sixth decades (7.5% loss, $p<.05$, between age group difference). The ANOVA test for significance of between group age differences reaches near-significance in females ($F=2.94$, $p<.10$). This non-linear bone core density aging pattern among females, reaching a peak after early adulthood before declining in old age, is similar to that reported for both sexes by several other authors who have included younger adults in their samples (Arnold et al., 1966; Mueller et al., 1966; Atkinson and Weatherall, 1967).

A 4 age group analysis, dividing the 40+ age group into 40-49 year and 50+ year groups, was also carried out on the bone core density data. Although the sample sizes here are small (n's for the fifth decade, and sixth decade and older individuals of 6 and 4, respectively, in males, and 4 and 5, respectively, in females), the results are quite interesting and indicate that males may also reach a "peak" core density in middle adulthood and begin declining thereafter, but at a later age than females. Means for the 20-29, 30-39, 40-49, and 50+ year age groups in males are 1.845, 1.850, 1.889, and 1.805 gm/cm^3. In females corresponding means are 1.814, 1.881, 1.714, and 1.776 gm/cm^3. The increase in density in the oldest female group is probably a result of sampling error, and in any event still represents a marked decrease from the 30-39 year age group. The decrease in the oldest male age group, averaging 4.6% below the "peak" 40-49 year values (statistically nonsignificant), is smaller than female age losses in the fifth and sixth decades, but may indicate that a maximum value for core density is reached among males as well, and that males also eventually decline in core density with aging, although at a later age, and to a lesser degree than females.

Although studies of bone loss in modern populations with sample sizes in the under 50 year age range comparable to those in the present study are not available, it is interesting that what limited results there are for this age range show basic similarities to the data presented here. Atkinson and Weatherall (1967) included several individuals between 20 and 60 years of age in their study of density changes in the femoral diaphysis with age (measured directly in a manner comparable to that of the present study). The mean difference in density between 20-39 year and 40-60 year ranges in their study (calculated by me as the average of anterior quadrant values at their 40% and 60% cross section levels) is 2.9% in females and 1.5% in males (both declines). If one individual in their sample with an anomalously high density value of 2.16 gm/cm^3 (.2 gm/cm^3 higher than any of the almost 400 other femoral quadrants tested)--an older female--is eliminated, the average female loss between age groups is 4.4%, quite similar to the 4.9% loss in core density among females found in the present study. The 1.5% average loss among males in Atkinson and Weatherall's study is greater than that among males in the present study (who show no age loss in 2 age group comparisons--

see Table 29, II), but as shown above, there are indications that males over 50 years in the present study sample begin to decline in core density, so this difference between studies may be a result of slightly different age characteristics of the two age groups (e.g., the average age among males in the Atkinson and Weatherall study older group is 51 years, in the present study older group 48 years (using the 6 age group midpoint technique described in chapter 3)), or to sampling error in general, due to the small samples involved. Most other studies of directly measured bone density changes with age in modern samples (e.g., Thompson, 1980; Amtmann, 1971) do not include individuals under 50 years of age, so are not strictly comparable to the present study data. However, they do support the general finding presented here that females decline more in density with aging than males. Comparisons of directly measured core densities in the present sample with bone mineral analyzer data are presented below.

A significant ($p<.05$) side difference in bone core density was found among males, with the right side averaging almost 5% denser than the left (Table 29, IV). No side difference in core density was observed among females. This directly contradicts the results of BMC/CA side comparisons presented above, where a right "dominance" in female femora and a variable pattern in male femora was found. However, these differences may be largely due to sampling error, given the small sample sizes for bone core density data in sex/side groups (see above). Also, as noted earlier, the sample of 39 individuals with core data does not correspond exactly to the 40 individual sample the above observations regarding BMC/CA are based on. If BMC/CA side differences at the femoral 50% cross section site are calculated for only those 34 individuals with bone core data, the side differences in BMC/CA are very similar to those in bone core density: 4.7% greater on the right side among females, and .9% greater on the left side in males. This also illustrates how important sampling error is in determining the nature of "side differences" using unpaired data and small samples, completely reversing results of a side comparison with the substitution and elimination of a few individuals in the smaller sample. As noted earlier (chapter 4), larger samples of at least 30 individuals in each sex/side group are necessary if side differences are to be considered representative of true bilateral differences <u>within</u> individuals. Thus, with the exception of the BMC/CA determinations for the full (n=119) sample at the femoral midshaft and neck (where the right side was found to be denser among both males and females), it is doubtful how much real significance can be attached to side differences in either BMC/CA or core densities in these smaller samples.

SUMMARY

A. Whole Bone Measurements

1. Two sex differences in whole bone dimensions were noted: a general size difference, reflected in bone length differences, and a difference in the knee axis angle and mechanical/ideal axes angle of the femur indicating a relatively more lateral position of the femoral head in females with respect to the body center of gravity. In general, whole bone angles for the Pecos sample are very similar to those reported for other population samples, with the exception of the femoral antetorsion angle, which averages larger than most other reported values, and may be a specific "racial" characteristic of Amerindians.
2. The most significant age difference in whole bone dimensions observed was a decrease in the femoral antetorsion angle, much larger in females (7°) than in males (2°). It was shown that measuring the cervico-diaphyseal angle as a projection into the frontal plane may overestimate age decreases in this feature if the antetorsion angle also decreases: no age differences in "corrected" cervico-diaphyseal angles were noted. Small age group differences in bone length (reaching 2% in males), probably a result of sampling error, were shown to have relatively little effect on age changes in cross-sectional geometric properties.
3. Both tibial distal torsion and femoral antetorsion angles were found to be smaller on the left side, i.e., left lower limb bones are less "twisted" at both proximal and distal ends. Also, using "corrected" cervico-diaphyseal angle values, this angle was found

to be larger on the left side, i.e., the femoral head-neck axis is closer to the longitudinal axis of the diaphysis on the left side.

B. Cross Section Geometry

1. The biomechanical significance of various cross section properties were reviewed. It was decided not to standardize geometric data by bone length, due to both theoretical and empirical difficulties in deriving an adequate index of "size" variation. Transformations in phi (orientation of greatest cross-sectional bending strength) were described which allow more meaningful interpretations of this property.
2. Relative 3-dimensional positions of cross section area centroids differed little between the sexes, except in the proximal femoral diaphysis, where females were more laterally placed, possibly indicating a thicker lateral cortex in this location. This sex difference was found to increase with age, which may reflect a progressive thickening (or less thinning) of the lateral cortex among older females. Age differences in the tibia indicated selective preservation and "drift" of the anterior crest to a more

medial position. Side differences in these properties generally reflected differences in phi values in individual cross sections on right and left sides.

3. Sex differences in cross section areas were generally greater in the tibia than in the femur, i.e., females exhibited relatively greater bone areas in the femur than the tibia. Within-bone differences in cortical area indicated relatively more cortical bone in the mid-tibia and distal femur among males, and in the proximal femur among females. Both sexes showed a "peak" in cortical bone area in the 25-40 year age range, after which CA declined slightly in males and markedly in females. Peak values occurred somewhat later and total bone loss was less in the femur than the tibia among females, while males showed no bone-specific differences. Both sexes increased in total subperiosteal area about equally throughout the adult age range, and increased in medullary cavity area due to endosteal resorption after age 35. Net reductions in CA among older females were due to relatively greater rates of endosteal resorption than males. Side differences in bone areas indicated a general left side "dominance" (i.e., greater left side values) in bone areas, except in the proximal femur among males. This "dominance" appeared to be greater in cross sections near the knee joint. With aging, males showed a reversal of "dominance" from left to right sides, while females increased in left "dominance".
4. Angles of orientation of the major axis, or axis of greatest bending strength (phi) showed a definite pattern of variation at

different cross section sites in the tibia (an A-P oriented, sigmoid curve), and the femur (a counterclockwise rotation from proximal to distal ends). The femoral neck showed an angle of orientation of greatest bending strength close to the vertical, although slightly angled forward. Males and females were generally very similar with respect to phi values, with the only near-significant difference occurring in the femoral midshaft, where males were more A-P oriented than females. With aging, major axes in the tibia rotated closer to the A-P axis, especially among males and on the left side. The most significant change with age in femoral phi's was a rotation closer to the M-L axis in the proximal femur among females, paralleling a decrease in the antetorsion angle of the head and neck. Side differences in phi generally paralleled differences in the tibial distal torsion and femoral antetorsion angles.

5. Analysis of principal moments showed males to have relatively larger maximum second moments of area (I_{max}) in cross sections where the orientation of I_{max} was closer to the A-P axis, i.e., in the mid-proximal tibia and mid-femur. Females showed relatively larger values of I_{max} in the more mediolaterally oriented cross sections of the proximal and distal femur. With age both sexes increased in principal moments due to subperiosteal expansion and remodeling, with relatively greater increases in the tibia in males and in the femur in females. Both sexes increased in I_{max} relatively more in the middle tibial cross sections, while in the femur, males increased most in I_{max} in the midshaft, and

females in the proximal diaphysis. The femoral neck also showed an increase in I_{max}, relatively greater among females. The I_{max}/I_{min} ratio was also found to be a useful indicator of age and sex-related changes in bone "shape", with male bone areas in the femur generally becoming more A-P distributed with age, and female bone areas more M-L distributed. Side differences in principal moments indicated greater values of I_{max} on the left side, except in the mid-proximal femur in males. Asymmetry in this property increased with age in females, but not in males.

6. Use of section moduli rather than principal moments to estimate relative bending strength was found to reduce differences between sites, between the sexes, and between sides. Certain age trends in bending strength were also made more clear, e.g., greater increases in A-P bending strength with age in male tibiae than in female tibiae, and a more marked increase in superior-inferior bending strength of the femoral neck in females, with no age increase among males.

C. Bone Mineral Measurements

1. Sex, age, and site-related variation in BMC and BMI were demonstrated to be principally due to variation in bone volume, and not bone "density". Bone mineral measurements standardized for cortical area differences showed no significant differences between the sexes or between different age groups. Some evidence indicated slightly more dense right side bones, although this pattern varied by site and by sex.
2. Bone core densities were found to decrease somewhat (5%) in females with age, and showed no significant age change in males when 20-39 year and 40+ year age groups were compared. Further analysis of smaller age groupings indicated a "peak" in core densities among females in the fourth decade, and a possible peak in males in the fifth decade, followed by a small decline.

Chapter 6

Discussion and Conclusions

One of the fundamental assumptions underlying this investigation is that bone morphology ("whole bone" and cross-sectional geometry, bone density) will reflect *in vivo* bone function, i.e., biomechanical loadings. The concept of stress induced bone remodeling was first popularized by Julius Wolff (1870, 1892), and has come to be known as "Wolff's Law". If Wolff's Law is applicable to the present study material, then the variations observed here in structure of the femur and tibia can be used to investigate differences in stress and use of the lower limb among the inhabitants of Pecos Pueblo.

Three factors were investigated for their effects on bone structural variation in this sample: sex, age, and side of the body. The effects of other variables, such as nutritional status and genetic background, were minimized by selecting a sample from an essentially closed, homogeneous population (see chapter 3). It should be emphasized that structural differences in the lower limb due to sex, age, or side differences, although consistent with Wolff's Law, are influenced by genetic/hormonal effects as well as environmental effects. It is probably largely futile to attempt a clear separation between these two kinds of effects, since many aspects of structural form are most likely brought about by an interaction between environment and genetics. For example, sex differences in shape of the lower limb bones may be brought about by genetically programmed development of sexual dimorphism in pelvic form, hormonally mediated at adolescence, and by activity differences between males and females, brought about partly by culturally-prescribed sex roles, and partly by pelvic form itself (and possibly relative muscle mass differences between the sexes). As another example, bilateral asymmetry in human limb bones has been shown to be increased by exercise, an environmental effect, e.g., among professional atheletes, but is also present to some degree in unborn fetuses, and so may have a genetic component as well (see chapter 3). Perhaps of most importance in the context of the present study, changes in skeletal structure with aging can be viewed as an interaction between a hormonal, genetically programmed effect--a general loss of bone (or at least, endosteal resorption of bone) with aging, especially among females--and an activity-related effect, whereby bone is redistributed under the influence of loadings and stress distributions to minimize structural weakening due to bone loss (see chapter 2). In fact, it is these interactions, and the possible roles of activity-induced remodeling of bone in mediating the effects of bone loss with aging, which form some of the most interesting aspects of the present study results. Thus, while Wolff's Law states that a relationship between bone stress and bone architecture exists, differences in the type of bone stress developed will depend upon a number of factors, both environmental and genetic, all of which must be considered if an accurate assessment of form-function relationships is to be arrived at.

In order to better understand and interpret the significance of variation in the structural properties described in the preceding chapter in terms of bone stress, it is first necessary to briefly summarize what is known about *in vivo* loadings of the femur and tibia. Some of these findings have been mentioned or implied in the course of chapter 5 where relevant to geometrical properties under discussion, but at this point a more thorough and organized account of these data is in order.

In Vivo Loadings and Stress Distributions in the Femur and Tibia

A voluminous literature on the biomechanics of the lower limb during weight bearing and bipedal locomotion has been amassed in the eighty years since Braune and Fischer's (1898-1904) pioneering work in this field. These studies fall into three general categories. The first, and most common type, has been the estimation of joint loadings and stress distributions within bones given theoretically applied loads based on knowledge of body segment weights, 3-dimensional position of joints, and general types of muscle action involved, usually greatly simplified and limited to static weight support, e.g., the single support phase of gait, or one-legged stance (Koch, 1917; Pauwels, 1935; Inman, 1947; McLeish and Charnley, 1970; Rybicki et al., 1972). The second type, initiated by Elftmann (1939) and the University of California group in the 1950's (Bresler and Frankel, 1950; Eberhart et al., 1954), and exemplified more recently by the studies of Paul (1966, 1967, 1971) and Morrison (1968, 1969, 1970), utilizes direct measurement of ground force reactions during gait using force plates, analysis of acceleration of body segments through high speed cinematography, and electromyographic recording of muscle activity, to calculate the magnitudes and orientations of forces and moments through the joints of the lower limb. These data have been used to calculate stresses in the lower limb during walking (Minns et al., 1977). Although several simplifications, for example, the combination of muscles into groups and the estimation of their combined lines of force, are still necessary in this type of study (Paul, 1967), because more information is gathered during the analysis and fewer assumptions are involved, these investigations provide a more complete and accurate assessment of joint loadings, and are not limited to static or single support phase analyses of gait. The third type of study involves direct measurement of strains (deformations) using strain gages attached to the bone surface to estimate stresses developed under loading. Although _in vivo_ applications of this kind using experimental animals have been common (e.g., Evans, 1953; Carter et al., 1980), to date the only _in vivo_ studies of human limb bones using this technique are those of Lanyon et al. (1975) and Carter (1978--a reanalysis of Lanyon's data) for the anteromedial aspect of the tibial midshaft.[1] Other investigators (e.g., Kimura, 1974, Minns etal., 1977; Oh and Harris, 1978) have conducted _in vitro_ strain gage experiments on excised human long bones subjected to loadings simulating _in vivo_ conditions. Stress coat analyses of stress patterns under various types of loadings have also been conducted by many authors (see Evans, 1973 for review).

It is not the intent of this section to provide an exhaustive review of these various approaches to the study of the biomechanics

[1] Rydell (1966) measured _in vivo_ strains in an implanted femoral head prothesis in two patients.

of the lower limb. Rather, some of the major conclusions of these investigations relevant to the interpretation of the present study results will be presented. The application of these findings to the present study data is discussed in the following section of this chapter.

Loadings and stresses in the femur.

Koch (1917) presented a simplified, two-dimensional analysis of stresses on the medial and lateral sides of the femur under a force applied to the femoral head simulating body weight (minus the weight of one lower limb) in the single support position. He found that under this loading the entire medial surface of the femur is under compression, reaching a peak in the middle of the femoral neck (inferior surface) and the proximal diaphysis, while the lateral surface of the femur (superior surface femoral neck) is under tension except near the distal end, again reaching peaks in the neck and proximal diaphyseal regions (ibid.: Fig. 18). Maximum compressive stresses were about 35% greater than tensile stresses under this model.

A more realistic model of stresses in the femur in the one-legged stance, taking account of the action of the major hip abductors (gluteus medius and minimus) as well as the tensor fascia lata, was reported by Rybicki et al. (1972). Inclusion of the hip abductors was found to increase tensile stress in the lateral cortex by approximately four times over tensile stresses in Koch's simple body weight support model. Adding the action of the tensor fascia lata decreased this tensile stress substantially, as well as decreasing compressive stress along the medial side. Figure 26 (next page), adapted from

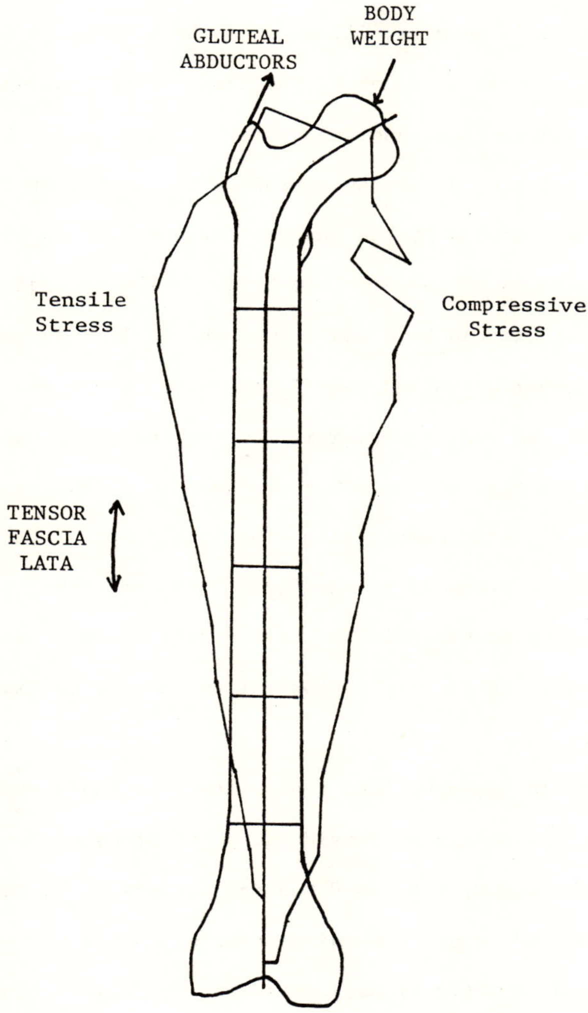

Figure 26. Stress magnitudes on the medial and lateral sides of the femur under axial loading (body weight) and loading by the gluteal abductors and tensor fascia lata (adapted from Rybicki et al., 1972--see text for details).

Rybicki et al.'s results, illustrates the pattern of stresses on the medial and lateral surfaces of the femur under loadings by body weight, hip abductors, and tensor fascia lata (note: this diagram was drawn assuming a tension of 135 lb in the tensor fascia lata, based on data quoted by the authors derived from Paul's studies of *in vivo* loadings of the hip--see below). The general pattern of stresses illustrated here conforms to the earlier observations of Koch, with both compressive and tensile stresses in the shaft increasing from the distal end to a peak in the proximal diaphysis, at a location approximating that of the 80% cross section in the present study. Rybicki et al. also demonstrated the inadequacy of the beam model for estimating stresses in the femoral neck-greater trochanter region, overestimating stress relative to a finite element analysis (thus, stresses shown in Figure 26 proximal to the 80% section are overestimated).

It should be emphasized that neither of these studies (Koch, Rybicki et al.) included consideration of stresses outside the frontal plane of the femur, i.e., A-P bending stresses. In fact, neither study exactly modeled even M-L bending stresses, since the entire femur (and all muscle actions) was considered to fall within the frontal plane, that is, no consideration was given to antetorsion of the femoral neck and head or to anterior curvature of the shaft. The significance of departures of the femur from the frontal plane in terms of cross-sectional geometry have already been touched upon to some extent in the preceding chapter (e.g., see Figure 21), and will be discussed further below.

The dynamic studies of human gait conducted by Paul (1966, 1967, 1971) and Morrison (1968, 1969, 1970) clearly show that A-P bending moments about the hip and knee joint axes are quite significant during several parts of the gait cycle, particularly just after heel strike and just before toe-off (Paul, 1971: Fig. 3; Morrison, 1968: Fig. 9). During these time periods, A-P bending moments applied to the femur may equal or exceed M-L bending moments (op. cit.). This results from the positions of lower limb segments relative to the body center of gravity, and to the action of muscle groups during these phases of gait. At heel-strike, the foot is situated some distance in front of the hip, giving rise to a negative bending moment at both hip and knee joints (i.e., concave surface of bending of the femur facing anteriorly--see Paul, 1971: Fig. 5). This is also a phase of gait where both the hamstring and quadriceps muscles are active, helping to decelerate the thigh (hamstrings) and stabilize both the knee and hip joints, further increasing A-P bending moments (Eberhart et al., 1954; also see Morrison, 1970, Paul, 1971). A-P bending moments about the hip and knee are relatively small during mid-stance (flat foot), when the body center of gravity is almost directly over the lower limb (in the sagittal plane), but increase again during the second half of the stance phase, prior to toe-off, when the foot is situated behind the body center of gravity, and the gastrocnemius and quadriceps muscles are active, creating positive A-P bending moments about the hip and knee (ibid.).

Although little empirical data are available regarding joint reaction forces and stresses in the lower limb during activities other than level walking, it may be hypothesized that in certain strenuous activities, A-P bending of the femur increases greatly, perhaps more so than M-L bending. Morrison (1969) presents some data on loadings of the knee during walking up and down a ramp or stairs. Maximum joint reaction forces on the knee during up stair walking averaged 25% greater than in level walking, and the force action of the quadriceps more than doubled. The force exerted by the quadriceps also increased greatly during down stair and down ramp walking. As Morrison notes, "The forces exerted by the quadriceps femoris in these activities better reflects the size of the muscle than do the forces calculated in level walking experiments", since strenuous extension (up stair walking) or controlled flexion (down stair or down ramp walking) of the knee joint is not normally required during level walking. Morrison also noted that during up ramp walking, the gastrocnemius showed a significant increase in force produced, which would also increase the A-P bending moment about the knee. It would also be expected that A-P bending about the hip would increase where the hip is actively flexed and extended, e.g., during up stair walking or running, where in addition to the action of other hip flexors and extensors the gluteus maximus becomes active.

Torsional loading of the femoral diaphysis during gait, i.e., twisting about its longitudinal axis, have been considered to be insignificant, due to the relatively small observed rotation of the thigh and second moments of mass about the longitudinal axis (Paul, 1967: 12). However, torsional loading of the femoral neck may be quite important, due to the rotation of the femoral head in the acetabulum during gait, and the lines of muscle action about the joint, coupled with its asymmetric cross-sectional outline (Backman, 1957). Shear stress due to torsion of the femoral neck has been investigated by Paul (1971) and Backman (*op. cit.*), who considered that torsional stress, in conjunction with increased pressure on the hip joint, e.g., resulting from a fall on the hip, may be a significant factor in causing certain types of femoral neck fractures.

In sum, *in vivo* loadings of the femur as a whole are a combination of A-P and M-L bending loads, with the probable addition of significant torsional loadings of the femoral neck. M-L bending loads predominate in the proximal femur and become less important in the distal femur (Figure 26). A-P bending loads are distributed throughout the femur, but become relatively more important in the distal femur, where M-L loading declines. In fact, a case can be made (at present unsupported by empirical evidence) that maximum A-P bending forces developed under *in vivo* loadings are *absolutely* greater in the mid-distal femur than in the proximal femoral shaft, due to the powerful action of the quadriceps muscle (mainly originating from the shaft of the femur) in extending the knee during vigorous physical activity (see above). Implications of these stress distributions in terms of sex and age variation in structural properties of femoral cross sections will be discussed below.

Loadings and stresses in the tibia.

As mentioned earlier, both theoretically derived stress distributions and direct *in vivo* measurements of strains under loading of the tibia are available. Loading configurations of the tibia and theoretical stress distributions will be described first, followed by strain gage measurements and resulting stress calculations. In these discussions, the role of the fibula is ignored. Inclusion of the fibula would probably have a relatively small effect on loading patterns, since it has been shown that this bone carries only about 1/6 of the total load on the leg (Lambert, 1971; Kimura, 1974).

The data of Paul and Morrison, already described above, indicate that the tibia undergoes a succession of negative and positive A-P bending moments throughout gait (i.e., concave and convex bendings of the tibial shaft). Minns et al. (1977) used these data and data on cross-sectional geometry to calculate stresses at six diaphyseal locations during four stages of the gait cycle. Maximum tensile stress occurred during the later part of stance phase, corresponding to the peak of activity of the calf plantarflexors prior to toe-off (Eberhart et al., 1954). Peak tensile stress at this time was located approximately 30% of the distance from the proximal to the distal end of the tibia (i.e., at about the 65% section of the tibia in the present study). These authors found the anteromedial surface of the tibia to be mainly under compression during the four stages of gait analyzed, with the anterolateral surface mainly in tension (except at heel-strike). As stated several times previously, the tibia is subjected to significant torsional loadings during gait. These are produced by the lateral rotation about the longitudinal axis which occurs during the later stages of the stance phase prior to and during toe-off (Eberhart et al., 1954: Fig. 15.20; see also Morrison, 1970: Fig. 15.20). Minns et al. (1977) found the peak shear stress due to torsional loading of the tibia to occur at about 70-80% of bone length, measured proximally-distally (i.e., between my 20% and 35% sections). Piziali et al. (1976: Fig. 6) using both a finite element model and a less precise beam model which used the polar moment of inertia, found peak shear stresses in the tibia under a torsional loading to occur at about 30-35% of bone length, measured from the distal end. Thus, a general pattern of greater A-P bending stress in the proximal tibia, and greater torsional stress in the distal tibia is indicated by these results.

Kimura (1974) conducted both theoretic tibial stress calculations similar to those of Minns et al. (but with much simpler loading configurations), and strain gage experiments on excised human tibiae under simulated *in vivo* loadings. A simple axial compression force through the centers of proximal and distal articulations produced tensile stresses along the anterior and anterolateral surfaces, and compressive stresses along the posterior and posteriomedial surfaces, peaking at midshaft. A loading configuration which included the action of the quadriceps and soleus muscles produced much the same stress distribution, although maximum stresses were located somewhat more proximally, between the midshaft and tibial tuberosities (i.e., my

50% to 80% sections of the tibia).

Lanyon et al. (1975) performed the only *in vivo* strain gage tests on human bone, attaching a strain gage to the anteromedial aspect of the tibial midshaft and recording bone deformations during walking and jogging. Carter (1978) reanalyzed these data using an anisotropic analysis, and presented results in terms of principal material directions (i.e., longitudinal and transverse bone axes). His findings parallel and confirm the theoretical and *in vitro* results described above, and extend these results to some interesting observations on differences in stress developed during walking and jogging. During walking, the longitudinal (axial) stress was compressive at heel-strike, tensile during stance phase, and compressive again at toe-off. Transverse stresses were very small throughout the cycle, while shear stress due to torsion became significant during the later part of the stance phase and at toe-off (*ibid.*: Fig. 2). During jogging (at 2.2 m/sec), a compressive stress at toe-strike was followed by a very high tensile stress which remained until toe-off. Both transverse and torsional shear stresses were small throughout the cycle (*ibid.*: Fig. 3). Thus, axial tensile stress may increase greatly during certain activities (jogging), while torsional stress actually decreases. While these data are strictly valid only for the specific location on the tibial diaphysis and the individual tested, in combination with other data presented above, they indicate a possible basis for differential remodeling of the tibial diaphysis in response to different habitual activities. The implications of these findings in terms of the present study data are presented below.

Relationship to cross section geometry.

In order to better visualize some of the loading and stress distribution concepts presented above in terms of variation in cross section geometry at different levels of the tibial and femoral diaphyses, the hypothetical locations of maximum tensile and compressive stress under a simplified axial loading are illustrated here using the cross section outlines for one individual (a 30-34 year old male) previously shown in chapter 5 (Figures 20a,b). The load considered here is a simple downwardly directed force on the femoral head representing body weight, with the load line passing through the femoral head and centers of the knee and ankle joints. In a sense, this is similar to the loading used by Koch (1917) in his analysis of stresses in the femur, discussed above. However, the analysis here is 3-dimensional, since the positions of articulation surface centers and all bone cross sections can be located with respect to each other in three planes (see chapter 4; also, illustration of 3-dimensional positions of area centroids in chapter 5). The addition of A-P coordinates of the hip, knee and ankle joints, and A-P eccentricities of the femoral and tibial shafts leads to some interesting results complementing and extending those of Koch and other more recent authors described above.

Complete stress distributions within cross sections were not determined here. However, to gain some idea of the approximate areas of maximum compressive and tensile stresses, the following procedure was carried out. A force line representing body weight was visualized as running through the hip, knee, and ankle joints, i.e., along the "mechanical" axis of the femur and the longitudinal axis of the tibia.[2] The A-P and M-L coordinates of the intersection of this line with the transverse plane of each cross section were then determined using previously recorded "whole bone" data points (see Table 3, chapter 4). Lines were then drawn from these points through each section centroid; these lines represent a rough approximation of the orientation of the plane of bending at each cross section under this loading (body weight). Maximum compressive stress should occur on the subperiosteal surface on the side nearest the force line, maximum tensile stress on the side farthest from the force line. Figures 27a,b illustrate the method and its results for femoral and tibial cross sections in one individual. The major axis of each section, as determined from SLICE, is also included for reference. (Because of the geometric complexities involved in determining the 3-dimensional position of the femoral neck cross section outline with respect to "whole bone" landmarks, e.g., accounting for angular displacement due to the antetorsion angle of the head and neck, only diaphyseal cross sections were included in this analysis. Also, no attempt was made to calculate relative _magnitudes_ of stress at each cross section, although certain observations regarding general levels of stress in different cross sections, based on other data reviewed earlier, will be made.)

As stated above, the force line through the three joints of the lower limb corresponds approximately to the mechanical and

---

[2] Macquet (1976:110) states that these two axes form a straight line _in vivo_ in the normal, unpathological condition.

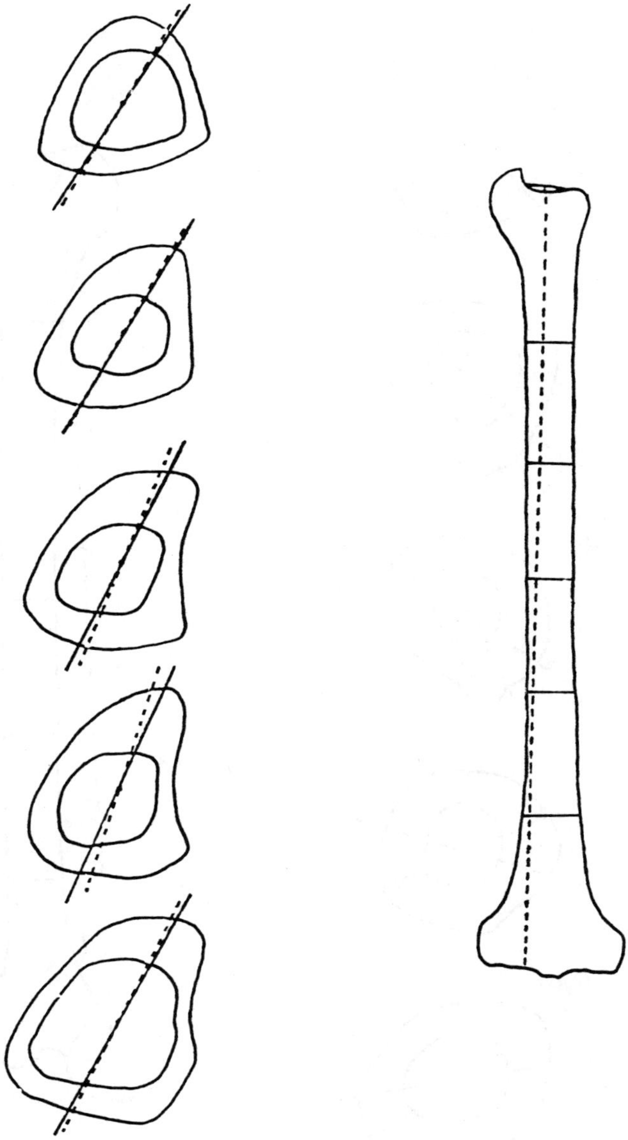

A B

Figure 27a. Planes of bending through tibial cross sections under axial loading. A. Cross section outlines, showing plane of bending (solid line) and major axis of section (dotted line). B. Diagram of tibia showing location of axial force line in frontal plane. See text for description of method.

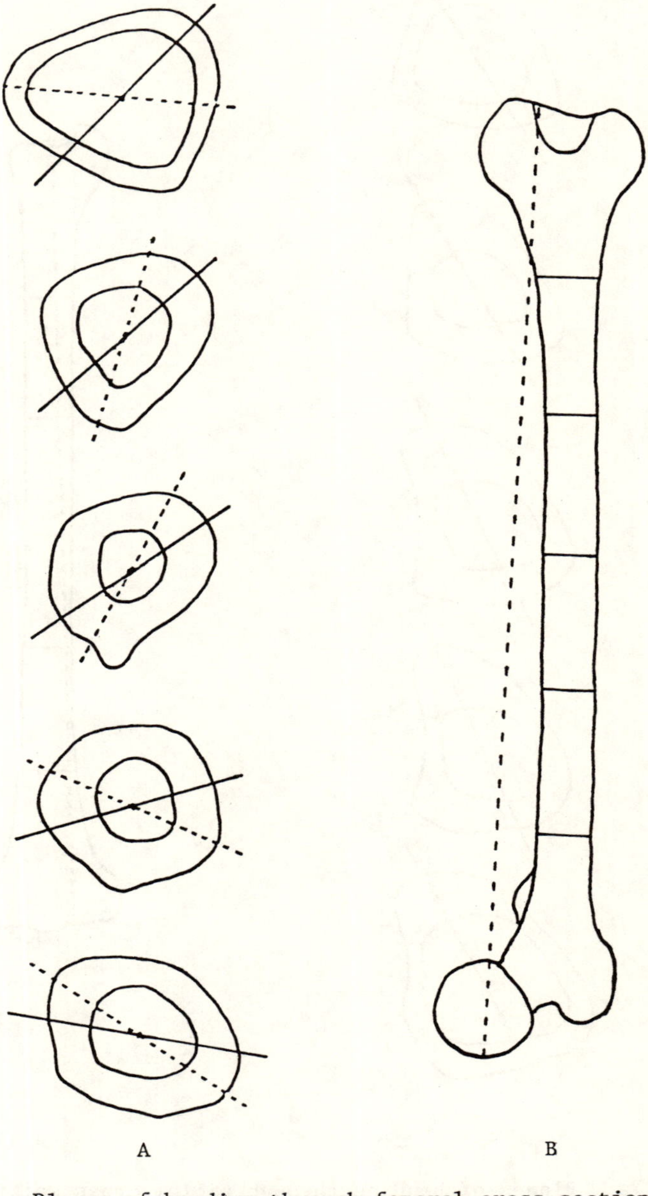

Figure 27b. Planes of bending through femoral cross sections under axial loading. A. Cross section outlines, showing plane of bending (solid line) and major axis of section (dotted line). B. Diagram of femur showing location of axial force line in frontal plane. See text for description of method.

longitudinal axes of the femur and tibia, respectively. However, slight modifications were made in the A-P and M-L coordinates of the knee joint center of articulation, based on data given by Morrison (1970) and Minns et al. (1977) regarding the in vivo loading of this joint. Morrison showed that during almost all of the weight-bearing, stance phase of gait, the center of pressure on the tibial condyles is medially displaced, i.e., over the medial condyle (op. cit.: Fig. 10). In his sample, this displacement averaged about .5 in, or 13 mm. In lieu of better data regarding displacement relative to bone length or body height (not reported by Morrison), this same figure was used in the present study to adjust the center of the tibial articulation (calculated as the midpoint of the two tibial plateau centers) medially.[3] Minns et al. (1977), using data from Steindler (1955) and Smidt (1973), estimated that the center of the position of contact between femur and tibia during gait is also displaced approximately 6 mm posteriorly from the center of the intercondylar eminence. This figure was also used in the present study to move the center of the proximal tibial articulation posteriorly. For consistency, these same adjustments in the position of the knee joint center were made with regard to the distal femoral articulation center (using point coordinates of the two femoral condyle centers determined as described

---

[3] Two out of three of the individuals in Morrison's sample were female, while the individual from Pecos is male. Thus, even given the shorter stature of the Pecos sample relative to most modern populations, Morrison's figure for medial displacement of the knee joint center should not be a significant overestimate relative to the height of the individual used here (calculated as 164 cm).

in chapter 4). It should be stressed that this analysis is only intended as a gross approximation of *in vivo* axial loading forces. The limitations involved, particularly in placement of the knee joint articulation center, should be borne in mind when interpreting the results presented below.

As a test of the method, a comparison was carried out between the present study results and results of an analysis of stress distribution in a femoral cross section reported by Amtmann (1971), following Pauwels (1950). This comparison is shown in Figure 28. Although there are some minor differences in the locations of maximum tensile and compressive stresses around the cortex (brought about partly by differences in the loading configurations and exact locations of cross sections used), the similarity between the two study results is striking. Thus, there is some evidence that the present study technique produces at least reasonable estimates of true stress distributions under an axial loading.

Results of the analysis of tibial cross sections, shown in Figure 27a, shows that simple axial loading will produce maximum compressive stress on the posteromedial surface, and maximum tensile stress on the anterolateral surface of the cortex. These results thus correspond well to those reported by Kimura (1974) and Minns et al. (1977) described above. The orientation of the major axis of each cross section (direction of greatest bending strength) is quite close to the hypothesized plane of bending under axial loading (and thus, the positions of maximum tensile and compressive stress). The general pattern of variation in the orientation of the bending

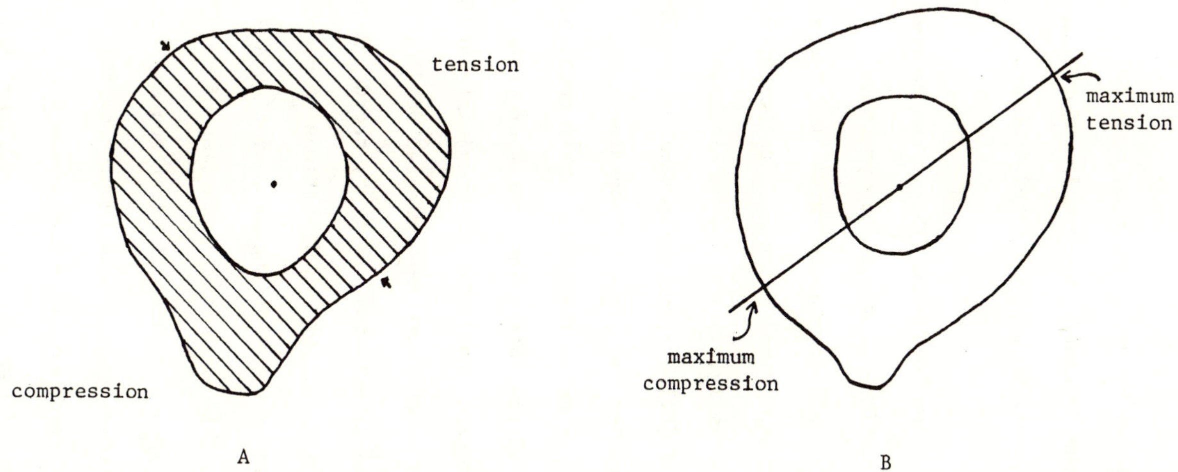

Figure 28. A. Stress distributions under axial loading in femoral cross section near midshaft as calculated by Amtmann (1971: Fig. 2b). Straight lines connect points of equal stress, with maximum compressive stress approximately 1.3 X maximum tensile stress. Arrows show location of neutral axis. B. Outline of femoral 50% cross section from Figure 27b, present report, showing hypothesized plane of bending under axial loading and locations of maximum tensile and compressive stress.

planes is similar to that of the major axes, becoming more M-L oriented in the distal tibia (the "sigmoid" pattern in major axis orientation noted earlier, however, is not reflected in these planes of bending).

In the femoral diaphysis, shown in Figure 27b, it can be seen that under axial loading the locations of maximum compressive and tensile stresses gradually rotate counterclockwise from medial and lateral, respectively, in the proximal diaphysis, to posteromedial and anterolateral, respectively, in the distal diaphysis. Although major axes orientations follow a generally similar pattern of variation, the correspondence between planes of bending and cross section major axes is not as close as in the tibia. This could be a result of a faulty placement of the knee joint center of pressure, described above. For example, moving the center of the joint laterally from its present position would create more A-P oriented bending planes, closer to the major axes directions. However, the deviations between predicted maximum stress locations and directions of greatest bending strength here could also result from the exclusion of A-P bending moments from the model (this would have much less of an effect on the tibia, since the planes of axial loading and A-P bending about the knee joint are much closer than in the femur--compare Figures 27a and 27b). Since in every cross section the position of the major axis is more A-P oriented than the predicted plane of axial bending, this can be seen as a combined resultant of the direction of an axial bending plane and an A-P bending plane. This hypothesis will need to be tested under more realistic loading configurations which include the forces of various muscle groups about the hip and knee joints (e.g., gluteal abductors and quadriceps femoris).

It is interesting that although the planes of bending under an axial loading in femoral cross sections do not correspond exactly to the major axes orientations, they do follow quite well, at least in the mid-distal femur, the "bulges" in the femoral cortex noted in the last chapter. This is especially apparent in the 20% and 35% sections, where the entire cross section outline appears asymmetrically "distorted", distributing more bone volume in the locations of hypothesized maximum tensile and compressive stress under axial loading. The slight bulge on the posteromedial endosteal surface of the 35% section shown in Figure 27b may also be significant in terms of adaptation to stress. This inward protrusion of the endosteal surface was observed in many individuals, and sometimes was developed enough to form a distinct internal ridge visible from the 20% through the 50% cross sections, always in the posteromedial location (note the intersection with the axial loading plane of bending).

The linea aspera of the femur may act as a kind of posterior strut in the middle region of the femoral diaphysis, contributing significantly to A-P bending strength. As can be seen subjectively from the cross section outlines in Figure 27b, it is primarily this feature of cross-sectional morphology which is responsible for the rotation of the major axes of the 50% and 35% sections to a more A-P orientation than the axial loading bending plane. The relatively circular shape of the 65% cross section (I_{max}/I_{min} ratio in this individual is only 1.13) may be a result of its "transitional" location between regions of relatively high M-L bending stress (80% section) and A-P bending stress (50% and 35% sections). In the same way, the transition from a more A-P distributed cross-sectional volume in the proximal tibia to a rounder cross section in the distal tibia, may be viewed as an adaptation to changes in the relative magnitudes of stresses at different levels, from high A-P bending to high torsional stress. These various observations will aid in interpreting the results of the geometrical analyses of femoral and tibial cross sections presented below.

Interpretation of Results

Sex differences.

Although the primary focus of the present investigation is the study of aging and its effects on the human skeleton, it is first necessary to consider the relationship between sex differences in loadings and structural properties, since these may to a large extent determine the type of age-related remodeling observed in different individuals. Bilateral asymmetry in use of the lower limb could also affect patterns of aging in the femur and tibia, but since equal numbers of right and left sides were included in each sex/age group, this factor may be ignored in preliminary analyses of sex and age differences. Side effects on structural remodeling will be discussed in a following section. Also, discussion here will focus on diaphyseal cross sections, with the femoral neck section considered along with other proximal femoral characteristics in a further section.

The most significant and general finding regarding sex-related differences in structure of the femur and tibia was an overall pattern of relatively greater anteroposterior bending strength among males, and greater mediolateral bending strength (and in the tibia, torsional strength) among females. This interpretation is supported by many observed features of cross-sectional geometry reported in the previous chapter. In terms of bone areas, males show relatively greater values of CA in the mid-proximal tibia and mid-distal femur, the sites of probable greatest A-P bending stress (see above), while females show relatively greater CA in the proximal femur and distal tibia, sites of high M-L and torsional stress, respectively. In general, sexual dimorphism in bone areas of the femur is less than in the tibia, due probably to overall relatively greater stresses in the female femur brought about by high M-L bending loads.

Even more important than bone volume differences however, are differences between the sexes in bone <u>shape</u>, as indicated by comparisons of relative magnitudes and orientations of principal moments. Again, males show adaptation for high A-P bending stress in the mid-proximal tibia, with sexual dimorphism in the A-P oriented maximum principal moments even greater than those observed in cortical area. This "shape" differences is made even more apparent by comparing I_{max}/I_{min} ratios in the mid-proximal tibia, clearly showing males to be more "platycnemic" (i.e., more A-P distributed) than females. On the other hand, females show relatively higher polar moments and "polar moment moduli" (J/c_{maj}) in the distal tibia, a region of high torsional stress. The rounder shape of the tibial diaphysis as a whole among females is more adaptive for torsional stress and not as "specialized" for A-P bending stress as the male tibia (Frankel and Burstein, 1965; Lovejoy et al., 1976). In the femur, males exhibit relatively greater values of I_{max} and I_{max}/I_{min} in the 35% and 50% sections, both subjected to relatively high A-P bending loads. The orientation of maximum bending strength in at least one of these cross sections (50%) may also be more anteroposterior among males. Females exhibit relatively higher values of I_{max} and I_{max}/I_{min} in the proximal diaphyseal sections of the femur, a region of high M-L bending stress. Sex-related differences in the 3-dimensional position of area centroids may also indicate a relatively thicker lateral cortex in the proximal femur among females, an area of particularly high tensile stress (see above). Again, sexual dimorphism in principal moments in the femur in general is less than in the tibia, indicating relatively greater bending strength in the female femur (and the male tibia).

Two possible factors may explain this pattern of sexual dimorphism in lower limb structure in the Pecos sample. Neither explanation is exclusive of each other, and in fact probably work in combination to produce the characteristic sex differences in femoral and tibial geometry described above.

First, sexual dimorphism in pelvic structure would be expected to produce higher M-L bending loads on the lower limb among females, particularly in the proximal femur. As was previously noted (chapter 5), "whole bone" measurements of femoral and tibial angles in this sample indicate a relatively more laterally placed femoral head among females.[4] This conforms to general observations regarding pelvic form, both in the Pecos sample and other samples, whereby the pelvic inlet is relatively broader among females as an adaptation for childbirth (e.g., see Krogman, 1962). Burr et al. (1977) analyzed sexual dimorphism in the biomechanics of the hip in a sample of 57 pelves and femora from a late prehistoric Amerindian site in Illinois. Using multivariate discriminant function analysis, they were able to show a sex-related difference in "abductor torque" (relative mediolateral position of the hip joint center with respect to the body center of gravity and the greater trochanter, multiplied by a body weight factor (Inman, 1947)), the moment arm of the gluteus medius about the hip joint, and the femoral head pressure (greater in females). They concluded:

> Because of the greater breadth of the female lesser pelvis due to adaptation to child bearing, the weight of the individual acting through the body's center of gravity lies at a greater distance from the center of the hip joint. This longer resistance arm, coupled with the shorter moment arm of the gluteus medius reduces the mechanical advantage of the abductor musculature. At the same time the shape of the female pelvis reduces the advantage of the abductors; greater abductor force is required to control pelvic tilt. (p. 277)

Lovejoy et al. (1973) came to similar conclusions using another sample. Thus, the combination of relatively greater M-L moments about the female hip joint, due to its relative distance from the body median plane, and relatively greater force required by the abductor mechanism, due to its relatively short moment arm (note the smaller

---

[4] That is, relative to the M-L position of the tibia, and thus presumably the median plane of the body.

"mechanical"/ideal axis angle among females in the present sample--Table 4, Chapter 5),should produce relatively greater mediolateral bending stresses in the femur, particularly in the proximal half of the shaft (see Figure 26 above). This factor could account for structural remodeling of the femur to increase M-L bending strength, as described above. It would seem to have less of an effect, however (if any at all) on the structure of the tibia. Also, one would expect sex differences in bone geometry to increase as one approached the proximal femur, while in fact, an index of bone "shape", I_{max}/I_{min}, actually shows greater sexual dimorphism in the distal half of the femur (see Table 16b, Chapter 5). Thus, it appears that an additional factor besides sexual dimorphism in pelvic structure may be necessary to explain all of the sex-related differences in structure observed in the lower limb.

This second factor may be differences in habitual activities engaged in by males and females of Pecos Pueblo. There is both ethnographic and historic evidence from populations living in this region that males traditionally participated more frequently in activities which would place much higher A-P bending stresses on the lower limb, in particular, long distance running. There are abundant references attesting to the importance and frequency of long distance foot-racing among many of the populations of the American Southwest (Stevenson, 1901-2: 317-328; Ortiz, 1979: 273, 385, 568, 425), including the inhabitants of Jemez Pueblo, considered to be most closely allied to Pecos Pueblo of all Southwest groups (Sando, 1979; also see below). Racing occurred in a variety of social situations ranging from informal races to institutionalized ceremonial events and formal betting races between individuals, kin groups, or tribes (ibid.). Stevenson (op. cit.) described several types of racing among the Zuni, including a series of 4 mile races run as part of a rain ceremony, and betting races of 25 miles run cross country. Only men of the villages took part in actual running, although whole villages were involved in preliminary preparations (which could take up to a week) and supervision of the race: "There is nothing more exciting to the Zunis, except for the scalp dance, than this game of ti'kwane [racing]. Those on horseback urge their ponies onward to keep pace with the racers, who run southward over the road of the Gods of War for a distance, then to the east... The most prominent religious positions do not debar men from taking part in these races" (p. 326-327). Stevenson also notes that racing among boys for wagers starts at an early age: "It is interesting to see the very young boys engaged in their footraces... Beginning at so early an age, these people develop naturally into swift runners" (p. 328). Ethnographic evidence indicates that this kind of running in the Southwest has persisted until the very recent past. Quoting from D. Tedlock (personal communication):

> My consultants made it very clear to me that until very recently it was common for males in roughly a 14-into-the-20's age bracket to spend vast quantities of time, beginning early in the spring (March) and going until planting time (mid-May), practicing for and competing in kick-stick races over a course that was at least ten miles long. To this day, long-distance high school track competitions at the state level in New Mexico are usually won by runners from the Pueblos--Jemez is a specific Pueblo I remember hearing about in this connection ... Back around the turn of the century there was at least one race (not a kick-stick race, but a straight race) from Zuni to Gallop (40 miles).

Thus, it is evident that long distance racing has played an important, even institutionalized role among young men in native Southwest societies. To what extent running was involved in other predominantly male activities, such as hunting on the eastern planes (Sando, 1979: 425; Schroeder, 1979: 434) or traveling long distances for certain staples such as salt (Stevenson, 1901-2: 354) or during warfare (Kidder, 1924: 4-15), is difficult to determine. However, it seems almost certain that males in these populations habitually ran more than females.

All of the above is not in any way meant to imply that males of the Pecos population were more active or engaged in more vigorous exercise in general than females. It is abundantly clear from ethnographic and historic records that women in Southwest society were responsible for many of the heaviest tasks performed, such as building houses, carrying water, and grinding corn (Stevenson, 1901-2: 349 ff; also see discussion of age differences below). Also, many of the daily activities, such as preparing fields and irrigation ditches, and planting and harvesting crops, were shared by both men and women (_ibid._). Thus, _total_ biomechanical stress on the skeletal system cannot be said to have been less among females than among males of this population. However, the particular _types_ of biomechanical stress each sex was subject to may have differed. As a review of the literature on the biomechanics of the human lower limb above indicated, running and increased action of the knee and hip flexors produces relatively high A-P bending stress in the lower limb. This is particularly apparent from the studies of Lanyon et al. (1970) and Carter(1978) on _in vivo_ strains and stresses in the tibia, which showed a very high tensile stress at the midshaft during jogging, while torsional stress actually declined from levels observed during walking. Thus, habitual running, beginning at at early age and continuing into at least early adulthood, could explain the observed remodeling of the tibia and femur to a more A-P distributed shape among males of this sample, particularly in cross sections near the knee. Torsional stress on the tibia would remain high during walking (and presumably, other daily activities which did not involve running or climbing), and thus a rounder, less platycnemic tibia would be more adaptive for other, less "specialized" forms of loading. Lovejoy et al. (1976) have made the same argument in their structural analysis of platycnemic and eurycnemic (rounder) tibiae.

The combination of two effects, then--sexual dimorphism in pelvic form, and sex-specific activity patterns--may explain the basic sex-related differences in structure of the lower limb bones observed in the present study sample. It is also possible that shapes of the femoral and tibial shafts in males and females are partly genetically determined and not wholly dependent on different _in vivo_ loading patterns throughout life, although the extent to which this factor is operable is not known. It would be interesting to compare degrees of development of these sex-related features in the femur and tibia in a sample of juveniles from Pecos to determine at what age they develop, since some sex differences in pelvic shape have been reported as early as fetal life (Krogman, 1962). It would also be of interest to compare these results to those derived from other population samples with different activity patterns, including sex-related activity differences, from those of native Southwest populations.

Age differences.

One of the most important results of the present study was the confirmation of several previous studies that subperiosteal expansion of long bones occurs throughout the adult lifespan. This is particularly significant in that to date, the present study is the only investigation to a) measure these changes using true bone areas, not extrapolations from radiographs, and b) include sufficient numbers of individuals in the young and middle adult age groups to clearly show age trends in this range (see chapter 2). The increase in tibial and femoral subperiosteal areas from the third and fourth through the fifth and sixth decades of life in the present sample averaged overall 9%. As first noted by Smith and Walker (1964) almost 20 years ago, the biomechanical significance of this phenomenon is that long bone cortices may become thinner through endosteal resorption, yet still maintain or even increase bending strength, i.e., second moments of area. Smith and Walker's study, however, is hampered by the fact that they had to rely on radiographs to estimate geometric properties, a technique which has been shown to give very misleading results (see chapter 2), and by the possible presence of secular trends in skeletal structure in their sample, which present difficulties in interpretation of age trends in cross-sectional design studies (Trotter et al., 1968). Also, all individuals in their sample were 45 years of age or older. In the present study, increases with age in I_{max} and I_{min} of tibial and femoral cross sections, measured directly, averaged 8%. It was shown previously that age differences in the present sample are unlikely to be caused by secular or other sampling error effects, thus, these results appear to indicate true changes with age in this population. Subperiosteal expansion and increase in principal moments is progressive and continuous for both males and females over the entire adult age range sampled, i.e., from 20 years through 50+ years (see Figure 24, Chapter 5).

This finding stands in direct contrast to that of Martin and Atkinson (1977), who carried out the only other direct measurement study of bone cross section geometric variation large enough to determine sex or age trends. These authors reported increases with age in second moments of area in the femoral shaft only among the males in their sample, with females showing a decline after about age 35 (ibid.: Fig. 7). Compact cortical bone density and calculated bone material strength declined with age in both sexes (Atkinson and Weatherall, 1967; Martin and Atkinson, 1977: Fig. 6). From these results the authors concluded that a sex-specific difference in structurally "compensatory" mechanisms exists:

> In a pilot study... one of us (B.M.) [had] observed that the increased bone porosity of old age was compensated by an increased structural strength (bending moment of inertia). Subsequent work reported here has shown that this occurs only in males. The results for females are very different, and possibly explain the special problems that women have with osteoporosis. Whether or not the compensatory changes observed in male femora are biomechanically induced, in keeping with Wolff's Law, is open to question. In any case female femora are apparently denied this advantage. (ibid.: 229-230)

It is uncertain to what precise extent bone tissue density declined with age in the Pecos sample, due to problems in the interpretation of bone mineral analyzer data gathered (see previous chapter). Bone mineral content standardized for volumetric differences showed no significant change with age in either sex, however, directly measured bone densities from a femoral midshaft core showed a small but significant decline in density among females, and possibly a small decline in males of the oldest (50+ years) age group. The decrease in bone core density among females from "peak" values in the 30-39 year age group averaged 7.7%, while males decreased 4.6% from "peak" values in the 40-49 year age group. Assuming that these figures are roughly representative of bone density loss through the femoral midshaft, and using the empirical relationship between bone density and compressive strength reported by Carter and Hayes (1976) (strength $\propto$ density^2), this would indicate a maximum decline with age in material strength at this site of approximately 15% among females, and 9% among males. However, <u>gains</u> with age in principal moments of area at this site averaged 12% in females, and 10% in males (see Tables 18a,b, Chapter 5). Granted the inaccuracies and limitations of these calculations (particularly the application of the density-strength formula, which was derived using a much wider range of density values than that in the present sample), still it appears that geometric changes with age do effectively "compensate" for losses in bone density in this cross section, about equally in both sexes. While it cannot be assumed that the femoral midshaft is representative of all other tibial and femoral sites, particularly largely trabecular bone sites such as the femoral neck, this pattern of age change is unlike that of any cross section site reported by Martin and Atkinson for the female members of their sample.

Several possible factors might explain this difference in findings between the two studies. Differences in the age ranges sampled, with Martin and Atkinson including more older individuals in their sample, could possibly explain why females were observed to decline in geometric parameters of bone strength. However, even if consideration is limited to only the younger and middle-aged adults in their sample (20-60 years), no increase with age in second moments of area beyond age 35 can be demonstrated among females (ibid.: Table 1, Figure 7). In contrast, females of the Pecos sample show continuous increases in these properties throughout the entire adult age range. It is also possible that sampling error might account for differences in results between the studies, particularly since Martin and Atkinson were able to include relatively few individuals in certain sex/age ranges (e.g., only 4 females between the ages of 40 and 60 years). However, age changes in properties other than second moments of area, such as average cortical area or cortical thickness, appear to be fairly similar in the two studies (ibid.: Table 1, my calculations; my values for cortical thicknesses extrapolated as described in Chapter 5). The principal differences between the two studies is that in their female sample, no increase with age in total subperiosteal area appears to have occurred (ibid.: Table 1, my extrapolations from their breadth data), thus, with cortical thinning second moments of area declined, while in the present sample, as described above, TA continues to increase throughout adulthood among females, leading to increases in second moments of area even with cortical thinning.

Thus, there may be a true population-specific difference between structural changes with aging among females in a modern, Western population sample (Martin and Atkinson's study), and the Pecos sample. This difference could be brought about by various genetic and environmental effects. While genetic differences certainly must contribute to structural variation, both within and between populations, it is probably most fruitful to investigate structural variation in terms of environmental or environmental-genetic interaction differences (also see discussion at beginning of this chapter). (In addition, hypothesizing a purely genetic mechanism would necessitate postulating different sex-specific genetic differences in the two populations, since males of both samples exhibited similar patterns of aging). Nutritional differences between populations might possibly account for the observed differences in aging patterns, since dietary deficiencies have been shown to produce loss of bone and alterations in skeletal structure (Jowsey, 1977). This would involve postulating different sex-specific nutritional intakes in the two populations, or differences in other nutritional demands, such as those associated with childbirth and lactation. However, consideration of these factors would seem to indicate greater losses of bone and bone strength among Pecos women with aging, rather than increases in structural strength. Also, as pointed out above, rates of cortical thinning and changes in cortical volume were not too different in females of the two samples, indicating more of a remodeling, rather than total bone loss or gain difference between the two populations.

One environmental variable which very likely differed in Martin and Atkinson's sample and the Pecos sample, and which could significantly affect the pattern of structural remodeling with aging, is activity level. While precise figures regarding general activity levels are not available for either sample, it is almost certain that the inhabitants of Pecos Pueblo, particularly women, were more physically active than their modern counterparts. Reference has already been made to the fact that Pueblo women frequently engaged in heavy manual labor such as house construction, water carrying, and agriculturally-related chores. Hooton, in his study of the Pecos skeletal material, noted this fact: "Indian women from the Southwest often, even usually, show a stronger development of muscle markings of the upper extremity than the men, perhaps because of the corn-grinding activities which fall to the lot of the housewife" (1930: 26; but see also my comments regarding Hooton's sexing of the Pecos material, Chapter 3). As discussed earlier, the _type_ of physical activities customarily engaged in by males and females of Pecos may have differed to some degree, however, the total stress on the skeleton resulting from exercise was probably not too different in the two sexes. This may contrast with modern Western populations, where manual subsistence tasks have largely been taken over by mechanized labor, and males may be more encouraged than females to participate in other forms of strenuous physical exercise, such as sports (at least at the time when the subjects of Martin and Atkinson's study were growing up).

The evidence for a physical activity-bone gain response has been reviewed in chapter 2. As noted there, geographical variation among living populations in the rate of proximal femoral fractures has been interpreted in terms of a general activity level difference, with more physically active populations suffering fewer fractures (Chalmers and Ho, 1970: age-specific rates). It is also of interest that only in European populations do the rates of hip fractures differ significantly between the sexes, with age-adjusted rates among females more than twice those among males (ibid.; also see Lewinnek et al., 1980). While fracture incidence rates for Pecos are not available, some idea of relative frequency of fractures in different sex and age groups can be gained by using Hooton's listing of observed fractures in a sample of 503 individuals from the site (1930: Table X-6; approximate sex and age distributions for the individuals inspected for fractures derived from data in Tables II-3, II-5). In this sample, 21 individuals with a long bone or pelvic fracture were recorded. Of these, 63% were male, 37% female. Since Hooton arrived at an approximately 60/40 male/female ratio for Pecos, it can be seen that the fracture incidence among both sexes was about equal.[5] A general increase with age in fracture incidence was noted by Hooton: age-specific fracture rates for the total sample (my calculations) rise from 3.3% and 2.6% in "younger" and "middle-aged" adults, to

---

[5] Hooton's probable inaccuracies in sexing and aging of the Pecos Collection have been previously noted (chapter 3). For the purposes of the present comparisons, these inaccuracies can be ignored without affecting the results to any large degree.

6.0% among "older" adults (aged as over 55 years by Hooton, but as shown in chapter 3, probably an overestimate). Interestingly, if these results are broken down by sex, males are seen to be responsible for this age trend (0%, 2.7%, and 8.5%, respectively, in the three age groups), while females show no age-related increases (5.8%, 2.2%, 3.7%). Another survey of fractures in this sample by G.D. Williams (_ibid._: Table X-8), produced essentially the same results.

Of course, there are many inaccuracies and problems in interpretation inherent in the data quoted above--small numbers of affected individuals, sexing and aging problems, possibility of missing some healed fractures, inability to distinguish traumatic fractures from nontraumatic fractures (e.g., Bauer, 1960). Also, there is no way of telling from these observations exactly when fractures occurred, i.e., evidence of healed fractures occurring in younger adulthood may persist into old age (however, this would tend to _overestimate_ frequency of fracture among older adults). Still, these data do _not_ support a theory of relatively increased fracture incidence among older females relative to males. It is also of interest that Hooton recorded no femoral neck or intertrochanteric fractures in his survey sample (Table X-11). This may be partly a function of the probable limited age range of the Pecos sample, since most proximal femoral fractures in modern populations occur after the seventh decade (Chalmers and Ho, 1970; Lewinnek et al., 1980). However, even by the age of 55-60 years, an appreciable increase in the incidence of hip fractures among females in European populations has been observed (_ibid._). Thus, based on the limited data avaiable, Pecos appears to be more representative of living non-Western populations with low hip fracture rates and no difference in fracture rates between the sexes (e.g., Bantu, Southeast Asian: ibid.) than European populations. In order to further investigate the relationship between age-related fracturing and structural age changes, additional data similar to those collected in the present study will need to be obtained for other populations with known fracture incidence rates.

Besides a general increase in subperiosteal dimensions and second moments of area, other more specific changes with age in geometric properties were observed in the Pecos study sample. These included site-specific changes in bone volume and magnitudes of principal moments at different levels of the tibial and femoral diaphyses, changes in the orientation of greatest bending strength and relative bending strength in different planes (i.e., cross-sectional "shape"), and changes in "whole bone" 3-dimensional morphology. If age changes in bone geometry are to reflect Wolff's Law, one would expect remodeling of bone in such a way that stress increases brought about by endosteal resorption and cortical thinning are kept at a minimum throughout the cross section. Generally this should entail preferential "protection" of bone in highly stressed areas, and greater losses of bone in less stressed areas, reflected in site and sex-specific changes in various geometric parameters. The general age-related increase in subperiosteal dimensions observed is one example of this type of remodeling, since maximum tensile and compressive stresses will occur on the outer surface of long bone cross sections (see Chapter 5). Given this hypothesis, one would expect the sex-related differences in bone geometry noted in the previous section to increase with age, since different loadings in the two sexes would produce continuous differential resorption and deposition of bone throughout adulthood. For most of the cross section properties examined here, this hypothesis is confirmed.

Examining age changes in the femur first, it was noted in chapter 5 that males show the greatest increase with age in I_{max} (maximum bending strength) in the 50% cross section, followed by the 35% section (Table 18a). As discussed above, both of these sections are subjected to relatively high A-P bending stress _in vivo_, particularly when engaging in strenuous physical activities which require powerful flexion and extension of the lower limb segments, such as in running. Interestingly, males show the greatest increase in I_{min} in the 65% section, which, using phi values for this section and age group (Table 15a), can be seen to lie in a generally A-P direction. Also, the I_{max}/I_{min} ratio for femoral sections shows an increase with age among males in those sections where I_{max} is more A-P oriented (35%, 50%), and a decrease where I_{max} is more M-L oriented (20%, 65%; the 80% section shows essentially no change). Thus, at every level of the femoral diaphysis, with the exception of the 80% section (which may be under the relatively lowest A-P bending loadings), males show age-related increases in A-P bending strength.

Female femora show a different pattern of age change in principal moments (Table 18b). By far the greatest increases in both I_{max} and I_{min} occur in the proximal half of the shaft, in particular, the 65% and 80% sections, those sections shown to be under the highest M-L bending stress. There is also some evidence from I_{max}/I_{min} ratios that the 50% cross section becomes more circular (i.e., more M-L distributed) with age, although no shape changes were observed in the two proximal sections. Why I_{min} increases proportionally to I_{max} in the proximal femur among females is unclear, since high M-L bending stress in this area would seem to favor a more M-L distributed shape. This may be related in some way to the decrease in the antetorsion angle of the femur with aging in females, described in another section below, or to a general requirement to maintain or increase overall strength in this region in females.

In the tibia, males show maximum age increases in both principal moments in the 35%, 50%, and 65% sections, while the greatest female increases in these properties occur in the 35% and 50% sections (Tables 17a,b). Similar aging patterns in the two sexes are characteristic of the polar moment of inertia as well. Thus, while sex-related differences in relative bending strength changes with age are not as distinct in the tibia as in the femur (possibly because loadings are not as distinctly different in the tibia), some evidence for a slightly more proximal maximal increase in bending strength is indicated for males; and, as discussed earlier, this is the location of relatively highest A-P bending stress in the tibia.

However, with regard to "shape" (I_{max}/I_{min}) changes with age in the tibia, unlike in the femur, males and females actually become more similar in shape with age. This is brought about by relatively greater increases (or smaller decreases) in I_{max} than in I_{min} among females, while except for the most proximal section, males show about equal increases in the two principal moments or a greater increase in I_{min}. It is hypothesized here that this apparently paradoxical finding is the result of age changes in other aspects of cross-sectional geometry and 3-dimensional morphology of the tibia and femur. One of the most interesting findings of the study was that the direction of greatest bending strength--phi--increased with age in the tibia, i.e., became more A-P oriented. This age change, although occurring in both sexes, was much more marked among males, particularly in the proximal tibia (Tables 14a,b). If, as described earlier, major axes orientations are interpreted as the combined resultants of different loadings applied, or planes of bending, this would indicate a more marked adaptation to A-P bending stresses with aging in the male tibia (see Figure 27a above). Thus, rather than changing the cross-sectional "shape" of the tibia (I_{max}/I_{min}), the orientation of I_{max} is rotated so that more bone is in the plane of A-P bending. Female tibiae may not be able to rotate the direction of greatest bending strength as far anteroposteriorly as males, due to changes at the proximal end of the lower limb with aging, i.e., the femoral antetorsion angle. The significance of the decrease with age in the antetorsion angle among females in terms of stresses in the proximal femur will be discussed in another section below; it will be noted here, however, that it can be shown that this will produce a more mediolateral axial bending plane through the tibia. Thus, following the same line of reasoning as that presented above, the combined resultant of an A-P bending moment about the knee and axial loading of the femoral head will produce more M-L oriented tibial major axes in older females than in older males (who show only a small age-related decrease in the antetorsion angle), particularly in the proximal tibia, which is closest to the femoral head and thus is affected more by changes in the position of this loading point (see Figures 27a,b above). It is unlikely that older females would engage more in activities which would place high A-P bending loads on the tibia than younger females. Rather, the "shape" change in the tibial diaphysis among females may be a kind of compensatory remodeling effect to preserve A-P bending strength, not necessary in males because of their greater rotation of the direction of maximum bending strength.

Thus, it can be seen that there may be several different ways of adapting to stress in the lower limb through bone remodeling. In general, sex-related differences in the type of age changes observed can be related to sex differences in loading patterns. When the 3-dimensional geometry of the "whole bone" (e.g., the femoral neck angle) changes, effects may be produced throughout the lower limb. All of this is hypothetical, of course, and must be further tested using simulated *in vivo* loadings and calculation of complete stress distributions (see below). However, it is clear that remodeling and changes in the cross-sectional shape of the lower limb bones do occur with aging, that these are not limited to a general expansion of subperiosteal dimensions, and that some may be sex-specific. These results confirm some earlier, non-quantitative observations reported by other authors, for example, Van Gerven et al. (1969), who suggested from visual inspection that male femoral midshaft sections underwent "cortical drift" with aging to a more A-P distributed cortical volume (ibid.: Fig. 17). Some remodeling changes in a more proximal femoral section involving selective preservation of the medial and lateral cortices were described by Amtmann (1971). Carlson et al. (1976) reported greater increases with age in average diaphyseal diameter in the distal femur among males, and the proximal femur among females (ibid.: Fig. 5). It is hoped that more investigations of the present kind will be carried out to further elucidate both intra- and inter-populational differences in the pattern of structural changes with aging.

Side differences.

Analyses of side differences in geometric properties showed left lower limb bones to be generally larger in cortical area and in second moments of area than right bones. This left "dominance" in size and bending strength was much more marked in females than males, however, and in fact males showed a slight right "dominance" in I_{max} in most of the femur. The greatest differences between right and left sides were found in the mid-distal femur among females, where CA reached an asymmetry of more than 10%, and principal moments more than 15% (Tables 11a,b; 19a,b). Side differences in I_{max} also showed an increase from the distal to the proximal tibia among females, and in both sexes the I_{max}/I_{min} ratio in tibial sections was much higher on the left side. Finally, the orientation of greatest bending strength in the tibia, and in the distal femur among males, was found to be more A-P oriented on the left side than on the right (Table 15). In fact, the entire left lower limb was observed to be less "twisted" (i.e., smaller femoral antetorsion and tibial distal torsion angles) than the right lower limb, especially among females (Table 6).

All of these data strongly suggest that the left lower limb was more stressed *in vivo*, particularly by greater A-P bending moments, than the right lower limb. These results also suggest that females had a more pronounced "left-leggedness" than males, especially with regard to loadings of the femur. This last observation might imply that females were subjected to greater M-L loadings of the left lower limb (since the femur is subjected to relatively high M-L loadings) except that the greatest differences between sides among females occurred distal to the regions of highest M-L bending stress, in fact reaching a peak in both femoral and tibial sections around the knee joint. Also, the I_{max}/I_{min} ratio in females indicated a somewhat more A-P distributed bone volume in both the left femur and left tibia. This suggests that differential use of the left and right lower limbs for activities producing high A-P bending stress was more marked among females. This observation is supported by the fact that females show an increase with age in side asymmetry in bone areas and bending strength in the A-P plane, and males do not (see Chapter 5).

While favoring of one of the lower limbs during running might conceivably produce such a remodeling effect, it seems more reasonable, given earlier observations on the probable prevalence of running among *males* in this sample, that these greater left side A-P bending moments among females were developed during some other activity, perhaps one involving pushing off of the lower limb, occurring, for example, during hoeing of fields or other agriculturally-related tasks. Thus, the interesting possibility exists that different types of asymmetry in lower limb bone structure can be used to reconstruct specific behavioral differences between males and females of the Pecos sample. It would be quite interesting to compare these data with those gathered from other populations with different customary patterns of activity, to see whether these features are generally characteristic of non-Western, more highly physically-stressed groups, only agriculturalist groups and not gatherer-hunters, etc. As noted earlier, other previous studies of various populations, both living and past, have generally indicated a left "dominance" in lower limb bone dimensions (e.g., Munter, 1936; Schultz, 1937; Latimer and Lowrance, 1965). An earlier radiographic study carried out by the present author on tibial and humeral asymmetry in cortical bone areas and "shape" (breadth indices) in an archaeological sample from a California shellmound also showed greater left bone areas and relatively greater A-P dimensions in the tibia, again more marked among females (Ruff and Jones, 1981).

There is other, independent evidence that the left lower limb was more stressed overall than the right lower limb among the inhabitants of Pecos. Jurmain (1977) carried out a study of degenerative joint disease in which he reported greater incidence of osteoarthritis at the knee joint on the left than on the right side in a sample from Pecos. Interestingly, closer examination of his data (<u>ibid.</u>: Table 1) indicates that this pattern of greater left involvement at the knee was more marked among females than among males. No other joint examined (hip, shoulder, elbow) showed significant side-related differences in osteoarthritis. In addition, other population samples examined by Jurmain (Eskimo, modern US Blacks and Whites) showed no evidence of a similar side difference in osteoarthritis in the lower limb. If degenerative joint disease is at least partly a function of increased stress (ibid.), these results indicate that asymmetry in stresses on the lower limb may have been particularly marked at Pecos in comparison with these other populations, especially among women.

Hooton's (1930) reported fracture incidence for right and left sides in the Pecos sample is also quite interesting with respect to hypothesized side-related differences in bone stress (and thus strength) of the lower limb. Of the 8 fractures observed by Hooton in the femur and tibia, 7 were on the right side (ibid.: Table X-6). While the number of affected individuals is small, and other variables (such as likelihood of particularly stressing right or left sides during activities which might cause fractures) may possibly play a part, this ratio of right-to-left fracture incidence in the lower limb does suggest that left bones were relatively stronger than right bones. In this regard, it is interesting that the incidence of hip fractures in modern populations has been reported to be higher on the left side (Chalmers and Ho, 1970; Lewinnek et al., 1980). In the present study, the femoral neck cross section was the only location in the lower limb among females which did not show a left "dominance" in bone areas or second moments of area (Tables 11b, 19b). In fact, right femoral necks showed greater I_{max}/I_{min} ratios, indicating greater superior-inferior bending strength. Since the femoral neck is probably subjected to higher M-L bending loads (i.e., in the superior-inferior plane of the neck) than A-P bending loads, due to the transfer of body weight from the femoral head to the shaft, this may indicate that during use of the left lower limb for movements producing high A-P bending moments (e.g., pushing off or forward on the limb), the right lower limb is being used more as a stabilizing, weight-bearing member. Since at least some femoral neck fractures are probably produced by loadings which create in effect a very high mediolateral bending moment through the superior-inferior plane of the neck (Backman, 1957), this may explain why hip fractures are generally less prevalent on the right side. It might be possible to test this hypothesis using clinical proximal femoral radiographs of individuals who have fractured right or left hips (e.g., do <u>types</u> of hip fractures differ on the two sides, between males and females, or between individuals with different activity patterns?) Analysis of side differences in trabecular patterns in the proximal femur, using the radiographs taken of the Pecos sample before sectioning (see Chapter 4), are also planned for the future.

<u>The proximal femur.</u>

Age changes in bone areas of the femoral neck section are fairly representative of bone area changes in the rest of the lower limb, i.e., an increase in TA and a somewhat greater increase in MA, leading to a relatively small decline in CA, larger among females. With regard to principal moments, whether calculated on the basis of the area of compact cortical bone or the total subperiosteal area, females show relatively greater increases with age in I_{max} and smaller increases (or larger decreases) in I_{min}. leading to a significantly higher I_{max}/I_{min} ratio among older females (Table 18b). Males show a similar pattern, but to a lesser extent than in females (Table 18 a). Thus, there is evidence that the femoral neck becomes relatively more adapted with aging for superior-inferior bending stress, particularly among females, where higher stresses of this kind would be expected. Interestingly, though, females as a whole do not show more superior-inferiorly distributed bone volume (i.e., higher I_{max}/I_{min} ratios) in the femoral neck than males (Table 16b), although <u>older</u> females do. As suggested earlier, torsional loading of the femoral neck may be an important form of <u>in vivo</u> stressing, and may account for the more circular cross sections in younger females. If this is true, it would suggest that older females become more "specialized" for superior-inferior bending, perhaps to preserve strength in the direction of most critical bending stress. Males may not show the same degree of remodeling of the neck a) because less overall bone is lost in males, and b) superior-inferior bending stress is relatively lower in males. Although direct tissue density (or apparent density) measurements for the femoral neck section are not available, BMC/CA and BMC/TA indices show age changes comparable to other femoral cross sections, i.e., no significant change in BMC/CA, and a slight decline in BMC/TA in both sexes (Tables 24-28).

The explanation for the age-related decline in the antetorsion angle of the femur, much greater in females, may involve a "protective" mechanism in the proximal femur, reducing tensile stresses in the femoral neck. It can be shown trigonometrically that the age increase in the "mechanical"/ideal axes angle of the femur in females will result in an increase in the mediolateral distance between the femoral head and the longitudinal axis of the shaft of 3.5mm. This increased M-L distance will result in an increased lever arm for the gluteal abductors (Radin, 1980; also see Lovejoy et al., 1973). This in turn could decrease the loading on the femoral neck, particularly tensile stresses on the superior surface (see Figure 26), which as stated before, are the most critical stresses in determining bone failure under loading. The fact that this change in 3-dimensional morphology is much more apparent in females also suggests its role in decreasing stress in the femoral neck, since as noted several times previously, this is a relatively more highly stressed area among females. Stresses in the proximal femoral <u>diaphysis</u>, however, may increase under such an increased loading of the gluteal abductors (Rybicki et al., 1972). It is suggested here that they <u>did</u> increase with age among females in this sample, and that this in part explains the very large increases in bending strength parameters in the proximal femoral cross sections observed among females (see above). Viewed in this way, changes in both the femoral neck antetorsion angle and morphology of the proximal femoral shaft serve as "protective" devices for the highly stressed and fracture vulnerable femoral neck.

Conclusions

Perhaps the most important conclusion to be drawn from the results presented in the present study is that biomechanical theory can make an important contribution to the study of skeletal structural variation. Many long bone morphological features of traditional anthropological interest, such as relative "robusticity" or "platycnemia" (Bass, 1971), can be more precisely quantified with respect to their specific effects on bone strength and adaptation to different loading configurations. Ultimately, using these data it may be possible to reconstruct past behavioral patterns and behavioral differences, both within and between populations. Several preliminary interpretations of this kind were made in the present study, including hypothesized sex-specific patterns of activity at Pecos Pueblo, and differential use of the right and left lower limbs, again possibly differing between the sexes. It is expected that extension of this kind of research to between-population differences will yield very interesting and useful results, both to the anthropologist interested in the origin and adaptiveness of structural characteristics, and the bioengineer interested in the general relationship between structure and function in living systems.

The major finding of clinical importance was the documentation of continued subperiosteal expansion and increase in geometrical parameters of bone strength throughout adult life in both males and females. This contrasts with a study of a modern Western population, where only males were found to increase in geometrical bending strength, compensating for a decline in bone density with aging (Martin and Atkinson, 1977). It is suggested here that the different results obtained in the two studies are due to general activity level differences between the two population samples, with Pecos females subjected to significantly more biomechanical stress throughout their lives than their modern counterparts. The implication of this finding is that osteoporosis preventive programs should include exercise as part of their therapeutic regimens. Fracture incidence among adults at Pecos, particularly age-related fractures such as those of the proximal femur, appears to have been relatively low, although there are problems in interpreting the data available for this sample. Comparisons of structural differences in populations with known differences in fracture rates, activity levels, and other characteristics such as nutritional levels, would greatly aid in elucidating the causative factors involved in age-related fracturing.

The results of the present study could be extended in several other ways. Analysis of stress distributions within cross sections from the Pecos sample under simulated loadings, using a computer program (STRESS) developed in the Beth Israel Biomechanics Laboratory, where much of the work involved in present study was carried out, is planned for the near future. This will allow much more precise testing of the various hypotheses put forward in the present study regarding specific loadings, resultant stresses, and bone remodeling. Other geometric and material properties of the study sample, including cortical thicknesses (which can be calculated using the perimeter point data stored by SLICE), and better estimates of density, measured directly at several different sites, should be determined to further clarify some of the age-related changes observed here. Inclusion of juveniles from Pecos could help to determine the age at which certain sex and side-related characteristics appear. Finally, analysis of trabecular systems and other structural features of interest using techniques such as finite element analysis, will allow a better quantification of stresses and strength in particularly important areas such as the proximal femur.

References

Aaron, J.E., Gallagher, J.C., Anderson, J., Stasiak, L., Langton, E.B., Nordin, B.E.C., and Nicholson, M. (1974) Frequency of osteomalacia and osteoporosis in fractures of the proximal femur. Lancet, 1: 229-233.

Acsadi, G., and Nemeskeri, J. (1970) History of Human Life Span and Mortality. Budapest, Akademiai Kiado.

Adams, P., Davies, G.T., and Sweetnam, P. (1970) Osteoporosis and the effects of ageing on bone mass in elderly men and women. Quart. J. Med., 39: 601-615.

Albright, F., and Reifenstein, E.C. Jr. (1948) The Parathyroid Glands and Metabolic Disease. Bailliere, Tindall and Cox, London.

Albright, F., Smith, P.H., and Richardson, A.M. (1941) Postmenopausal osteoporosis. Its clinical features. JAMA, 116: 2465-2474.

Alffram, P.A. (1964) An epidemiological study of cervical and trochanteric fractures of the femur in an urban population. Acta Orthop. Scand. Suppl., 65.

Alffram, P.A., and Bauer, G.C.H. (1962) Epidemiology of fractures of the forearm. A biomechanical investigation of bone strength. J. Bn. Jt. Surg., 44A: 105-114.

Amtmann, V.E. (1968) The distribution of breaking strength in the human femur shaft. J. Biomech., 1: 271-277.

Amtmann, V.E. (1971) Mechanical stress, functional adaptation and the variation structure of the human femur diaphysis. Adv. Anat. Emb. Cell Biol., 44 (3). Springer-Verlag, Berlin.

Arnold, J.S. (1960) Quantification of mineralization of bone as an organ and tissue in osteoporosis. Clin. Orthop., 17: 167-175.

Arnold, J.S., Bartley, M.H., Tonot, S.A., and Jenkins, D.P. (1966) Skeletal changes in aging and disease. Clin. Orthop., 49: 17-38.

Atkinson, P.J., and Weatherall, J.A. (1967) Variation in the density of the femoral diaphysis with age. J. Bn. Jt. Surg. 49B: 781-788.

Backman, S. (1957) The proximal end of the femur. Acta Radiol. Suppl., 146.

Barnett, E., and Nordin, B.E.C. (1960) The radiological diagnosis of osteoporosis--a new approach. Clin. Radiol., 11: 166-174.

Bass, W.M. (1971) Human Osteology. Missouri Arch. Soc., Columbia, Missouri.

Bauer, G.C.H. (1960) Epidemiology of fracture in aged persons. A preliminary investigation in fracture etiology. Clin. Orthop., 17: 219-225.

Berry, F.R. (1952) Angle variation patterns of normal hip, knee and ankle in different operations. Univ. Calif. (Berkeley) Pros. Dev. Res. Proj.

Billing, L. (1954) Roentgen examination of the proximal end of the femur in children and adolescents. Acta Radiol. Suppl., 110.

Bloom, R.A., and Laws, J.W. (1970) Humeral cortical thickness as an index of osteoporosis in women. Br. J. Radiol., 43: 522-527.

Blumenbach, J.F. (1776) De Generis Humani Varietate Nativa. Vandenhoeck, Gottingen.

Bohler, L. (1958) The Treatment of Fractures. Grune and Stratton, New York.

Boyd, R.M., Cameron, E.C., McIntosh, H.W., and Walker, V.R. (1974) Measurement of bone mineral content in vivo using photon absorptiometry. Canad. Med. Assoc. J., 111: 1201-1205.

Bresler, B., and Frankel, J.P. (1950) The forces and moments in the leg during level walking. Trans. ASME, 72: 27-36.

Brooks, S.T. (1955) Skeletal age at death: The reliability of cranial and pubic age indicators. Am. J. Phys. Anthrop., 13: 567-597.

Brothwell, D.R. (1972) Digging Up Bones. British Museum, London.

Buhr, A.J., and Cooke, A.M. (1959) Fracture patterns. Lancet, 1: 531-534.

Burr, D.B., VanGerven, D.P., and Gustav, B.L. (1977) Sexual dimorphism and mechanics of the human hip: a multivariate assessment. Am. J. Phys. Anthrop., 47: 273-278.

Cadien, J.D., Harris, E.F., Jones, W.P., and Mandarino, L.J. (1976) Biological lineages, skeletal populations, and microevolution. Yrbk. Phys. Anthrop., 1974: 194-201.

Cameron, J.R., and Sorenson, J. (1963) Measurement of bone mineral in vivo, an improved method. Science, 142: 230-232.

Cameron, J.R., Mazess, R.B., and Sorenson, J.A. (1968) Precision and accuracy of bone mineral determination by direct photon absorptiometry. Invest. Radiol., 3: 141-150.

Carlson, D.S., Armelagos, G.J., and VanGerven, D.P. (1976) Patterns of age-related cortical bone loss (osteoporosis) within the femoral diaphysis. Hum. Biol., 48: 295-314.

Carter, D.R. (1978) Anisotropic analysis of strain rosette information from cortical bone. J. Biomech., 11: 199-202.

Carter, D.R., and Hayes, W.C. (1977) The compressive behavior of bone as a two-phase porous material. J. Bn. Jt. Surg., 59A: 954-962.

Carter, D.R., and Hayes, W.C., (1976) Bone compressive strength: The influence of density and strain rate. Science, 194: 1174-1176.

Carter, D.R., Hayes, W.C., and Schurman, D.J. (1976) Fatigue life of compact bone II. Effects of microstructure and density. J. Biomech., 9: 211-218.

Carter, D.R., Smith, D.J., Spengler, D.M., Daly, C.H., and Frankel, V.H. (1980) Measurement and analysis of in vivo bone strains on the canine radius and ulna. J. Biomech., 13: 27-38.

Carter, D.R., and Vasu, R. (1981) Plate and bone stresses for single- and double-plated femoral fractures. J.Biomech., 14: 55-62.

Chalmers, J., and Ho, K.C. (1970) Geographical variations in senile osteoporosis. J. Bn. Jt. Surg., 52B: 667-675.

Churches, A.E., Howlett, C.R., and Waldron, K.J. (1979) The response of living bone to controlled time-varying loading: Method and preliminary results. J. Biomech., 12: 35-45.

Crane, L. (1959) Femoral torsion and its relation to toeing-in and toeing-out. J. Bn. Jt. Surg., 41A: 421-428.

Currey, J.D. (1968) The adaptation of bone to stress. J. Theoret. Biol., 20: 91-106.

Currey, J.D. (1969) The mechanical consequences of variation in the mineral content of bone. J. Biomech., 2: 1-11.

Currey, J.D. (1970) The mechanical properties of bone. Clin. Orthop., 73: 210-231.

Dalen, N., and Olsson, K.E. (1974) Bone mineral content and physical activity. Acta Orthop. Scand., 45: 170-174.

Dequeker, J. (1975) Bone and ageing. Ann. Rheum. Dis., 34: 100-115.

Dequeker, J., and Merlevede, W. (1971) Collagen content and collagen extractability pattern of adult human trabecular bone according to age, sex and amount of bone mass. Biochem. Biophys. Acta, 244: 410.

Dewey, J.R., Armelagos, C.J., and Bartley, M.H. (1969) Femoral cortical involution in three Nubian populations. Hum. Biol., 41: 13-28.

Dixon, W.J., and Brown, M.B. (eds.) (1979) BMDP-79 Biomedical Computer P-Series. Univ. Calif. Press, Berkeley, Calif.

Doyle, F. (1972) Involutional osteoporosis. In: Clinics in Endocrinology and Metabolism, McIntyre, J. (ed.).

Doyle, F., Brown, J., and Lachance, C. (1970) Relation between bone mass and muscle weight. Lancet, 1: 391-393.

Dupertuis, C.W., and Haddon, J.A. Jr. (1951) On the reconstruction of stature from long bones. Am. J. Phys. Anthrop., 9: 15-54.

Eberhart, H.D., Inman, V.T., and Bresler, B. (1954) The principal elements in human locomotion. In: Human Limbs and Their Substitutes, Klopsteg, P.E., and Wilson, P.D. (eds.). McGraw-Hill, New York.

Elftmann, H. (1939) Forces and energy changes in the leg during walking. Am. J. Physiol., 125: 339.

Enlow, D.H. (1963) Principles of Bone Remodeling. Charles C. Thomas, Springfield, Ill.

Epker, B.N., and Frost, H.M. (1966) Periosteal appositional bone growth from age two to age seventy in man. A tetracycline evaluation. Anat. Rec., 154: 573-577.

Ericksen, M.F. (1976) Cortical bone loss with age in three Native American populations. Am. J. Phys. Anthrop., 45: 443-452.

Evans, F.G. (1953) Methods of studying the biomechanical significance of bone form. Am J. Phys. Anthrop., 11: 413-435.

Evans, F.G. (1973) Mechanical Properties of Bone. Charles C. Thomas, Springfield, Ill.

Eveleth, P., and Tanner, J.M. (1976) Worldwide Variation in Human Growth. Cambridge Univ. Press, London.

Exton-Smith, A.N., Millard, P.H., Payne, P.R., and Wheeler, E.F. (1969) Pattern of development and loss of bone with age. Lancet, 2: 1154-1157.

Foltz, A.S., Relnick, D.J., and Jurisy, J.M. (1976) The relationships between bending stiffness, mineral content, geometry, and strength of the human ulna. Am. J. Roentgen., 126: 1270-1271.

Frankel, V.F., amd Burstein, A.H. (1965) Load capacity of tubular bone. In: Biomechanics and Related Bio-engineering Topics, Kenedi, R.M. (ed.). Pergamon, London.

Frankel, V.H., and Burstein, A.H. (1970) Orthopaedic Biomechanics. Lea and Febiger, Philadelphia.

Frazer, J.E. (1948) The Anatomy of the Human Skeleton. Churchill Ltd., London.

Fredensborg, N., and Nilsson, B.E. (1977) The bone mineral content and cortical thickness in young women with femoral neck fracture. J. Bn. Jt. Surg., 44B: 520-527.

Frisancho, A.R., Garn, S.M., and Ascoli, W. (1970) Subperiosteal and endosteal bone apposition during adolescence. Hum. Biol., 42: 639-664.

Frost, H.M. (1963) Bone Remodeling Dynamics. Charles C. Thomas, Springfield, Ill.

Garn, S.M. (1970) The Earlier Gain and the Later Loss of Cortical Bone. Charles C. Thomas, Springfield, Ill.

Garn, S.M., Mayer, G.H., and Shaw, H.A. (1976) Paradoxical bilateral asymmetry in bone size and bone mass in the hand. Am. J. Phys. Anthrop., 45: 209-210.

Garn, S.M., Rohmann, C.G., and Wagner, B. (1967a) Bone loss as a general phenomenon in man. Fed. Proc., 26: 1729-1736.

Garn, S.M., Rohmann, C.G., Wagner, B., and Ascoli, W. (1967b) Continuing bone growth throughout life: a general phenomenon. Am. J. Phys. Anthrop., 26: 313-318.

Gilbert, B.M. (1976) Anterior femoral curvature: Its probable basis and utility as a criterion of racial assessment. Am. J. Phys. Anthrop., 45: 601-604.

Gilbert, B.M., and Mckern, T.W. (1973) A method for aging the female os pubis. Am. J. Phys. Anthrop., 38: 31-38.

Goldsmith, N.F., Johnston, J.O., Ury, H., Vose, G., and Colbert, C. (1971) Bone mineral estimation in normal and osteoporotic women: A comparability trial of four methods and seven bone sites. J. Bn. Jt. Surg., 53A: 83-100.

Goldstein, M.S. (1953) Some vital statistics based on skeletal material. Hum. Biol., 25: 3-12.

Hall, M.C. (1961) The trabecular patterns of the neck of the femur with particular reference to changes in osteoporosis. Canad. Med. Assoc. J., 85: 1141-1144.

Hanihara, K. (1952) Age changes in the male Japanese pubic bone. J. Anthrop. Soc. Nippon, 62: 245-260.

Hanna, R.E., and Washburn, S.L. (1953) The determination of sex of skeletons, as illustrated by a study of the Eskimo pelvis. Hum. Biol., 25: 21-27.

Harris, W.H., and Heaney, R.P. (1969) Skeletal renewal and metabolic bone disease. New Engl. J. Med., 280: 193-253.

Hayes, W.C., Swenson, L.M., and Schurman, D.J. (1978) An axisymmetric finite element model of the lateral tibial plateau. J. Biomech., 11: 21-23.

Hays, W.L., and Winkler, R.L. (1970) Statistics: Probability, Inference, and Decision. Holt, Rinehart and Winston, New York.

Hooton, E.A. (1930) The Indians of Pecos Pueblo. A Study of Their Skeletal Remains. Pap. Phil. Acad. SW Exped., 4. Yale Univ. Press, New Haven.

Horsman, A., and Simpson, M. (1975) The measurement of sequential changes in cortical bone geometry. Brit. J. Radiol., 48: 470-476.

Howells, W.W. (1960) Estimating population numbers through archaeological and skeletal remains. In: The Application of Quantitative Methods in Archaeology, Heizer, R.F., and Cook, S.F. (eds.). Viking Fund Pub. Anth, No. 28. Wenner-Gren, New York.

Hutter, C.G., and Scott, W. (1949) Tibial Torsion. J. Bn. Jt. Surg., 31A: 511-518.

Ingalls, N.W. (1931) Observations on bone weights. Am. J. Anat., 48: 45-98.

Inman, V.T. (1947) Functional aspects of the abductor muscles of the hip. J. Bn. Jt. Surg., 29: 607-619.

Iskrant, A.P., and Smith, R.W. (1969) Osteoporosis in women 45 years and over related to subsequent fractures. Publ. Health Rep., 84: 33-38.

Israel, H. (1973) Age factor and the pattern of change in craniofacial structures. Am. J. Phys. Anthrop., 39: 111-128.

Johnston, C.C., Smith, D.M., Yu, P.L., and Deiss, W.P. (1968) In vivo measurement of bone mass in the radius. Metabolism, 17: 1140-1153.

Johnston, F.E., and Watts, E.S. (1969) Endosteal deposition of bone at the midshaft of the second metacarpals of adolescent females. Anat. Rec., 163: 67-70.

Jones, H.H., Priest, J.D., Hayes, W.C., and Tichenor, C.C (1977) Humeral hypertrophy in response to exercise. J. Bn. Jt. Surg., 59A: 204-208.

Jowsey, J. (1960) Age changes in human bone. Clin Orthop., 17: 210-218.

Jowsey, J. (1977) Metabolic Diseases of Bone. Saunders, Philadelphia.

Jowsey, J., Riggs, B.L., Kelly, P.J., and Hoffman, D.L. (1972) Effect of combined therapy with sodium fluoride, vitamin D and calcium in osteoporosis. Amer. J. Med., 53: 43.

Jungers, W.L., and Minns, R.J. (1979) Computed tomography and biomechanical analysis of fossil long bones. Am. J. Phys. Anthrop., 50: 285-290.

Jurmain, R.D. (1975) Distribution of Degenerative Joint Disease in Skeletal Populations. PhD. Thesis, Harvard Univ.

Jurmain, R.D. (1977) Stress and the etiology of osteoarthritis. Am J. Phys. Anthrop., 46: 353-366.

Kenedi, R.M. (ed.) (1965) Biomechanics and Related Bio-Engineering Topics. Pergamon Press, London.

Kidder, A.V. (1924) An Introduction to the Study of Southwestern Archaeology with a Preliminary Account of the Excavations at Pecos. Pap. Phil. Acad. SW Exped., 1. Yale Univ. Press, New Haven.

Kidder, A.V. (1931) The Pottery of Pecos, Vol. 1. Dull Paint Wares. Pap. Phil. Acad. SW Exped., 5. Yale Univ. Press, New Haven.

Kidder, A.V. (1932) The Artifacts of Pecos. Pap. Phil. Acad. SW Exped., 6. Yale Univ. Press, New Haven.

Kidder, A.V. (1958) Pecos, New Mexico: Archaeological Notes. Pap. Peabody Found. Arch., 5: Phillips Academy, Andover, Mass.

Kidder, M.A., and Kidder, A.V (1917) Notes on the pottery of Pecos. Am. Anthrop., 19: 325-360.

Kidder, A.V., and Shepard, A.O. (1936) The Pottery of Pecos, Vol. 2. The Glaze-Paint, Culinary, and Other Wares. Pap. Phil. Acad. SW Exped., 7. Yale Univ. Press, New Haven.

Kimura, T. (1974) Mechanical characteristics of human lower leg bones. J. Fac. Sci., Univ. Tokyo, Sect. 5, 4: 319-393.

King, J.W., Brelsford, H.J., and Tullos, H.S. (1969) Analysis of the pithching arm of the professional baseball pitcher. Clin. Orthop., 67: 116-123.

Kingsley, P.C., and Olmsted, K.L. (1948) A study to determine the angle of anteversion of the femoral neck of the femur. J. Bn. Jt, Surg., 30A: 745-751.

Klenerman, I., Swanson, S.A.V., and Freeman, M.A.R. (1967) A method for the clinical estimation of the strength of a bone. Proc. Roy. Soc. Med., 60: 10-14.

Koch, J.C. (1917) The laws of bone architecture. Am. J. Anat., 21: 177-298.

Krogman, W.H. (1962) The Human Skeleton in Forensic Medicine. Charles C. Thomas, Springfield, Ill.

Lambert, K.L. (1971) The weight-bearing function of the fibula. J. Bn. Jt. Surg., 53A: 507-513.

Lanyon, L.E., Hampson, W.G.J., Goodship, A.G., and Shah, J.S. (1975) Bone deformation recorded in vivo from strain gages attached to the human tibial shaft. Acta Orthop. Scand., 46: 256-268.

Lanz, von T. (1949) Anatomische undentwicklungsgeschichtliche probleme am huftgelenk. Verhandl.d. Dtsch. Orthop. Ges., 37: 7-40.

Lanz, von T., and Wachsmuth, W. (1935) Praktische Anatomie. Bd. I, Teil 4. Bein und Statik, Berlin.

Latimer, H.B., and Lowrance, E.W. (1965) Bilateral asymmetry in weight and in length of human bones. Anat. Rec., 152: 217-224.

Le Damany, P. (1909) La torsion du tibia normale, pathologique, experimentale. J. Anat. Physiol., 45: 598-615.

Ledley, R.S., DiChiro, G., Luessenhop, A.J., and Twigg, H.L. (1974) Computerized transaxial x-ray tomography of the human body. Science, 186: 207-212.

Legoux, P. (1966) Determination de l'Age Dentaire de Fossiles de la Lignee Humaine. Libraire Maloine, Paris.

Lewinnek, G.E., Kelsey, J., White, A.A. III, and Kreiger, N.J. (1980) The significance and a comparative analysis of the epidemiology of hip fractures. Clin. Orthop., 152: 35-43.

Liskova, M., and Hert, J. (1971) Reaction of bone to mechanical stimuli. Part 2. Periosteal and endosteal reaction of tibial diaphysis in rabbit to intermittent loading. Fol. Morph., 29: 301-317.

Lovejoy, C.O. (1970) Biomechanical Methods for the Analysis of Skeletal Variation with an Application by Comparison of the Theoretical Diaphyseal Strength of Platycnemic and Euricnemic Tibias. PhD. Thesis, Univ. Mass.

Lovejoy, C.O., Heiple, K.G., and Burstein, A.H. (1973) The gait of Australopithecus. Am. J. Phys. Anthrop., 38: 757-780.

Lovejoy, C.O., and Barton, T.J. (1980) A simple, rapid method of obtaining geometrical properties from sections or laminograms of long bones. J. Biomech., 13: 65-67.

Lovejoy, C.O., and Burstein, A.H. (1977) Geometrical properties of bone sections determined by laminography and physical section. J. Biomech., 10: 527-528.

Lovejoy, C.O., Burstein, A.H., and Heiple, K.G. (1976) The biomechanical analysis of bone strength: A method and its application to platycnemia. Am. J. Phys. Anthrop., 44: 489-506.

Lovejoy, C.O., Meindl, R.S., Pryzbeck, T.R., Barton, T.S., Heiple, K.G., and Kotting, D. (1977) Paleodemography of the Libben site, Ottowa County, Ohio. Science, 198: 291-293.

Lovejoy, C.O., and Trinkaus, E. (1980) Strength and robusticity of the neanderthal tibia. Am J. Phys. Anthrop., 53: 465-470.

Macquet, P.G.J. (1976) Biomechanics of the Knee. Springer-Verlag, Berlin.

Martin, R. (1928) Lehrbuch der Anthropologie. Gustav Fischer, Jena.

Martin, R.B., and Atkinson, P.J. (1977) Age and sex-related changes in the structure and strength of the human femoral shaft. J. Biomech., 10: 223-231.

Mazess, R.B. (1971) Estimation of bone and skeletal weight by direct photon absorptiometry. Invest. Radiol., 6: 52-60.

Mazess, R.B., and Cameron, J.R. (1974) Bone mineral content in normal US Whites. In: Proceedings of International Conference on Bone Mineral Measurements, Mazess, R.B. (ed.). Gov. Printing Office, Washington, D.C.

Mazess, R.B., Cameron, J.R., O'Connor, R., and Knutzen, D. (1964) Accuracy of bone mineral measurement. Science, 145: 388-389.

Mazess, R.B., and Jones, R. (1972) Weight and density of Sadlermuit Eskimo long bones. Hum. Biol., 44: 537-548.

McHenry, H.M. (1975) Biomechanical interpretation of the early hominid hip. J. Hum. Evol., 4: 343-356.

McKern, T.W., and Stewart, T.D. (1957) Skeletal Changes in Young American Males, Analyzed from the Standpoint of Identification. Headquarters Quartermaster Res. Dev. Command, Tech. Rep. EP-45. Natick, Mass.

McLeish, R.D., and Charnley, J. (1970) Abduction forces in the one-legged stance. J. Biomech., 3: 191-209.

Meema, H.E., Bunker, M.L., and Meema, S. (1965) Loss of compact bone due to menopause. Obstet. Gynec., 26: 333-343.

Meema, H.E., and Meema, S. (1969) Cortical bone mineral density versus cortical thickness in the diagnosis of osteoporosis: A roentgenologic-densitometric study. J. Am. Geriat. Soc., 17: 120-141.

Meema, H.E., and Meema, S. (1974) Involutional (physiologic) bone loss in women and the feasibility of preventing structural failure. J. Am. Geriat. Soc., 22: 443-452.

Melick, R.A., and Miller, D.R. (1966) Variations of tensile strength of human cortical bone with age. Clin. Sci., 30: 243-248.

Mensforth, R.P., and Lovejoy, C.O. (1978) Femoral cortical involution in a Late Woodland skeletal population: epidemiological and biomechanical observations. (abstract) Am. J. Phys. Anthrop., 48: 419.

Mikulicz, J. (1878) Uber die individuellen formdifferenzen am femur und der tibia des menschen. Arch. Anat. Physiol., Anat. Abt., S.: 351-404.

Miles, A.E.W. (1963) The dentition in the assessment of individual age. In: Dental Anthropology, Brothwell, D.R. (ed.) Pergamon, London.

Miller, G.J., and Purkey, W.W. (1980) The geometric properties of paired human tibiae. J. Biomech., 13: 1-8.

Minns, R.J., Bremble, G.R., and Campbell, J. (1975) The geometric properties of the human tibia. J. Biomech., 8: 253-255.

Minns, R.J., Bremble, G.R., and Campbell, J. (1977) A biomechanical study of internal fixation of the tibial shaft. J. Biomech., 10: 569-579.

Morgan, D.B. (1973) The metacarpal bone: A comparison of the various indices for the assessment of the amount of bone and for the detection of loss of bone. Clin. Radiol., 24: 77-82.

Morrison, J.B. (1968) Bioengineering analysis of force actions transmitted by the knee joint. Bio-Med. Engin., 3: 164-170.

Morrison, J.B. (1969) Function of the knee joint in various activities. Bio-Med. Engin., 3: 573-580.

Morrison, J.B. (1970) The mechanics of the knee joint in relation to normal walking. J. Biomech., 3: 51-61.

Mueller, K.H., Trias, A., and Ray, R.D. (1966) Bone density and composition. Age-related and pathological changes in water and mineral content. J. Bn. Jt. Surg., 48A: 140-148.

Munter, A.H. (1936) A study of the lengths of the long bones of the arms and legs in man, with special reference to Anglo-Saxon skeletons. Biometrika, 28: 258-294.

Nagurka, M.L., and Hayes, W.C. (1980) An interactive graphics package for calculating cross-sectional properties of complex shapes. J. Biomech., 13: 59-64.

Nelson, C.T. (1938) The teeth of the Indians of Pecos Pueblo. Am J. Phys. Anthrop., 23: 261-298.

Newton-John, M.B., and Morgan, D.B. (1970) The loss of bone with age, osteoporosis, and fractures. Clin. Orthop., 71: 229-252.

Nilsson, B.E., and Westlin, N.E. (1971) Bone density in atheletes. Clin. Orthop., 77: 179-182.

Nilsson, B.E., and Westlin, N.E. (1977) Bone mineral content and fragility fractures. Clin. Orthop., 124: 161-164.

Nordin, B.E.C., MacGregor, J., and Smith, D.A. (1966) The incidence of osteoporosis in normal women: Its relation to age and the menopause. Quart. J. Med., 35: 25-37.

Oh, I., and Harris, W.H. (1978) Proximal strain distribution in the loaded femur. J. Bn. Jt. Surg., 60A: 75.

Ortiz, A. (ed.) (1979) Handbook of North American Indians. Vol. 9 (Southwest). Smith. Inst., Washington, D.C.

Owsley, D.W., and Bass, W.M. (1979) A demographic analysis of skeletons from the Larson site (39WW2) Walworth County, South Dakota: Vital statistics. Am. J. Phys. Anthrop., 51: 145-154.

Parsons, F.G. (1914) The characters of the English thigh bone. J. Anat. Physiol., 48: 238-267.

Paul, J.P. (1966) The biomechanics of the hip-joint and its clinical relevance. Proc. Roy. Soc. Med., 59: 943-948.

Paul, J.P. (1967) Forces transmitted by joints in the human body. Proc. Inst. Mech. Engin., 181: 8-15.

Paul, J.P. (1971) Load actions on the human femur in walking and some resultant stresses. Exp. Mech., 11: 121-125.

Pauwels, F. (1935) Der Schenkelhalsbruch Ein mechanisches Problem. Ferninand Enke, Stuttgart.

Pauwels, F. (1950) Die bedeutung der muskelkrafte fur die regelung der beansprunchung des rohrenknochens wahrend der bewegung der glieder. Dritter beitrag zur funktionellen anatomie und kausalen morphologie des stutzapparates. K. Anat. Entwickl.-Gesch., 115: 327-351.

Pearson, (1898) On the reconstruction of the statur of prehistoric races. Phil. Trans. Roy. Soc., series A, 192: 169-244.

Pearson, K., and Bell, J. (1919) A Study of the Long Bones of the English Skeleton. Pt. I. The Femur. Draper's Co. Res. Mem., Biom. Series X, Cambridge Univ. Press, London.

Per Linton, G. (1949) Types of displacement in fractures of the femoral neck. J. Bn. Jt. Surg., 31B: 184-187.

Perzigion, A.J. (1973) Osteoporotic bone loss in two prehistoric Indian populations. Am. J. Phys. Anthrop., 39: 87-95.

Piotrowski, G., and Wilcox, G.A. (1971) The STRESS program: A computer program for analysis of stresses in long bones. J. Biomech., 4: 497-506.

Piziali, R.L., Hight, T.K., and Nagel, D.A. (1976) An extended structural analysis of long bones--application to the human tibia. J. Biomech., 9: 695-701.

Plato, C.C., Wood, J.L., and Norris, A.H. (1980) Bilateral asymmetry in bone measurements of the hand and lateral hand dominance. Am. J. Phys. Anthrop., 52: 27-31.

Radin, E.L. (1980) Biomechanics of the human hip. Clin Orthop., 152: 28-34.

Reilly, D.T., and Burstein, A.H. (1974) The mechanical properties of cortical bone. J. Bn. Jt. Surg., 56A: 1001-1022.

Roberts, W.H. (1962) Femoral torsion in normal human development and as related to dyplasia. Anat. Rec., 143: 369-374.

Rockwood, C.A., and Green, D.P. (eds.) (1975) Fractures. Lippincott, Philadelphia.

Ruff, C.B. (1977) Compact Cortical Bone Loss with Age--Methodological Issues and Application to an Archaeological Sample from California. Masters Thesis, Dept. Anthrop., Univ. Penn.

Ruff, C.B. (1980) Age differences in craniofacial dimensions among adults from Indian Knoll, Kentucky. Am. J. Phys. Anthrop., 53: 101-108.

Ruff, C.B. (n.d.) CT scanning, radiographic and direct measurement of the biomechanical properties of long bone cross sections--a comparability trial. Unpublished manuscript.

Ruff, C.B., and Jones, H.H. (1981) Bilateral asymmetry in cortical bone of the humerus and tibia--sex and age factors. Hum. Biol., 53: in press.

Rybicki, E.F., Simonen, F.A., and Weis, E.B. Jr. (1972) On the mathematical analysis of stress in the human femur. J. Biomech., 5: 203-215.

Rydell, N.W. (1966) Forces acting on the femoral head prothesis. Acta Orthop. Scand. Suppl., 88.

Sando, J.S. (1979) Jemez Pueblo. In: Handbook of North American Indians. Vol. 9 (Southwest), Ortiz, A. (ed.). Smith. Inst., Washington, D.C.

Saville, P.D. (1972) Osteoporosis: An overview. In: Clinical Aspects of Metabolic Bone Disease, Proceedings of the International Symposium on Clinical Aspects of Metabolic Bone Disease, Frame, B., Parfitt, A.M., and Duncan, H. (eds.). Excerpta Medica, Amsterdam.

Schroeder, A.H. (1979) Pecos Pueblo. In: Handbook of North American Indians. Vol. 9 (Southwest), Ortiz, A (ed.). Smith. Inst., Washington, D.C.

Schultz, A.H. (1926) Fetal growth of man and other primates. Quart. Rev. Biol., 1: 465-521.

Schultz, A.H. (1937) Proportions, variability and asymmetries of the long bones of the limbs and the clavicles in man and apes. Hum. Biol., 9: 281-328.

Shands, A.R. Jr., and Steele, M.K. (1958) Torsion of the femur. J. Bn. Jt, Surg., 40A: 803-816.

Shighley, J.F. (1976) Applied Mechanics of Materials. McGraw-Hill, New York.

Singer, R. (1953) Estimation of age from cranial suture closure. J. For. Med., 1: 52-59.

Smidt, G.L. (1973) Biomechanical analysis of knee flexion and extension. J. Biomech., 6: 31-39.

Smith, D.M., Johnston, C.C, and Yu, P-L. (1972) In vivo measurement of bone mass. JAMA, 219: 325-329.

Smith, E.L., and Babcock, S.W. (1973) Effects of physical activity on bone loss in the aged. (abstract) Med. Sci. Sports, 5: 68.

Smith, E.L., and Reddan, W.G. (1975) The effects of physical activity on bone in the aged. (abstract) Med. Sci. Sports, 7: 84.

Smith, R.W., and Walker, R.R. (1964) Femoral expansion in aging women: Implications for osteoporosis and fractures. Science, 145: 156-157.

Steinbock, R.T. (1976) Paleopathological Diagnosis and Interpretation. Bone Diseases in Ancient Populations. Charles C. Thomas, Springfield, Ill.

Steindler, A. (1955) Kinesiology. Charles C. Thomas, Springfield, Ill.

Stevens, J., Freeman, P.A., Nordin, B.E.C., and Barnett, E. (1962) The incidence of osteoporosis in patients with femoral neck fracture. J. Bn. Jt. Surg., 44B: 520-527.

Stevenson, M.C. (1901-2) The Zuni Indians: Their Mythology, Esoteric Fraternities, and Ceremonies. 23rd Ann. Rep. Am. Bur. Ethn. Smith. Inst., Washington, D.C.

Stevenson, P.H. (1929) On racial differences in stature long bone regression formulae, with special reference to stature reconstruction formulae for the Chinese. Biometricka, 21: 303-318.

Stewart, I.M. (1955) Fractures of neck of femur: Incidence and implications. Br. Med. J., 1: 698-701.

Stewart, T.D. (1954) Evaluation of evidence from the skeleton. In: Legal Medicine, Gradwohl, R.B.H. (ed.). Mosby, St. Louis.

Stewart, T.D. (1957) Distortion of the pubic symphyseal surface in females and its effects on age determination. Am. J. Phys. Anthrop., 15: 9-18.

Stewart, T.D. (1958) The rate of development of vertebral osteoarthritis in American Whites and its Significance in Skeletal Age Identification. The Leech, 28: 144-151.

Suchey, J.M. (1979) Problems in the aging of females using the os pubis. Am. J. Phys. Anthrop., 51: 467-470.

Thompson, D.D. (1980) Age-related changes in bone mineralization, cortical thickness, and haversian canal area. Submitted for publication.

Timoshenko, S.P., and Gere, J.M. (1972) Mechanics of Materials. Van Nostrand Reinhold, New York.

Todd, T.W. (1920) Age changes in the pubic bone: I. The male White pubis. Am. J. Phys. Anthrop., 3 (o.s.): 285-334.

Todd, T.W. (1921) Age Changes in the pubic bone: II. Pubis of male Negro-White hybrid. III. Pubis of White female. IV. Pubis of female Negro-White hybrid. Am. J. Phys. Anthrop., 4 (o.s.): 1-70.

Todd, T.W., and Lyon, D.W. (1924) Endocranial suture closure, its progress and age relationship. Part I. Adult males of White stock. Am. J. Phys. Anthrop., 7 (o.s.): 325-384.

Trotter, M., Broman, G.E., and Peterson, R.R. (1959) Density of cervical vertebrae and comparison with densities of other bones. Am. J. Phys. Anthrop., 17: 19-25.

Trotter, M., Broman, G.E., and Peterson, R.R. (1960) Densities of bones of White and Negro skeletons. J. Bn. Jt. Surg., 42A: 50-58.

Trotter, M., and Gleser, G.C. (1952) Estimation of stature from long bones of American Whites and Negroes. Am. J. Phys. Anthrop., 10: 463-514.

Trotter, M., and Gleser, G.C. (1958) A re-evaluation of estimation of stature based on measurements of stature taken during life and of long bones after death. Am. J. Phys. Anthrop., 16: 79-123.

Trotter, M., Peterson, R.R., and Wetter, R. (1968) The secular trend in the diameter of the femur of American Whites and Negroes. Am. J. Phys. Anthrop., 28: 65-74.

Ubelaker, D.H. (1974) Reconstruction of Demographic Profiles from Ossuary Skeletal Samples. A Case Study from the Tidewater Potomac. Smith. Inst., Washington, D.C.

VanGerven, D.P. (1973) Thickness and area measurements as parameters of skeletal involution of the humerus, femur, and tibia. J. Geront., 28: 40-45.

VanGerven, D.P., and Armelagos, G.J. (1970) Cortical involution in prehistoric Mississippian femora. J. Geront., 25: 20-22.

VanGerven, D.P., Armelagos, G.J., and Bartley, M.H. (1969) Roentgenographic and direct measurement of femoral cortical involution in a prehistoric Mississippian population. Am. J. Phys. Anthrop., 31: 23-38.

Vilensky, J.A. (1979) Masses, centers-of-gravity, and moments of inertia of the body segments of the rhesus monkey (Macaca mulatta). Am. J. Phys. Anthrop., 50: 57-66.

Walensky, N.A. (1965) A study of anterior femoral curvature in man. Anat. Rec., 151: 559-570.

Washburn, S.L. (1948) Sex differences in the pubic bone. Am. J. Phys, Anthrop., 6: 199-208.

Washburn, S.L. (1951) The new physical anthropology. Trans. N.Y. Acad. Sci., 13: 298-304.

Watson, R.C. (1973) Bone growth and physical activity in young males. In: Proceedings of International Conference on Bone Mineral Measurement. Gov. Printing Office, Washington, D.C.

Watson-Jones, R. (1976) Fractures and Joint Injuries. Churchill Livingstone, London.

Weiss, K.M. (1972) On the systematic bias in skeletal sexing. Am. J. Phys. Anthrop., 37: 239-250.

Wells, C. (1967) A new approach to paleopathology: Harris's lines. In: Diseases in Antiquity, Brothwell, D. (ed.). Thomas and Sandison, Springfield, Ill.

Wells, J.P., and Wood, G.A. (1975) The application of biomechanical motion analysis to aspects of green monkey (Cercopithecus a. sabaeus) locomotion. Am. J. Phys. Anthrop., 43: 217-226.

Willey, G.R. (1966) An Introduction to American Archaeology, Vol. 1. North and Middle America. Prentice-Hall, Englewood Cliffs, N.J.

Wolff, J. (1870) Ueber die innere architectur der knochen und'ihre bedeutung fur die frage von knochenwachstum. Uirchows Arch., 50: 389-450.

Wolff, J. (1892) Das Gesetz der Transformation der Knochen. A. Hirchwild, Berlin.

Woo, S. L-Y., Kuei, S.C., Dillon, W.A., Amiel, D., White, F.C., and Akeson, W.H. (1981) The effect of prolonged physical training on the properties of long bone--a study of Wolff's Law. J. Bn. Jt. Surg.: in press.